Analysis of Geometrically Nonlinear Structures

Analysis of Geometrically Nonlinear Structures

Second Edition

by

Robert Levy
Technion – Israel Institute of Technology,
Haifa, Israel

and

William R. Spillers
New Jersey Institute of Technology,
Newark, New Jersey, U.S.A.

KLUWER ACADEMIC PUBLISHERS
DORDRECHT / BOSTON / LONDON

A C.I.P. Catalogue record for this book is available from the Library of Congress.

ISBN 1-4020-1654-9

Published by Kluwer Academic Publishers,
P.O. Box 17, 3300 AA Dordrecht, The Netherlands.

Sold and distributed in North, Central and South America
by Kluwer Academic Publishers,
101 Philip Drive, Norwell, MA 02061, U.S.A.

In all other countries, sold and distributed
by Kluwer Academic Publishers,
P.O. Box 322, 3300 AH Dordrecht, The Netherlands.

Printed on acid-free paper

Printed in the Netherlands.

והארץ שקטה ממלחמה
יהושע יד טו

And the land had rest from war
Joshua 14 15

To Lydia, Itai and Shelly
R.L.

To the late Jewell Garrelts of
Columbia University
W.R.S.

BOOK CONTENTS

PREFACE

The availability of computers has, in real terms, moved forward the practice of structural engineering. Where it was once enough to have any analysis given a complex configuration, the profession today is much more demanding. How engineers should be more demanding is the subject of this book.

In terms of the theory of structures, the importance of geometric nonlinearities is explained by the theorem which states that

"*In the presence of prestress, geometric nonlinearities are of the same order of magnitude as linear elastic effects in structures*."

This theorem implies that in most cases (in all cases of incremental analysis) geometric nonlinearities should be considered. And it is well known that problems of buckling, cable nets, fabric structures, ... REQUIRE the inclusion of geometric nonlinearities.

What is offered in the book which follows is a unified approach (for both discrete and continuous systems) to geometric nonlinearities which incidentally does not require a discussion of large strain. What makes this all work is perturbation theory. Let the equations of equilibrium for a system be written as

$$\mathbf{N}^{\mathrm{T}}\mathbf{F} = \mathbf{P}$$

where **P** represents the applied loads, **F** represents the member forces or stresses, and **N** represents the operator which describes system equilibrium. (This equation can also be thought of as the matrix equation of node equilibrium for a discrete system.) Under a load perturbation d**P** this system responds as

$$\mathrm{d}\mathbf{N}^{\mathrm{T}}\mathbf{F} + \mathbf{N}^{\mathrm{T}}\mathrm{d}\mathbf{F} = \mathrm{d}\mathbf{P}$$

It is the first term in the above equation which describes so-called geometric nonlinearity and it is the second term which returns linear theory. For a discrete system it turns out to be a relatively simple matter to convert this equation into the usual

$$(\mathbf{K}_{\mathrm{E}} + \mathbf{K}_{\mathrm{G}})\delta = \mathrm{d}\mathbf{P}$$

Here $\mathbf{K}_E$ and $\mathbf{K}_G$ are the geometric and elastic stiffness matrices and δ represents the system displacements. With this approach, nonlinear structural analysis becomes simply an application of Newton's method.

In terms of organization, this book begins with an introduction which cites many practical phenomena involving geometric nonlinearities. We then procede through a hierarchy of discrete and continuous systems ... from trusses to frames and from beams to a membrane and shell finite elements. In all cases computer programs are a central part of the discussion. Finally in a slightly different spirit, a chapter on fabric structures is included. Fabric structures of course require nonlinear analysis but this chapter goes outside the other material of this book when talking about shape finding and patterning. And we leave out much of the beautiful work which comprises much of what is now called nonlinear elasticity primarily on the grounds that it depends on issues of large strain.

Finally, we hope that the material which follows will have appeal to the practicing engineer. In fact, this work appears in its present form largely because of a comment made by Horst Berger (Horst Berger Partners, New York City) in the 1980's who suggested that since fabric structures were nonlinear anyway, their analysis should include the nonlinear effects (such as buckling) in the supporting frame. He might well amend this today to include more general classes of structures. That is, there are many reasons today to include nonlinear effects in structures rather than simply settling for linear analysis. And the technology to do so is available and inexpensive.

Any modern discussion of the incremental analysis of structures takes much from the now classic work of Biot (1965). We would be happy to be regarded to be in his company.

Robert Levy
William R. Spillers

Haifa and Newark, August, 1994

Preface to the Second Edition

With the exception of new material on three-dimensional beams (Chapter 8) and shells (Chapter 9) this text retains most of the material of the first edition.

Robert Levy
William R. Spillers

Haifa and Newark, July, 2003

USING THE CD

This book is sold with a CD which contains FORTRAN source code and data files to be used with these programs. It is presumed that the reader has access to a personal computer (PC) and a FORTRAN compiler. The authors have attempted to make computer programs an integral part of this text and see the reader with a PC at his or her side. This approach probably derives from the old saw which tells us that you don't really understand something until you code it for a computer. Having said this, it seems to the authors that the issue of how best to combine computer code and text remains largely unsettled.

The following comments attempt to deal with some preliminary items that the reader should understand at the outset:

PCs Versus Workstations. The computer programs associated with this text all run on a PC running DOS. This is an expression of the authors' view that PCs not workstations will be computers of choice in both the workplace and the university. But the reader may note that with the exception of computer graphics features, these programs move easily to a Sun workstation. (We run them on the SUN under X-windows which is used to supply the graphics routines.)

Computer Graphics. Computer graphics is fun but also a serious component analysis programs, particularly with regard to checking data and visualizing three-dimensional problems. The reader who wishes to take advantage of computer graphics features should go directly to Appendix 6.

Hardware. The programs were run on a Pentium 4, 1.9GHz with Windows® XP.

Practical Applications. This book is written more for use as a learning tool than for production. That is particularly true of the computer programs which do not use sparse matrix methods and are thus limited in terms of the size of problems which can be treated. But 1) adding sparse matrix methods to a computer algorithm is by now a fairly routine matter and 2) the code used in this text has served as the basis of many commercial applications.

Contact With the Authors. We would be pleased to receive E-mail from the readers. We can be reached at

cvrlevy@tx.technion.ac.il
spillers@admin1.njit.edu

CHAPTER 1

OVERVIEW

1.1 Introduction

There are a number of practical phenomena in structures which simply can not be described using a linear formulation. Among these are buckling, the behavior of cable nets and fabric structures, the formation of plastic hinges, the nonlinear material behavior of concrete, structural collapse,... In all these cases (and many others) what has come to be known as the elementary theory of structures does not work or at least needs to be augmented. The question is of course how.

Since the elementary theory of structures is the low end - the most simple attempt to describe the response of structures - what is to be added to it is an open-ended question. In the case of the work which follows what is added to elementary theory is in some sense the "next step". To make this point it is convenient to introduce the terms "geometric" and "material" nonlinearities. Over the years effects due to constitutive equations (stress-strain relations) which are nonlinear, such as those describing the yielding of mild steel and the behavior of reinforced concrete, have been referred to as material nonlinearities. These effects are NOT considered in the work which follows.

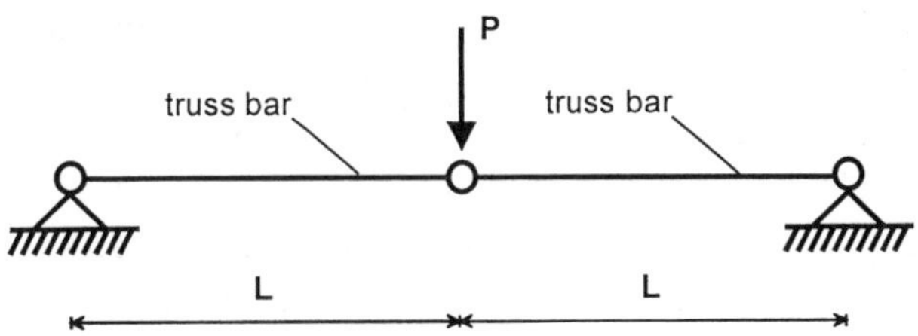

Figure 1.1 Geometrically unstable structure.

This book addresses the low end of what is then left of structural theory when material nonlinearities are not considered. Elementary structural theory is of course restricted to issues of small strain. With the exception of the truss problem, this book is also restricted to problems of small strain. (The truss is an anomaly in which it is a simple matter to work with large strain.) What is left is the effect of deformed geometry.

Elementary theory is a strange mix in which behavior is described in the "undeformed configuration". That is, in elementary theory the effect of deformations - even though they are computed - are neglected when writing the equations of equilibrium and motion. Biot (1965) has used the structure of Figure 1.1 to characterize these "geometric" difficulties which are inherent in elementary theory. This structure is said to be (geometrically) unstable, but it can function quite well in carrying load when sufficiently prestressed. In fact, it is not even necessary for this structure to undergo large deformations to work well, but its behavior is not included in the linear theory of structures.

The central idea of this book is quite simple: Real structures are in equilibrium in their deformed configuration, not their undeformed configuration as implied by elementary theory. Using the deformed configuration implies nonlinear analysis in the work below which will typically involve applying Newton's method to compute the effect of a load perturbation upon some given initial state. Buckling occurs when the system matrix becomes singular. This is not only easy to say. It is easy to do for many common structural systems.

This chapter attempts an overview by discussing familiar if eclectic problems of structures with the idea of setting the stage for a consistent approach to be developed in the later chapters.

1.2 Newton's Method

Central to the solution process used in this book is Newton's method. It is utilized whenever need arises to solve nonlinear equations. It has the advantage of rapid convergence and solutions are obtained with relative ease. Newton's method is usually described for a scalar system (Figure 1.2). Here an attempt is being made to find the value x^* for which $f(x^*)=0$. The typical step of Newton's method involves starting at some given point x_0 and linearizing the system to find a new point x_1. Happily, Newton's method works for an $n \times n$ system of n equations in n unknowns, $\mathbf{f}(\mathbf{x})=\mathbf{0}$. Formally,

$$\mathbf{f}(\mathbf{x})=\mathbf{0} \Rightarrow \mathbf{f}(\mathbf{x}_0)+\nabla\mathbf{f}(\mathbf{x}_0)\cdot d\mathbf{x} \cong \mathbf{0} \tag{1.1}$$

or

$$d\mathbf{x}=-\nabla\mathbf{f}(\mathbf{x}_0)^{-1}\mathbf{f}(\mathbf{x}_0) \tag{1.2}$$

with

$$d\mathbf{x}=\mathbf{x}_1-\mathbf{x}_0 \tag{1.3}$$

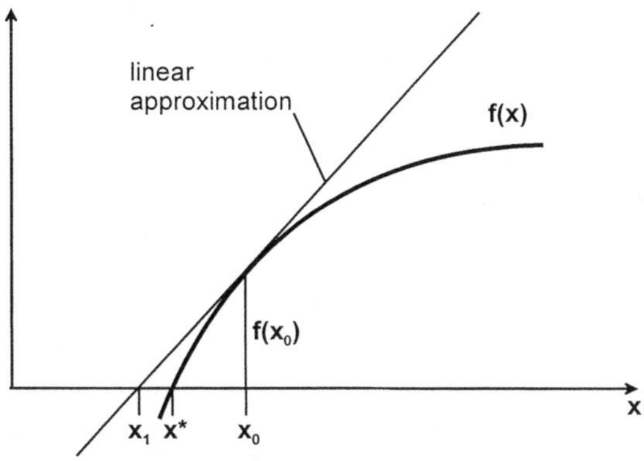

Figure 1.2 Newton's method.

Equation (1.1) only involves keeping the linear term in a Taylor series expansion. $\nabla\mathbf{f}$ is of course the gradient matrix of the n-vector $\mathbf{f}$. Newton's method linearizes the nonlinear system $\mathbf{f} = \mathbf{0}$ at a point and requires the solution of a system of linear equations for the next point. But solving linear equations is something structural engineers are very good at. It might also be noted that Newton's method requires the gradient matrix which will be computed explicitly in the work which follows. In structures, the gradient matrix of the equations of equilibrium is the tangential stiffness matrix containing within both the usual elastic stiffness matrix, $\mathbf{K}_{\mathrm{E}}$ and the geometric stiffness matrix, $\mathbf{K}_{\mathrm{G}}$ which will be discussed below.

One of the remarkable facets of Newton's method is that it possesses the property of "quadratic convergence". The remainder of this section discusses quadratic convergence for the simple case of a single variable x and a single equation $f(x)$. For the n-dimensional case, the reader can refer to any of the standard texts on numerical analysis. The Taylor series expansion of $f(x)$ about x_0 with remainder, is

$$f(x) = f(x_0) + f'(x_0)\cdot(x - x_0) + \frac{1}{2}f''(\xi)\cdot(x - x_0)^2 \tag{1.4}$$

where ξ is a point in the interval (x, x_0). For a single variable and a scalar function Eq. 1.1 reduces to

$$x_1 - x_0 = -f'(x_0)^{-1}\cdot f(x_0) \tag{1.5}$$

For $x \to x^*$ so that $f(x^*) = 0$ the substitution of Eq. 1.4 into Eq. 1.5 yields

$$x_1 - x^* = (x^* - x_0)^2 \frac{f''(\xi)}{2f'(x_0)} \tag{1.6}$$

For $\left|\frac{f''(\xi)}{2f'(x_0)}\right| \leq 1$ it follows that

$$|x_1 - x^*| \leq |x^* - x_0|^2 \tag{1.7}$$

which implies quadratic convergence in the neighborhood of x^*. That is when you are sufficiently close to a root so that $|x^* - x_0| << 1$, the next step is much closer.

1.3 Restrictions of Small Strain

With the exception again of the truss, this book deals with the special case of large rotations (deformations) but small strain. There are several reasons for doing so but the principal one is simplicity. Using the restrictions of small strain it is possible, in terms of mechanics, to keep much of classical theory. Another reason for dealing with this special case has to do with materials. The common structural components, with the notable exception of mild steel, cannot tolerate large strains without fracture. This is true of cables, fabrics commonly used in permanent structures, wood, aluminum,...

When dealing with nonlinear problems, it is particularly important to be careful with regard to the semantics used. A case in point is the implications of the restrictions of "small strain" used in linear theory (see Figure 1.3). Given a bar and a displacement, linear theory takes the member length change to be the projection of the displacement in the initial direction of the bar. The actual length change is, on the other hand, obtained by

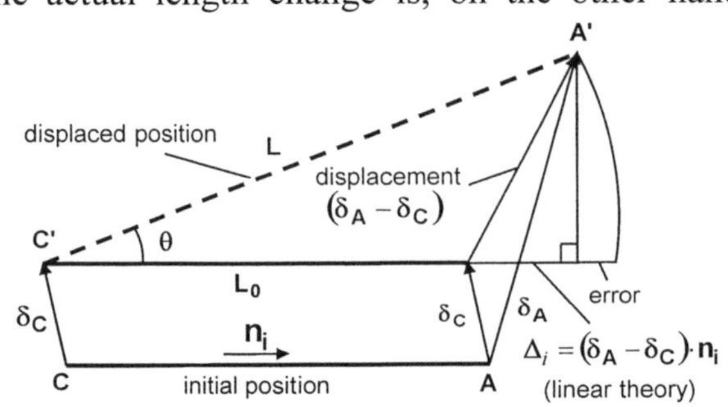

Figure 1.3 A truss bar. Approximation of linear theory.

drawing an arc of a circle from the displaced end of the bar, in this case, to the horizontal. The error is indicated in the figure. For the case of small rotations this error can be shown to be small.

The point is that linear theory, if misused, can lead to the idea of "fictitious" or "spurious" strain - strain when there really isn't any. In the work below, steps will be taken to prevent the appearance of spurious strain. This will typically be done by removing the effect of rigid bodymotion prior to computing strain. (This step is fundamental to the concept of large rotations but small strain.)

1.4 Stress Stiffening

Figure 1.1 shows the classic example of stress stiffening, the phenomenon in which prestressing a structure increases its stiffness. This simple structure can not work within linear theory because equilibrium is impossible in the undeformed configuration as indicated in Figure 1.4. Clearly, since the undeformed structure implies horizontal bar forces there is no mechanism to carry an applied vertical load. The kinematic version of the impossibility of equilibrium is the fact that a small vertical displacement of the loaded point produces no length change in the bars (within elementary theory).

Another look at this structure comes from the theory of strings (string vibrations from partial differential equations). Under the assumptions of string theory, the structure may displace without changing the string tension. If that is the case (Figure 1.5), deformed equilibrium may be used to compute a joint stiffness. Here, vertical equilibrium implies that $P = 2T \sin\theta$ where T is the prestress (string tension). But $\sin\theta \cong \tan\theta \cong \theta \cong d / L$ so that

$$P = \left(\frac{2T}{L}\right) \cdot d \tag{1.8}$$

and the "joint stiffness" is, thus, equal to *2T/L* with each member contributing an equal amount of *T/L* to the stiffness.

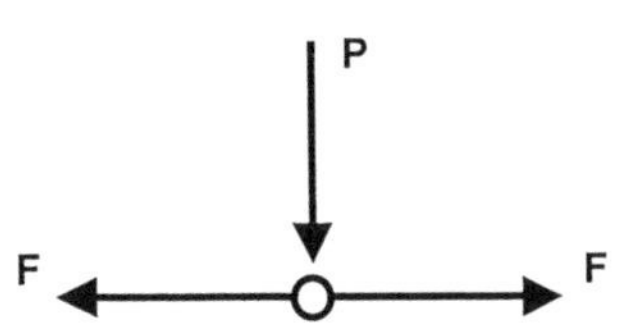

Figure 1.4 Undeformed joint equilibrium.

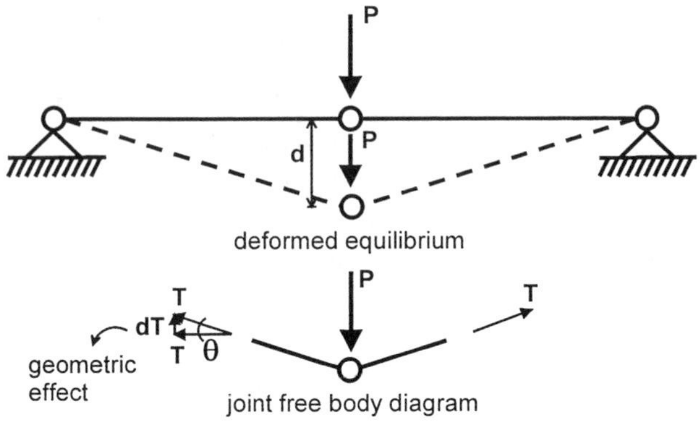

Figure 1.5 Joint equilibrium.

The "geometric effect" is the change dT in the internal force in response to the change in geometry of the structure.

The "geometric stiffness" T/L of the members is their ability to carry load normal to their centerline. Clearly this stiffness is a function of the tension T and increases with increasing T. The geometric stiffness matrix discussed below will include this effect of stress stiffening.

1.5 Buckling

Figure 1.6 shows the classic buckling example used by Timoshenko (1936). In this case a rigid bar is in equilibrium under the applied force P. Within linear theory, the only solution available to this problem is the one indicated in the schematic in which the spring is undeformed. In order to study buckling, it is common to apply a "perturbation", in this case a displacement d, to joint A and look for a position of deformed equilibrium. For deformed equilibrium the spring force is kd; for a small angle θ the vertical component of the force P becomes $P\theta$. For deformed equilibrium it follows that

$$P\theta = kd \tag{1.9}$$

Here k is of course the spring constant. If $\theta = d / L$, Eq. 1.9 becomes

$$\frac{Pd}{L} = kd \quad \Rightarrow \quad P_{\mathrm{CR}} = kL \tag{1.10}$$

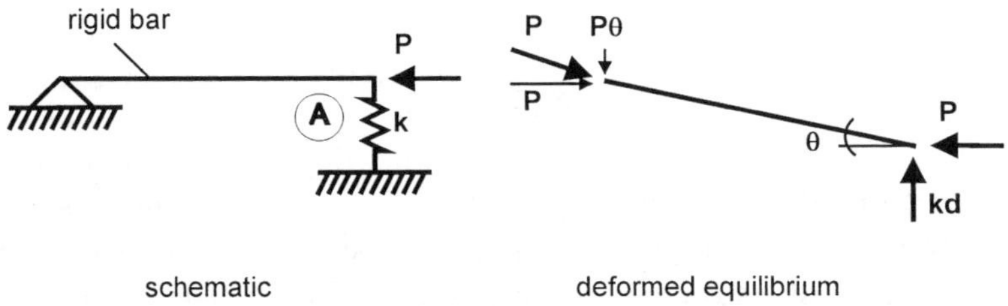

Figure 1.6 Buckling example.

It is said that the buckling load is equal to kL implying that for $P \le P_{CR}$ the system will return to its unperturbed configuration when disturbed and that $P \ge P_{CR}$ implies catastrophic failure. Again, this solution is quite intuitive but offers little help in terms of a general approach for problems of buckling.

1.5.1 Snap Through

Figure 1.7 shows the simple case of a symmetric two bar truss and its load deflection curve for large displacements. This curve can be produced in the following manner. First of all, it is necessary to use a nonlinear strain-displacement relationship such as requiring the force to be proportional to the "real" bar length change rather than the approximate length change of elementary theory. Then the load P can be computed directly from the displacement d:

- Assume any d
- Compute the real length change ΔL
- Compute the bar forces $F=k\ \Delta L$

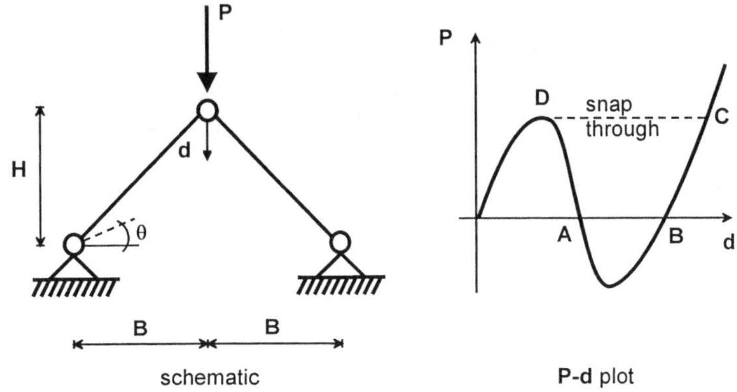

Figure 1.7 Snap-through buckling.

- Compute P using the deformed equilibrium ($P = 2F \sin\theta$)

Here again k is the linear spring constant, θ is the angle of inclination of the bar in its deformed position, and F is the bar force.

There are several points of interest in Figure 1.7. First of all, point D is typically what would be referred to as buckling; point A corresponds to the truss deformed to the point at which it becomes flat; and point B corresponds to the structure hanging down below the horizontal without load. After point B the structure continues to stiffen as it is loaded.

In order to generate this behavior in a laboratory, it would be necessary to control the displacement and measure the load. Otherwise, as the load P is increased from the value zero, the structure wants to "snap through" when point D is reached and jump to point C since following the curve downward to point A requires unloading.

1.5.2. Thermal Buckling

The solution of the problem of the preceding section implies the solution of the thermal buckling problem of Figure 1.8. (This problem looks a little like pavement buckling.) Here the bars are heated. If they are heated sufficiently the structure will buckle.

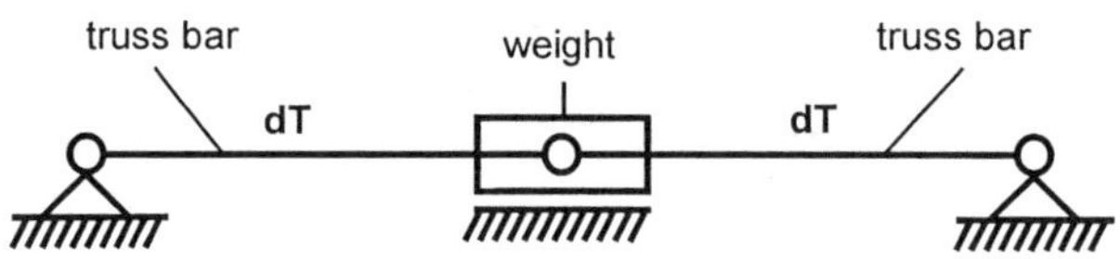

Figure 1.8 Thermal buckling.

The analysis of this buckling problem can be carried out in the following manner. For any temperature change, dT, a "free bar length" can be computed. (This is the length each bar would have if it were free to expand.) For this bar length the analysis of Section 1.5.1 can be carried out. If the attached weight is greater than the buckling load computed, thermal buckling will not occur; otherwise it is unstable for the given dT. That is, the structure will not buckle as long as the attached weight is sufficient to hold it down.

1.5.3. Euler's column

There is of course a commonalty between discrete and continuous systems. This can be seen in the classical problem of the buckling of a pin-ended

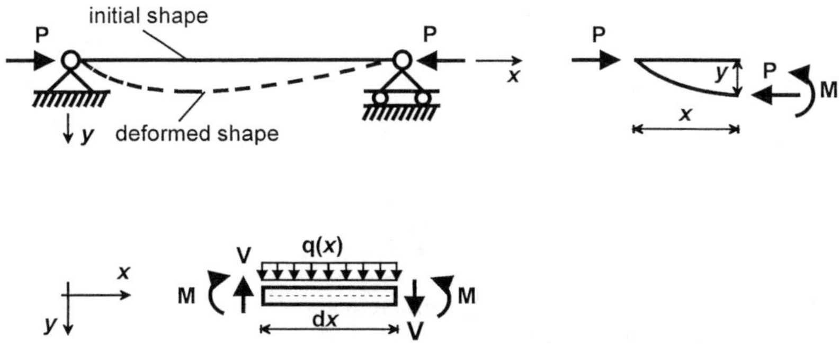

Figure 1.9 Euler's column.

column (Figure 1.9). In this simple case, deformed equilibrium implies that $M = Py$. That is, the bending moment M is equal to the axial load P times the lateral displacement y. If the constitutive equation $M = -EIy''$ is invoked (E=Young's modulus and I is the moment of inertia of the beam), it follows that,

$$Py = -EIy'' \quad or \quad y''+k^2 y = 0 \tag{1.11}$$

with $k^2 = P / EI$. Clearly,

$$y = A \sin kx + B \cos kx \tag{1.12}$$

Using the boundary condition $y = 0 @ x = 0$ it follows that $B = 0$; if $y = 0 @ x = L$ it follows that $\sin kL = 0$. This implies that $kL = n\pi \ (n = 1,2,...)$. By picking the smallest buckling load it is argued that $kL = \pi$ or $k^2 L^2 = \pi^2$ and $P_{CR} = \pi^2 EI / L^2$.

Note that the approach used here is again one of perturbation. The initially straight column is given a small lateral deformation and conditions for deformed equilibrium used to identify the buckling load.

1.5.4. Moment Distribution

Gere (1963) gives an example of the use of the method of "moment distribution" to determine the buckling load of a frame. His discussion includes the simple frame of Figure 1.10. Using Gere's analysis this frame buckles when the rotational stiffness of the joint becomes zero. Since this frame has no side sway and if member length change is neglected, this frame becomes a single degree of freedom system. Since the horizontal member

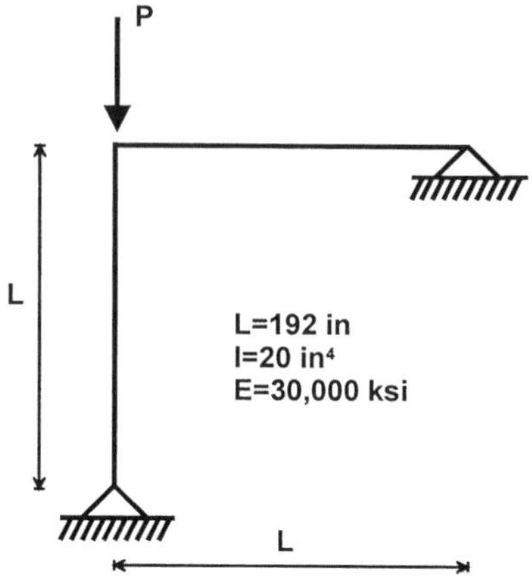

Figure 1.10 Frame buckling.

has no axial load, its stiffness is simply $3EI/L$. This is also the stiffness of the vertical frame member when axial load effects are neglected. In fact the linear elastic solution of this problem simply has a compressive load P in the vertical member.

Gere introduces buckling by considering "beam-column" effects. (See Appendix 1.) Within beam-column theory, the stiffness of a member (the end moment required to produce unit rotation) decreases as the compressive load in a member is increased and increases as the tension in a member increases. This is another example of stress-softening/stiffening. Consider a specific case of the frame in Figure 1.10. As the load increases from zero in this problem, at some point the stiffness of the vertical member goes to zero, ($kL \approx 3.125$), (see Figure 1.11). The structure is at this point not yet unstable because the horizontal member can still provide stiffness to the joint. If the load P is increased further the stiffness of the vertical member becomes increasingly negative. At some point the stiffness of the vertical member decreases to $-3EI/L$ and thus the total rotational stiffness of the joint becomes zero. The condition for zero stiffness for this case can be written according to Appendix 1, as

$$1+\frac{1}{\alpha}=0 \tag{1.13}$$

This is an implicit trigonometric equation whose solution was obtained using a fixed point iterative scheme to yield a critical value, P_{CR}, of 226 kips. Returning to Figure 1.11 the stiffness of $-3EI/L$ occurs at kL=3.725 which translates to:

$$P_{CR} = \frac{EI(kL)^2}{L^2} = \frac{30{,}000 \times 20 \times (3.725)^2}{(192)^2} \approx 226 \text{ kips}$$

It may be noted here that this example illustrates the importance of "member stiffness". That is, without beam-column effects it is impossible to discuss buckling in this case. The example is interesting from another point of view. When dealing with nonlinear problems it is often difficult

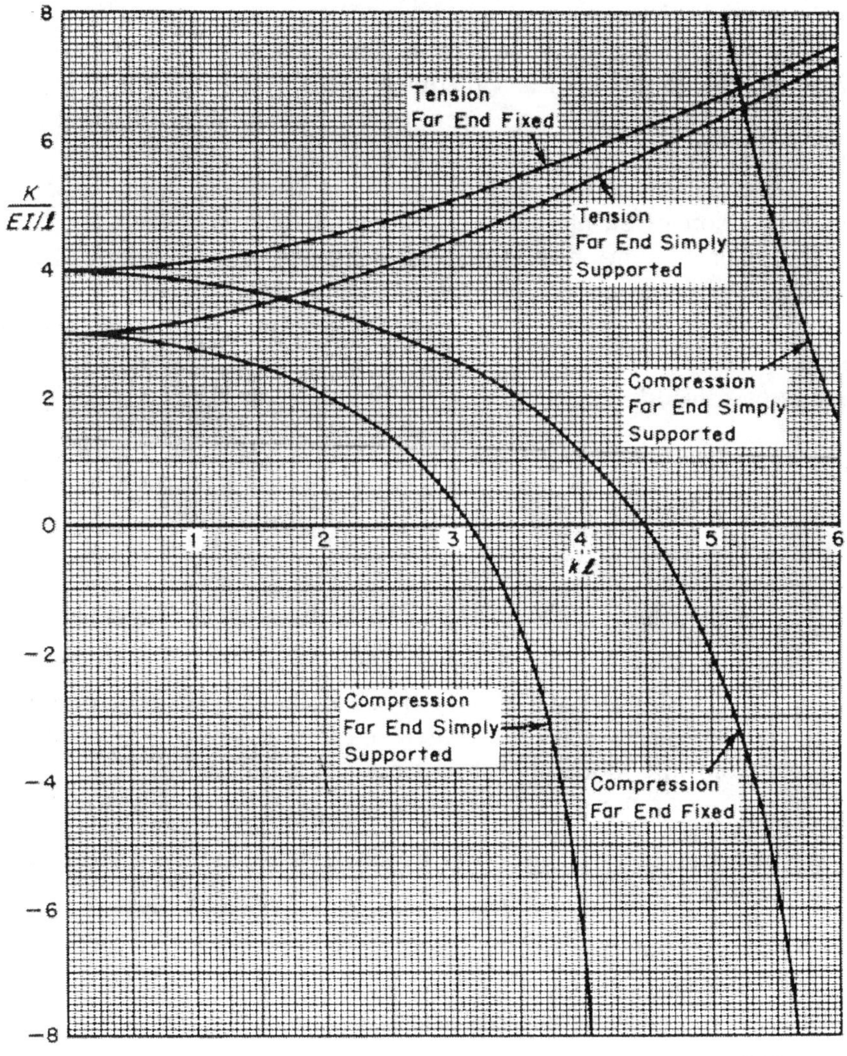

Stiffness factors for beams with axial loads. K – stiffness; $k = \sqrt{\frac{P}{EI}}$

Figure 1.11 (after Gere (1963).)

to keep track of the level of approximation involved. But in this case it is only member length change which is neglected. It remains to determine how sensitive the buckling load is to member length change.

1.5.5 Eigenvalue Approach

This book argues that buckling is most appropriately described by incrementing the load on a structure to failure - that is until its stiffness matrix becomes singular. This section describes a simplified, alternative approach to buckling which is sometimes found in the literature. (Note that this approach depends on unit vectors in the undeformed configuration and thus does not properly represent deformed equilibrium.)

Consider the structure of Figure 1.12. It is a simple two-bar truss under the action of an applied load P for which a deformed equilibrium position exists in addition to the undeformed equilibrium described above. Linear theory, of course, does not give a buckling load.

If the "string effects" described in Section 1.4 are included it is possible to compute a buckling load as follows. First of all note that linear theory gives a load/deformation relationship, $\mathbf{K}_{\mathrm{E}}\boldsymbol{\delta} = \mathbf{P}$,

$$\frac{AE}{L}\begin{bmatrix} 1 & 0 \\ 0 & 1 \end{bmatrix}\begin{Bmatrix} \delta_x \\ \delta_y \end{Bmatrix} = \begin{Bmatrix} 0 \\ P \end{Bmatrix} \tag{1.14}$$

Here $\mathbf{K}_{\mathrm{E}}$ is the elastic stiffness matrix, δ is the node displacement matrix $(\delta_x = d)$, and P is the joint load matrix. The "string effects" can be included by noting that the horizontal displacement of the loaded joint produces a horizontal force component *Pd/L* (Figure 1.12) and that there is no effect of this type from the horizontal bar since it is unloaded. This type of phenomenon is commonly described using the so-called "geometric stiffness matrix" $\mathbf{K}_{\mathrm{G}}$.

Writing this string effect on the joint forces as $\mathbf{K}_{\mathrm{G}}\boldsymbol{\delta}$,

$$\begin{bmatrix} P/L & 0 \\ 0 & 0 \end{bmatrix}\begin{Bmatrix} \delta_x \\ \delta_y \end{Bmatrix} \tag{1.15}$$

and combining both effects will then give equilibrium as $(\mathbf{K}_{\mathrm{E}} + \mathbf{K}_{\mathrm{G}})\boldsymbol{\delta} = \mathbf{P}$. In general, given **P**, it is possible to solve this system. The singular case

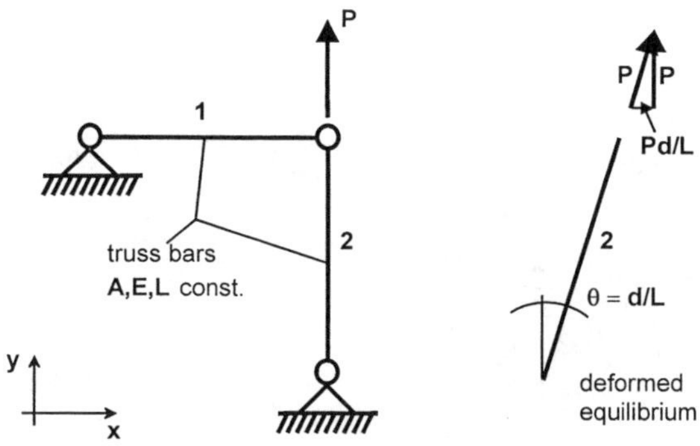

Figure 1.12 Eigenvalue buckling.

(buckling) corresponds to the problem of finding **P** so that the determinant of $\mathbf{K}_E + \mathbf{K}_G$ is zero. In this trivial case the determinant is zero only when

$$\frac{AE}{L} + \frac{P}{L} = 0 \quad \Rightarrow \quad P_{CR} = -AE \tag{1.16}$$

At buckling then a horizontal displacement, *d*, of the loaded joint produces an elastic force *AEd/L* in bar *1* which is balanced by the string effect force from bar *2* of *Pd/L.*

This example, of course, still leaves open the question of a general approach. But it is simple enough so its shortcomings are obvious. The problem here is that the effect of the deflections is not introduced in any consistent manner. (Both the elastic and the geometric stiffness matrices make use of the undeformed rather than the deformed configuration.) It will be seen later that a better approach is not difficult

Figure 1.13 is an attempt to introduce a more general perspective to geometric effects and the contribution of a generic member to the overall geometric stiffness matrix. A more rigorous derivation based on a perturbation of the equilibrium equations will be discussed later. It will be seen to produce identical results.

Member *i* connects nodes A and C with a positive incidence toward node A. If the bar force F_i remains constant in the deformed and undeformed configurations ($\theta << 1$), the string effect or geometric effect is a force $F_i\theta$ in the direction of vector **d** that can be written as

$$\mathbf{dF} = F_i\theta \frac{\mathbf{d}}{|\mathbf{d}|} \tag{1.17}$$

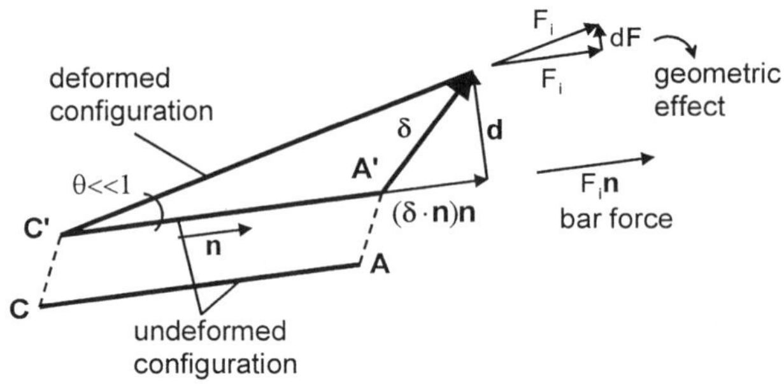

Figure 1.13 The geometric effect.

Now $\theta = \dfrac{|\mathbf{d}|}{L_i}$ and $\mathbf{d} = \boldsymbol{\delta} - (\boldsymbol{\delta}\cdot\mathbf{n})\mathbf{n}$ where **n** is a unit vector in the direction of the undeformed configuration of member i. **dF** (Eq. 1.17) now becomes

$$\mathbf{dF} = \frac{F_i}{L_i}\left[\boldsymbol{\delta} - (\boldsymbol{\delta}\cdot\mathbf{n})\mathbf{n}\right] \tag{1.18}$$

In matrix form Equation 1.18 becomes

$$\mathbf{dF} = \frac{F_i}{L_i}(\mathbf{I} - \mathbf{n}\mathbf{n}^{\mathrm{T}})\boldsymbol{\delta} \tag{1.19}$$

In the general case of a member for which both ends are displaced, so that $\boldsymbol{\delta} = \boldsymbol{\delta}_{\mathrm{A}} - \boldsymbol{\delta}_{\mathrm{C}}$ and $\mathrm{d}\mathbf{F}_{\mathrm{A}} = -\mathrm{d}\mathbf{F}_{\mathrm{C}}$ the contribution of member i to the geometric stiffness matrix associated with the appropriate degrees of freedom $\delta_{\mathrm{A}r} \rightarrow r = \mathrm{x}, y, z$ and $\delta_{\mathrm{C}r} \rightarrow r = \mathrm{x}, y, z$ is simply

$$\left[\mathbf{K}_{\mathrm{G}}\right]_i \begin{Bmatrix} (\boldsymbol{\delta}_{\mathrm{A}})_i \\ \vdots \\ (\boldsymbol{\delta}_{\mathrm{C}})_i \end{Bmatrix} = \frac{F_i}{L_i} \overset{\text{col A} \qquad\qquad\qquad \text{col C}}{\begin{bmatrix} (\mathbf{I} - \mathbf{n}_i\mathbf{n}_i^{\mathrm{T}}) & -(\mathbf{I} - \mathbf{n}_i\mathbf{n}_i^{\mathrm{T}}) \\ -(\mathbf{I} - \mathbf{n}_i\mathbf{n}_i^{\mathrm{T}}) & (\mathbf{I} - \mathbf{n}_i\mathbf{n}_i^{\mathrm{T}}) \end{bmatrix}} \begin{Bmatrix} \delta_{\mathrm{A}x} \\ \delta_{\mathrm{A}y} \\ \delta_{\mathrm{A}z} \\ \vdots \\ \delta_{\mathrm{C}x} \\ \delta_{\mathrm{C}y} \\ \delta_{\mathrm{C}z} \end{Bmatrix} \begin{matrix} \text{row A} \\ \\ \text{row C} \end{matrix} \tag{1.20}$$

1.5.6 An Exact Approach

This section introduces the notion of nonlinear buckling to describe overall stability when large rotations and equilibrium in the deformed state are considered. Moreover exact member length changes are used here to define member deformations. The eigenvalue formulation of the last section is restricted to small rotations and equilibrium in the initial state and is referred to as linear buckling. Both of these formulations naturally fall within nonlinear structural analysis since they describe buckling which is a nonlinear phenomenon. Differentiation between the two levels of complexity within buckling leads to the linear and nonlinear terminology. The discussion in section 1.5.1 relates, of course, to nonlinear buckling.

Consider the fundamental non symmetric two-bar truss of Figure 1.14. When treated "exactly" it no longer is a simple statically determinate structure but rather a complex system governed by implicit equations of equilibrium and conditions for nonlinear stability. This section will only discuss the symmetric case in which $B_1 = B_2 = B$ and v=0. For a discussion of the general case see Levy and Brill (1989).

Assuming that member stiffnesses, k, are constant and Hooke's law applies, equilibrium in the deformed state may be written in terms of the displacement, u, as

$$P - 2k(L_0 - L)\frac{H_1}{L} = 0 \qquad (1.21)$$

where $H_1 = H - u$ and L_0 and L are the initial and deformed lengths of the members respectively.

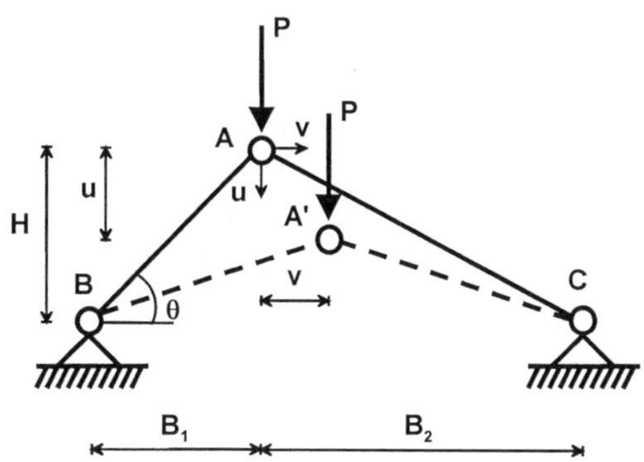

Figure 1.14 The non-symmetric two-bar truss.

As the loading is increased proportionally the derivative $\mathrm{d}P/\mathrm{d}u$ will vanish at a certain critical point along the equilibrium path and provide the following nonlinear buckling condition:

$$\frac{\mathrm{d}P}{\mathrm{d}u}=0 \quad \Rightarrow \quad \left(\frac{H_1}{L}\right)^2+\frac{L-L_0}{L_0}=0 \tag{1.22}$$

The critical height, H_1, is first obtained from Eq. 1.22. In terms of θ, the original angle of inclination ($\theta=\arctan(H/B)$), it can be written as

$$H_1=B(\sec^{2/3}\theta-1)^{1/2} \tag{1.23}$$

Then H_1 is substituted in Eq. 1.21 to yield the critical buckling load

$${}^{\text{nonlinear}}P_{\text{CR}}=2kL_0(1-\cos^{2/3}\theta)^{3/2} \tag{1.24}$$

When linear buckling, as described by the eigenvalue approach of section 1.5.5, is used to determine the buckling load of this truss, two modes are evident. Modes that interchange roles at $\theta=45°$. The critical loads for these buckling modes are

$$\begin{aligned}{}^{\text{linear}}P_{\text{CR}}&=2kL_0\left(\frac{\sin^3\theta}{\cos^2\theta}\right) \quad 0^o\le\theta\le 45^o\\ {}^{\text{linear}}P_{\text{CR}}&=2kL_0\left(\frac{\cos^2\theta}{\sin\theta}\right) \quad 45^o\le\theta\le 90^o\end{aligned} \tag{1.25}$$

Linear and nonlinear (exact) buckling can now be compared (Figure1.15) and the result is striking. Up to an angle of $\theta=62°$ the linear eigenvalue representation of stability overestimates the buckling load broadly. Defining the "linear buckling load factor", λ, as the ratio of the critical buckling load obtained via the eigenvalue representation to the critical nonlinear buckling load, a curve of λ versus θ can be generated. A peak of $\lambda=7.55$ at 45° is observed which implies that the eigenvalue representation is largely ineffective. This is an important result. It suggests that in most cases linear buckling analysis is highly inaccurate. The error of linear buckling here is reminiscent of shell theory where elementary buckling loads do not agree with experiment. In that case it is common to cite "imperfections" as the culprit. Here the difference between linear buckling and nonlinear buckling cannot be explained in such terms.

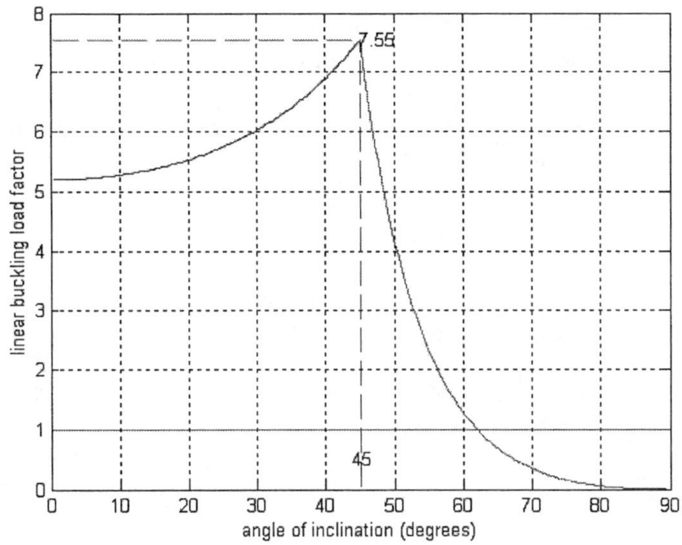

Figure 1.15 The linear buckling load factor.

1.6 Prestress

This section addresses the relative position of geometric stiffness effects in structures. It will be argued that "prestress" is the driving force and that quite generally geometrically nonlinear effects are of the same order of magnitude as classical linear effects and should be included in most analyses. This speaks to the question raised by Figure 1.1 concerning why a simple "unstable" truss can perform well under conditions of small deformation. Specifically, it will be shown using a uniform Taylor series expansion that:

Without prestress-

•Keeping terms of first order in displacement components leads to the classical linear theory of structures.

•Keeping second order terms leads to the "geometric stiffness" effect but also leads to a length change term due to rotation which is of the same order.

With prestress-

•A zeroth order term occurs which corresponds to the presence of the prestressing force.

•Keeping first order terms leads to the classical elastic response and also to the "geometric stiffness" effect.

•Length change due to rotation appears as a higher order term.

The point is that when there is prestress, nonlinear effects which are incorporated in the geometric stiffness matrix can be of the same order of magnitude as the effects ordinarily included in the linear theory of structures and must therefore be included in any analysis in order to develop a consistent analysis procedure. This type of conclusion appears in other areas of mechanics (Baumhauer and Tiersten, 1973).

1.6.1 A Fundamental Theorem

In Figure 1.16 a truss bar is shown in both the undeformed and deformed position. In general the length is

$$L=\sqrt{(L_0+\delta_x)^2+(\delta_y)^2}=\sqrt{(L_0)^2+2L_0\delta_x+(\delta_x)^2+(\delta_y)^2} \tag{1.26}$$

and the force components are

$$F_x=(L-L_0)k\frac{(L_0+\delta_x)}{L}=\left(1-\frac{L_0}{L}\right)k(L_0+\delta_x) \tag{1.27}$$

$$F_y=(L-L_0)k\frac{\delta_y}{L}=\left(1-\frac{L_0}{L}\right)k\delta_y \tag{1.28}$$

where k is the bar stiffness. If Eq. 1.26 is now written as,

$$L=L_0\sqrt{(1+x)} \tag{1.29}$$

where

$$x=2\frac{\delta_x}{L_0}+\left(\frac{\delta_x}{L_0}\right)^2+\left(\frac{\delta_y}{L_0}\right)^2 \tag{1.30}$$

a Taylor series expansion can be used to eliminate the radical in the expression for length and write Eqs. 1.27 and 1.28 in powers of δ_x and δ_y. Now

$$\left(1-\frac{L_0}{L}\right)=1-\frac{1}{\sqrt{1+x}}\cong\frac{1}{2}x-\frac{3}{8}x^2 \tag{1.31}$$

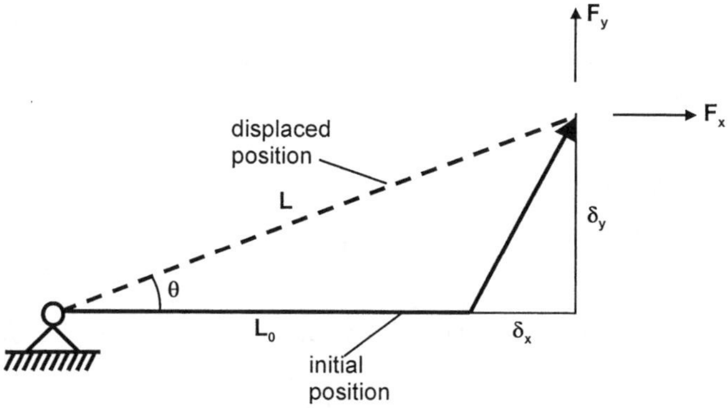

Figure 1.16 A truss bar.

since

$$\frac{1}{\sqrt{1+x}} = 1 - \frac{1}{2}x + \frac{3}{8}x^2 - \cdots \text{ for } |x| << 1 \tag{1.32}$$

Equation 1.27 thus becomes

$$F_x = k(L_0 + \delta_x)\left[\frac{1}{2}x - \frac{3}{8}x^2 \cdots\right] \tag{1.33}$$

or, after substitution of Eq. 1.30

$$F_x = k\left(\delta_x + \frac{\delta_y^2}{2L_0}\right) + \cdots \text{higher order terms} \tag{1.34}$$

Similarly Eq. 1.28 becomes

$$F_y = k\delta_y\left[\frac{1}{2}x - \frac{3}{8}^2 \cdots\right] \tag{1.35}$$

or, after substitution of Eq. 1.30

$$F_y = k\delta_x \frac{\delta_y}{L_0} + \cdots \text{higher order terms} \tag{1.36}$$

From Eqs. 1.34 and 1.36 it is clear that keeping terms uniformly in the first order of the joint displacements gives rise to the classical x - component of the first order or linear theory of structures,

$$F_x^{(1)} = k\delta_x \qquad F_y^{(1)} = 0 \tag{1.37}$$

Going to quadratic terms gives rise to two new components,

$$F_x^{(2)} = k\frac{\delta_y^2}{2L_0} \qquad F_y^{(2)} = k\frac{\delta_x\delta_y}{L_0} \tag{1.38}$$

If $k\delta_x$ is identified with the classical bar force and δ_y / L_0 with the angle θ in Fig. 1.16 (assuming small angle changes), the $F_y^{(2)}$ component which becomes $F_x^{(1)}\theta$ can be identified with the "string effect" (Eq. 1.8). But at the same time Eq. 1.38 indicates that there is a "length change due to rotation" which is of the same order in the displacement components and must also be included in order to be consistent. Specifically, to second order terms,

$$\sec\theta \approx 1 + \frac{1}{2}\theta^2 \tag{1.39}$$

and

$$F_x^{(2)} = k\left[L_0 \frac{1}{2}\left(\frac{\delta_y}{L_0}\right)^2\right] \tag{1.40}$$

which can be viewed as a second order correction to length change. Note that Eqs. 1.39 and 1.40 are identical since $\sec\theta = L / L_0$ and $\theta = \delta_y / L_0$.

These results leave open the question of the relationship of linear buckling theory to this Taylor series analysis since in the linear buckling of trusses it is common to neglect terms corresponding to length change due to rotation. That point will now be clarified by considering the effect of prestress F_0 in a member. In this case, it is only necessary to modify Eqs.

1.27 and 1.28 to read

$$F_x = [F_0 + (L - L_0)k]\frac{(L_0 + \delta_x)}{L} \tag{1.41}$$

$$F_y = [F_0 + (L - L_0)k]\frac{\delta_y}{L} \tag{1.42}$$

Then Eqs. 1.34 and 1.36 become

$$F_x = F_0 + k\delta_x + \cdots \text{ higher order terms} \tag{1.43}$$

$$F_y = F_0 \frac{\delta_y}{L_0} + \cdots \text{ higher order terms} \tag{1.44}$$

In Eqs. 1.43 and 1.44 there is now a zeroth order term due to the prestress. Keeping first order terms uniformly leads to the classical linear term and the geometric term with the length change due to rotation appearing as a higher order term. The right-hand-side term of Eq. 1.44 is the now familiar "geometric effect" that appeared in this chapter as the "string effect" (section 1.4) and later as $F_i\theta$ (Eq. 1.17), a more general term, that was used to construct the contribution of member i to the geometric stiffness matrix (Eq. 1.20). A fundamental theorem as depicted from Eqs. 1.43 and 1.44 may be stated as

In the presence of prestress, geometric nonlinearities are of the same order of magnitude as linear elastic effects in structures.

1.7 Problems

1. Redo the proof of geometric convergence for Newton's method given in section 1.2 for the case in which **f** is an n-vector.

2. Show geometrically the conditions under which the "error" indicated in Figure 1.3 is small. (Hint: $\theta \ll 1$)

3. Take a specific 2-bar truss and recreate the *P-d* relationship of Figure 1.7 numerically.

4. Redo Gere's example of Figure 1.10 for the case in which the frame has "fixed" supports rather than the "pinned" supports shown in the figure.

5. Show that Eqs. 1.18 and 1.19 are identical.

6. Take a specific 2-bar truss and verify the "linear buckling load factor" for the case in which $\theta = 45°$.

CHAPTER 2

LINEAR STRUCTURAL ANALYSIS

2.1 Introduction

This chapter is restricted to a review of the displacement or node method of linear structural analysis. Many good references on this topic are available to the reader. (See for example Spillers, 1972.)

Using the node method, the equilibrium equations of each structural joint are written in terms of the joint displacements giving a square, linear system of equations to be solved by the computer. These equations are commonly written in matrix form as

$$\mathbf{K}_{\mathrm{E}}\boldsymbol{\delta} = \mathbf{P} \tag{2.1}$$

where

$\mathbf{K}_{\mathrm{E}}$	-	elastic stiffness matrix (n × n)
$\boldsymbol{\delta}$	-	joint displacement matrix (n × 1)
$\mathbf{P}$	-	joint load matrix (n × 1)

Physically, an element of $\mathbf{K}_{\mathrm{E}}$, $(\mathbf{K}_{\mathrm{E}})_{ij}$, is the force at "$i$" due to a unit displacement at "j". That can be seen clearly by selecting $\boldsymbol{\delta}$ as indicated in Eq. 2.2 below and forming the product $\mathbf{K}_{\mathrm{E}}\boldsymbol{\delta}$,

$$\begin{bmatrix} & \cdots & (\mathbf{K}_{\mathrm{E}})_{1j} & \cdots \\ & & \vdots & \\ & \cdots & (\mathbf{K}_{\mathrm{E}})_{ij} & \cdots \\ & & \vdots & \\ & & \vdots & \end{bmatrix} \begin{Bmatrix} \\ 0 \\ \\ 1 \\ 0 \end{Bmatrix} \begin{matrix} \\ \\ \\ \text{row j} \\ \\ \end{matrix} = \begin{Bmatrix} (\mathbf{K}_{\mathrm{E}})_{1j} \\ \vdots \\ (\mathbf{K}_{\mathrm{E}})_{ij} \\ \vdots \\ \vdots \end{Bmatrix} \text{row i} \tag{2.2}$$

The elastic stiffness matrix is also symmetric. This may be argued from many points of view, for example, in terms of the conservation of energy but that issue will not be pursued here.

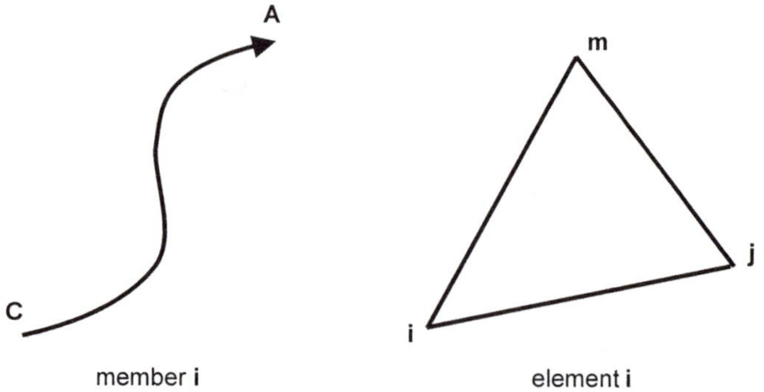

$$(\mathbf{K}_{\mathrm{E}})_i = \begin{bmatrix} [\times] & [\times] \\ [\times] & [\times] \end{bmatrix} \begin{matrix} \text{row A} \\ \text{row C} \end{matrix} \qquad \text{(columns A, C)}$$

$$(\mathbf{K}_{\mathrm{E}})_i = \begin{bmatrix} [\times] & [\times] & [\times] \\ [\times] & [\times] & [\times] \\ [\times] & [\times] & [\times] \end{bmatrix} \begin{matrix} \mathrm{i} \\ \mathrm{j} \\ \mathrm{m} \end{matrix} \qquad \text{(columns i, j, m)}$$

Figure 2.1 Contributions to the stiffness matrix.

In practical terms the matrix $\mathbf{K}_{\mathrm{E}}$ is usually constructed by summing the contribution of each bar or finite element to that matrix. Given the above physical interpretation of the elements $(K_{\mathrm{E}})_{ij}$, it follows that a member with two ends contributes four (matrix) terms to the stiffness matrix and that an element with three nodes contributes nine. (See Figure 2.1). The size of these matrix terms depends on the degrees of freedom of each node. The size of the matrix term for plane trusses, for example, is 2×2 whereas for plane frames it is 3×3 etc. This is because trusses have two degrees of freedom at each node and plane frames have three degrees of freedom at each node.

The equations of the node method can be written in many forms. The form used here is motivated by both network methods and the equations of a linear elastic continuum:

	<u>the node method</u>	<u>elasticity</u>
node equilibrium	$\mathbf{N}^{\mathrm{T}}\mathbf{F} = \mathbf{P}$	$\sigma_{ij,i} + f_j = 0$
constitutive eqs.	$\mathbf{F} = \mathbf{K}\boldsymbol{\Delta}$	$\sigma_{ij} = 2\mu\varepsilon_{ij} + \lambda\delta_{ij}\varepsilon_{kk}$
member node disp. eqs	$\boldsymbol{\Delta} = \mathbf{N}\boldsymbol{\delta}$	$\varepsilon_{ij} = (u_{i,j} + u_{j,i})/2$

Here

$\mathbf{F}$	-	member force matrix	~	stress tensor	σ_{ij}
$\mathbf{\Delta}$	-	member displacement matrix	~	strain tensor	ε_{ij}
$\mathbf{K}$	-	primitive stiffness matrix	~	Lamé constants	λ, μ
$\boldsymbol{\delta}$	-	node displacement matrix	~	displacement vector	u_i
$\mathbf{P}$	-	node load matrix	~	body force vector	f_i

and δ_{ij} is Kronecker's delta.

It follows directly from these equations that $\mathbf{N}^{\mathrm{T}}\mathbf{F} = \mathbf{P} \Rightarrow \mathbf{N}^{\mathrm{T}}\mathbf{K}\mathbf{\Delta} = \mathbf{P} \Rightarrow \mathbf{N}^{\mathrm{T}}\mathbf{K}\mathbf{N}\boldsymbol{\delta} = \mathbf{P} \Rightarrow \mathbf{K}_{\mathrm{E}} = \mathbf{N}^{\mathrm{T}}\mathbf{K}\mathbf{N}$ and that once $\mathbf{K}_{\mathrm{E}}\boldsymbol{\delta} = \mathbf{P}$ has been solved for $\boldsymbol{\delta}$, $\mathbf{F}$ and $\mathbf{\Delta}$ follow as $\mathbf{\Delta} = \mathbf{N}\boldsymbol{\delta}$ and $\mathbf{F} = \mathbf{K}\mathbf{\Delta}$. Furthermore, when $\mathbf{K}_{\mathrm{E}}$ is written as $\mathbf{K}_{\mathrm{E}} = \mathbf{N}^{\mathrm{T}}\mathbf{K}\mathbf{N}$, its symmetry only requires the symmetry of the matrix $\mathbf{K}$.

2.2 The Truss Problem

In order to be specific and anticipating some of the material to follow, the matrix formulation of the truss problem will now be described. Other types of structures will be discussed later as the need arises.

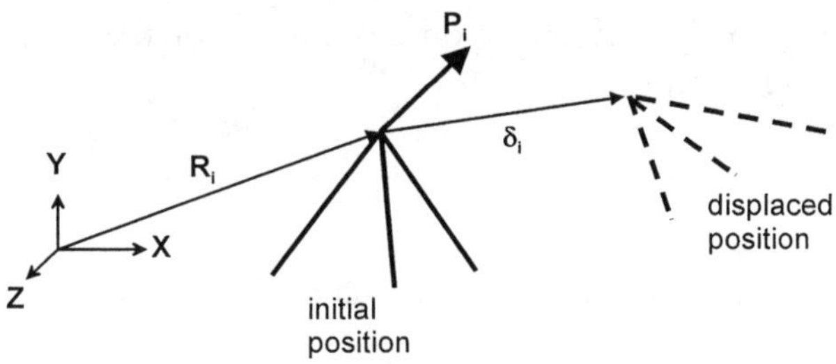

$$\mathbf{R} = \begin{Bmatrix} \mathbf{R}_1 \\ \mathbf{R}_2 \\ \vdots \end{Bmatrix} = \begin{Bmatrix} (\mathbf{R}_1)_x \\ (\mathbf{R}_1)_y \\ (\mathbf{R}_1)_z \\ (\mathbf{R}_2)_x \\ (\mathbf{R}_2)_y \\ (\mathbf{R}_2)_z \\ \vdots \end{Bmatrix}; \quad \mathbf{P} = \begin{Bmatrix} \mathbf{P}_1 \\ \mathbf{P}_2 \\ \vdots \end{Bmatrix} = \begin{Bmatrix} (\mathbf{P}_1)_x \\ (\mathbf{P}_1)_y \\ (\mathbf{P}_1)_z \\ (\mathbf{P}_2)_x \\ (\mathbf{P}_2)_y \\ (\mathbf{P}_2)_z \\ \vdots \end{Bmatrix}; \quad \boldsymbol{\delta} = \begin{Bmatrix} \boldsymbol{\delta}_1 \\ \boldsymbol{\delta}_2 \\ \vdots \end{Bmatrix} = \begin{Bmatrix} (\boldsymbol{\delta}_1)_x \\ (\boldsymbol{\delta}_1)_y \\ (\boldsymbol{\delta}_1)_z \\ (\boldsymbol{\delta}_2)_x \\ (\boldsymbol{\delta}_2)_y \\ (\boldsymbol{\delta}_2)_z \\ \vdots \end{Bmatrix};$$

Figure 2.2 Joint (node) quantities.

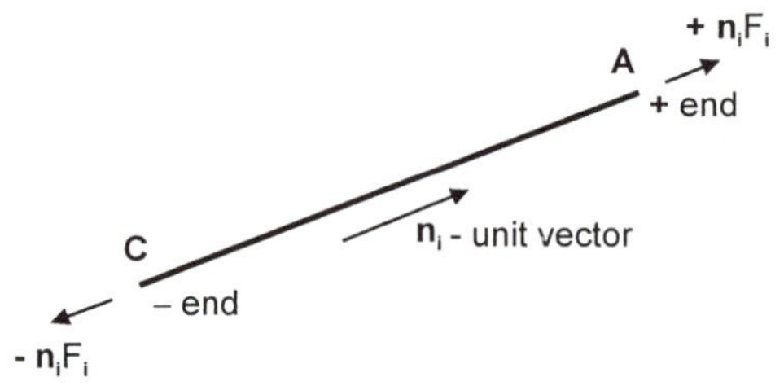

Figure 2.3 Typical bar i.

For the purposes here, a truss is a collection of pin-connected members in which loads are only applied to nodes. In a typical truss, associated with each node i is a position vector $\mathbf{R}_i$, a load vector $\mathbf{P}_i$, and a displacement vector $\boldsymbol{\delta}_i$. In matrix formulations these are collected in column matrices as indicated in Figure 2.2. (At this point, the use of matrices is only a matter of bookkeeping.)

Next, a typical truss *member* can be described in terms of its member force F_i, its member displacement (length change) Δ_i, and a unit vector $\mathbf{n}_i$ which describes the slope of the bar. (See Figure 2.3.)

Figure 2.4 can now be used to discuss the equations of node equilibrium. There are two cases to be considered here: a bar is either positively or negatively incident upon any node. For the case of Figure 2.4 node equilibrium is simply

$$-\mathbf{n}_\alpha F_\alpha + \mathbf{n}_\beta F_\beta + \mathbf{n}_\gamma F_\gamma - \mathbf{n}_\delta F_\delta + \mathbf{P}_i = \mathbf{0} \tag{2.3}$$

It can be shown that node equilibrium in matrix form can be written as

$$\mathbf{N}^{\mathrm{T}}\mathbf{F} = \mathbf{P} \tag{2.4}$$

where the matrix **N** is defined as

$$N_{ij} = \begin{cases} \mathbf{n}_i^{\mathrm{T}} & \text{if node } j \text{ is the + end of member } i \\ -\mathbf{n}_i^{\mathrm{T}} & \text{if node } j \text{ is the – end of member } i \\ \boldsymbol{0} & \text{otherwise} \end{cases} \tag{2.5}$$

Here $\mathbf{n}_i$ is the 3×1 column matrix of unit vector components.

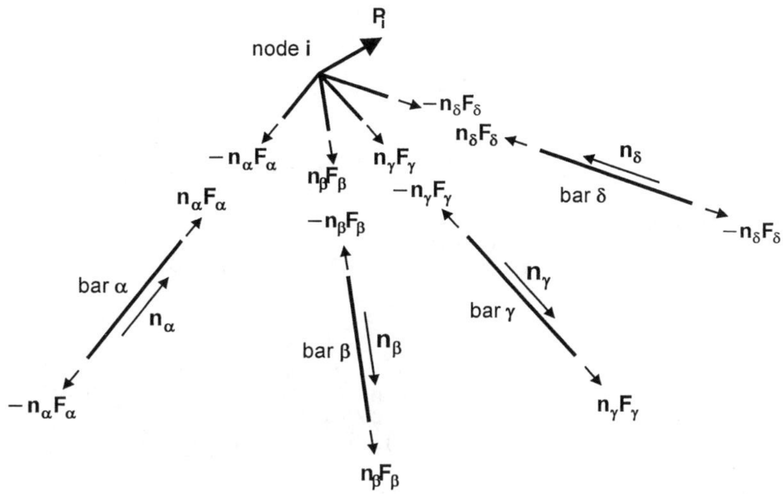

Figure 2.4 Equilibrium of node i.

Before the matrix formulation can be completed two more relations are needed. First, there is the approximate (linear) member/node displacement relationship which was discussed in Chapter 1

$$\Delta_i = \mathbf{n}_i \cdot (\boldsymbol{\delta}_A - \boldsymbol{\delta}_C) \tag{2.6}$$

It can be verified by direct calculation that the matrix equation

$$\boldsymbol{\Delta} = \mathbf{N}\boldsymbol{\delta} \tag{2.7}$$

reproduces Eq. 2.6 for each bar in the system.

Finally there is the constitutive equation $F_i = k_i \Delta_i$ where $k_i = A_i E / L_i$; A_i = cross sectional area of member *i*; L_i = length of member *i* and E = Young's modulus. The constitutive relations can be written in matrix form as

$$\mathbf{F} = \mathbf{K}\boldsymbol{\Delta} \tag{2.8}$$

where **K** is diagonal with elements $K_{ii} = k_i = A_i E / L_i$.

Given the matrices of the node method it is a simple matter to form the system matrix $\mathbf{K}_\mathrm{E} = \mathbf{N}^\mathrm{T}\mathbf{K}\mathbf{N}$. Rather than doing so directly, the contribution of member *i* to the system matrix, $(\mathbf{K}_\mathrm{E})_i$, is usually formed allowing the system matrix to be assembled as the sum of the contributions of all of the members. For the truss bar of Fig. 2.3 this contribution turns out to be

$$\begin{array}{c} \\ (\mathbf{K}_\mathrm{E})_i = \end{array} \begin{array}{c} \text{col A} \qquad\qquad \text{col C} \\ \begin{bmatrix} (\mathbf{K}_\mathrm{E})_i^{\mathrm{AA}} & -(\mathbf{K}_\mathrm{E})_i^{\mathrm{AA}} \\ -(\mathbf{K}_\mathrm{E})_i^{\mathrm{AA}} & (\mathbf{K}_\mathrm{E})_i^{\mathrm{AA}} \end{bmatrix} \end{array} \begin{array}{l} \\ \text{row A} \\ \text{row C} \end{array} \tag{2.6}$$

with

$$(\mathbf{K}_\mathrm{E})_i^{\mathrm{AA}} = \frac{A_i E}{L_i}\left(\mathbf{n}_i \mathbf{n}_i^\mathrm{T}\right) = \frac{A_i E}{L_i}\begin{bmatrix} (n_i)_x^2 & (n_i)_x (n_i)_y & (n_i)_x (n_i)_z \\ (n_i)_y (n_i)_x & (n_i)_y{}^2 & (n_i)_y (n_i)_z \\ (n_i)_z (n_i)_x & (n_i)_z (n_i)_y & (n_i)_z{}^2 \end{bmatrix} \tag{2.7}$$

and

$$\mathbf{n}_i = \{(n_i)_x \quad (n_i)_y \quad (n_i)_z\}^\mathrm{T}$$

This contribution can be computed in many ways. For example, Figure 2.5 describes how this can be done by applying the interpretation, that the elements of the stiffness matrix are forces due to unit displacements. In this figure $\mathbf{n}_i$ indicates the original direction of the undeformed bar and also (from linear theory considerations) the direction of the deformed bar. Let, $(\boldsymbol{\delta}_\mathrm{A})_x = 1$ with all other displacement components held at zero. The member length change is then:

$$\Delta_i = \mathbf{n}_i \cdot (\boldsymbol{\delta}_\mathrm{A} - \boldsymbol{\delta}_\mathrm{C}) = (\mathbf{n}_i)_x$$

the member force:

$$F_i = K_{ii}\Delta_i = (\mathbf{n}_i)_x A_i E / L_i$$

and the end-of-member forces are:

$$\pm (\mathbf{n}_i)_x \mathbf{n}_i A_i E / L_i$$

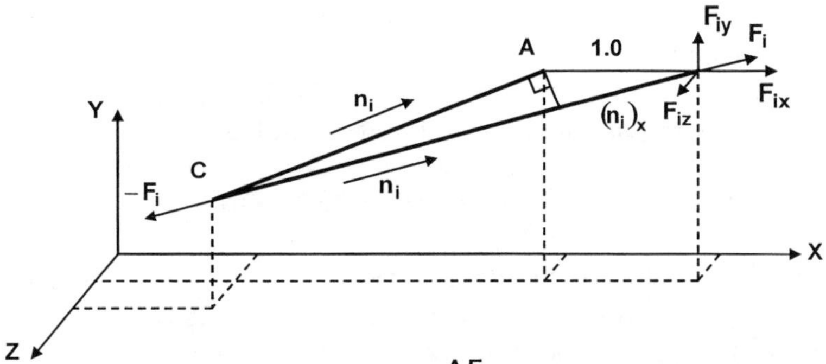

$$F_i = (n_i)_x \frac{A_i E}{L_i};$$

$$F_{ix} = (n_i)_x (n_i)_x \frac{A_i E}{L_i}; F_{iy} = (n_i)_x (n_i)_y \frac{A_i E}{L_i}; F_{iz} = (n_i)_x (n_i)_z \frac{A_i E}{L_i}$$

Figure 2.5 Truss element stiffness.

This step has produced the elements of $(\mathbf{K}_E)_i$ which lie in column (3*A-2) of the global stiffness matrix. The terms in the other columns follow in a similar fashion.

The computer programs that come with the CD which is included with this book show how these ideas can be implemented. Here it will simply be noted that all programs for linear structural analysis have three steps:

Step 1. Form the stiffness matrix $\mathbf{K}_E$ by summing the contributions to it of the various structural elements.

Step 2. Solve the system $\mathbf{K}_E \boldsymbol{\delta} = \mathbf{P}$ for $\boldsymbol{\delta}$

Step 3. Compute the forces $\mathbf{F}$ and the member displacements Δ in the structural elements.

2.3 Computer Programs

This section presents the program TR3D.FOR that performs linear analysis on three dimensional trusses. A two dimensional version, TR2D.FOR, is also discussed. A listing of TR3D.FOR is provided in Appendix 5. This program illustrates the concepts of linear structural analysis and serves as a basis for the nonlinear analysis computer programs to follow.

It should be noted that sparse matrix methods are not used in this book and that the solver is typically a Gaussian elimination program which does NOT pivot. All supports are required to be completely fixed with support joints numbered last. Fictitious fully fixed members with infinite areas of cross section may be added to partially fixed joints (roller supports) to meet the above requirement. All loads are joint loads.

2.3.1 Programs P1-TR3D.FOR and P2-TR2D.FOR

It is appropriate to start by noting that TR2D.FOR is directly obtained from TR3D.FOR by changing 3-D vectors to 2-D vectors wherever they are associated with joint entities. With regard to the coding:

1. `MAXC` is the dimension of array `C`.

2. Young's modulus E is set to 30,000,000 psi. (This is the only place units appear in the program. For example if E has units of psi then the load matrix `P` should have units of pounds and the coordinate matrix `R` should have units of inches.) Conversion of the program to other system of units is thus trivial.

3. Read as input and echoed as output are the number of bars `NB`, the number of nodes `NN`, the number of supports `NS`, the joint coordinate array `R(I)` and the joint load array `P(I)`.

4. The system matrix $\mathbf{K}_E$ is constructed as matrix `C(I,J)`. Its elements are first set to zero and then the contributions of all members (Eq. 2.7) are added. This step uses subroutine `UNITV` to generate the unit vector $\mathbf{n}_i$ of member i and subroutine `INSERT` to actually assemble `C`. Associated with each member the quantities `NP(I)`, `NM(I)`, and `S(I)` are read and echoed. These are the "+ end", the "- end" and the area of bar i respectively.

2.4 Examples

2.4.1 Example 2.1. A 3-Bar Truss

A statically indeterminate three bar plane truss is fully fixed at nodes 2, 3 and 4 and loaded at node 1 with a horizontal load of -10,000 lbs and a vertical load of -5,000 lbs. The connectivity of the members is shown in Figure 2.6 which was generated using PLOT3D.FOR with the data file TR2D21.DAT. Member lengths are 10, 10√2 and 20 inches. All member areas are .657 square inches and E=30,000,000 psi.

Listed below is the data file which is read by the linear analysis program TR2D.FOR. The first line provides the number of bars, the number of nodes and the number of supports. The next four lines supply information on each of the four nodes. Each line gives the x-coordinate, the y-coordinate,

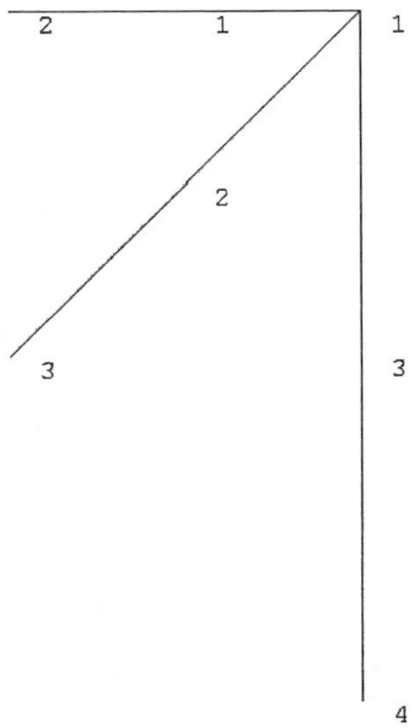

Figure 2.6 The three bar truss.

the horizontal load and the vertical load respectively. Finally each of the last three lines supplies connectivity details (node number of the + end and the node number of the - end) and the member area. Young's modulus is specified within the program.

Data file TR2D21.DAT

```
 3    4    3
       10.          20.        -10000.0    -5000.0
                    20.
                    10.
       10.0
 1    2            .657
 1    3            .657
 1    4            .657
```

The output file below first echoes the data and then lists the nodal displacements, member elongations, member forces and member stresses. This is also the order of their computation. The linear system $\mathbf{K}_E\boldsymbol{\delta} = \mathbf{P}$ is solved for the nodal displacements, $\boldsymbol{\delta}$, then the member deformations, $\boldsymbol{\Delta}$, are calculated from $\mathbf{N}\boldsymbol{\delta} = \boldsymbol{\Delta}$, the member forces, $\mathbf{F}$, are obtained from $\mathbf{F} = \mathbf{K}\boldsymbol{\Delta}$ and the member stresses, $\boldsymbol{\sigma}$, from $\sigma_i = F_i / A_i$, where A_i is the area of cross-section of member i.

Output file TR2D21.OUT

```
   3  NO. MEMBERS
   4  NO. NODES
   3  NO. SUPPORTS

                    COORDINATES                                LOADS
                       X                  Y                     PX                 PY
  1       .10000000D+02      .20000000D+02     -.10000000D+05     -.50000000D+04
  2       .00000000D+00      .20000000D+02      .00000000D+00      .00000000D+00
  3       .00000000D+00      .10000000D+02      .00000000D+00      .00000000D+00
  4       .10000000D+02      .00000000D+00      .00000000D+00      .00000000D+00

   MEMBER        + END       - END               AREA
       1           1           2         .65700000E+00
       2           1           3         .65700000E+00
       3           1           4         .65700000E+00

DISPLACEMENTS
                            X                     Y
       1         -.33325938D-02       -.15916210D-02
       2          .00000000D+00        .00000000D+00
       3          .00000000D+00        .00000000D+00
       4          .00000000D+00        .00000000D+00

   MEMBER                  DL                 FORCE               STRESS
       1         -.33325938D-02       -.65685425D+04       -.99977815D+04
       2         -.34819457D-02       -.48528137D+04       -.73863222D+04
       3         -.15916210D-02       -.15685425D+04       -.23874315D+04
Stop - Program terminated.
```

2.4.2 Example 2.2. A 24-Bar Space Truss Dome

A twenty four bar space truss which is fixed at its edges and loaded vertically downwards at its central node with a load of 220.46 lbs. is shown in Figure 2.7. *EA* is constant for all the members with *E*=30,000,000 psi and *A*=0.0155 square inches.

Here again PLOT3D.FOR was activated to generate the drawings of Fig. 2.7 with the analysis data in TR3D22.DAT. That data was read by the three dimensional truss analysis program, TR3D.FOR and results saved in TR3D22.OUT. This example falls in the category of single layered reticulated shallow shells that are apt to experience loss of stability (snap through) at rather unexpectedly low loads. The example will be further studied in the next chapter when the geometrically nonlinear analysis tools will have been introduced.

Data file TR3D22.DAT

```
  24    13     6
              .000000    .000000   3.234646                         -220.46
             4.921303   8.523848   2.447244
            -4.921246   8.523880   2.447244
```

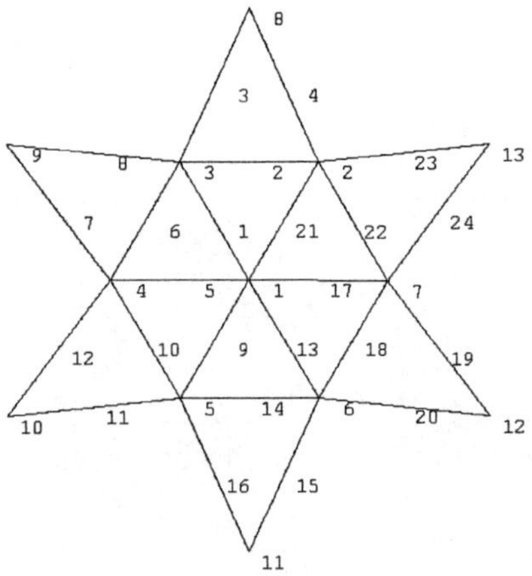

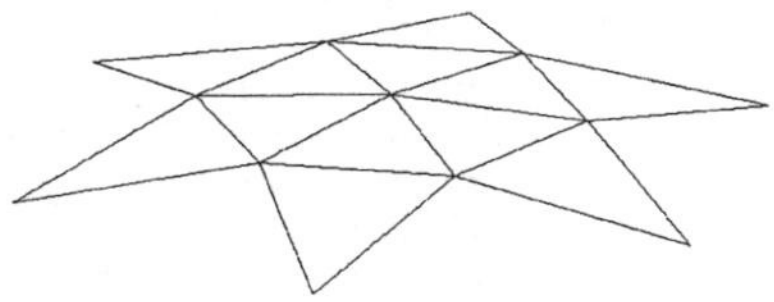

Figure 2.7 Plan and schematic views of the twenty four bar truss dome.

```
          -9.842520    .000023  2.447244
          -4.921286 -8.523857  2.447244
           4.921228 -8.523890  2.447244
           9.842520  -.000043  2.447244
            .000025 19.685040   .000000
         -17.047720  9.842553   .000000
         -17.047770 -9.842475   .000000
           -.000065-19.685040   .000000
          17.047700 -9.842588   .000000
          17.047790  9.842440   .000000
    1    3          .1550E-01
    2    3          .1550E-01
    3    8          .1550E-01
    2    8          .1550E-01
    1    4          .1550E-01
    3    4          .1550E-01
    4    9          .1550E-01
    3    9          .1550E-01
    1    5          .1550E-01
    4    5          .1550E-01
    5   10          .1550E-01
    4   10          .1550E-01
    1    6          .1550E-01
    5    6          .1550E-01
    6   11          .1550E-01
    5   11          .1550E-01
    1    7          .1550E-01
    6    7          .1550E-01
    7   12          .1550E-01
    6   12          .1550E-01
    1    2          .1550E-01
    7    2          .1550E-01
    2   13          .1550E-01
    7   13          .1550E-01
```

Output file TR3D22.OUT

```
   24  NO. MEMBERS
   13  NO. NODES
```

6 NO. SUPPORTS

COORDINATES

	X	Y	Z
1	.00000000D+00	.00000000D+00	.32346000D+01
2	.49212500D+01	.85239000D+01	.24472000D+01
3	-.49212500D+01	.85239000D+01	.24472000D+01
4	-.98425000D+01	.00000000D+00	.24472000D+01
5	-.49212500D+01	-.85239000D+01	.24472000D+01
6	.49212500D+01	-.85239000D+01	.24472000D+01
7	.98425000D+01	.00000000D+00	.24472000D+01
8	.00000000D+00	.19685000D+02	.00000000D+00
9	-.17047200D+02	.98425000D+01	.00000000D+00
10	-.17047200D+02	-.98425000D+01	.00000000D+00
11	.00000000D+00	-.19685000D+02	.00000000D+00
12	.17047200D+02	-.98425000D+01	.00000000D+00
13	.17047200D+02	.98425000D+01	.00000000D+00

LOADS

	PX	PY	PZ
1	.00000000D+00	.00000000D+00	-.22046000D+03
2	.00000000D+00	.00000000D+00	.00000000D+00
3	.00000000D+00	.00000000D+00	.00000000D+00
4	.00000000D+00	.00000000D+00	.00000000D+00
5	.00000000D+00	.00000000D+00	.00000000D+00
6	.00000000D+00	.00000000D+00	.00000000D+00
7	.00000000D+00	.00000000D+00	.00000000D+00
8	.00000000D+00	.00000000D+00	.00000000D+00
9	.00000000D+00	.00000000D+00	.00000000D+00
10	.00000000D+00	.00000000D+00	.00000000D+00
11	.00000000D+00	.00000000D+00	.00000000D+00
12	.00000000D+00	.00000000D+00	.00000000D+00
13	.00000000D+00	.00000000D+00	.00000000D+00

MEMBER	+ END	- END	AREA
1	1	3	.15500000E-01
2	2	3	.15500000E-01
3	3	8	.15500000E-01
4	2	8	.15500000E-01
5	1	4	.15500000E-01
6	3	4	.15500000E-01
7	4	9	.15500000E-01
8	3	9	.15500000E-01
9	1	5	.15500000E-01
10	4	5	.15500000E-01
11	5	10	.15500000E-01
12	4	10	.15500000E-01
13	1	6	.15500000E-01
14	5	6	.15500000E-01
15	6	11	.15500000E-01
16	5	11	.15500000E-01
17	1	7	.15500000E-01
18	6	7	.15500000E-01
19	7	12	.15500000E-01
20	6	12	.15500000E-01
21	1	2	.15500000E-01
22	7	2	.15500000E-01
23	2	13	.15500000E-01
24	7	13	.15500000E-01

DISPLACEMENTS

	X	Y	Z
1	-.48536751D-17	-.11946443D-17	-.20641184D+00
2	.37159444D-02	.64361834D-02	.91781931D-02
3	-.37159444D-02	.64361834D-02	.91781931D-02

```
          4       -.74321153D-02        -.29720790D-19         .91780850D-02
          5       -.37159444D-02        -.64361834D-02         .91781931D-02
          6        .37159444D-02        -.64361834D-02         .91781931D-02
          7        .74321153D-02        -.96950537D-18         .91780850D-02

     MEMBER                   DL                 FORCE                STRESS
          1       -.97840054D-02        -.46076256D+03        -.29726617D+05
          2        .74318888D-02         .35111285D+03         .22652442D+05
          3       -.24987545D-02        -.93394741D+02        -.60254673D+04
          4       -.24987545D-02        -.93394741D+02        -.60254673D+04
          5       -.97838205D-02        -.46075567D+03        -.29726172D+05
          6        .74319838D-02         .35111595D+03         .22652642D+05
          7       -.24986957D-02        -.93394458D+02        -.60254490D+04
          8       -.24987116D-02        -.93396601D+02        -.60255873D+04
          9       -.97840054D-02        -.46076256D+03        -.29726617D+05
         10        .74319838D-02         .35111595D+03         .22652642D+05
         11       -.24987116D-02        -.93396601D+02        -.60255873D+04
         12       -.24986957D-02        -.93394458D+02        -.60254490D+04
         13       -.97840054D-02        -.46076256D+03        -.29726617D+05
         14        .74318888D-02         .35111285D+03         .22652442D+05
         15       -.24987545D-02        -.93394741D+02        -.60254673D+04
         16       -.24987545D-02        -.93394741D+02        -.60254673D+04
         17       -.97838205D-02        -.46075567D+03        -.29726172D+05
         18        .74319838D-02         .35111595D+03         .22652642D+05
         19       -.24986957D-02        -.93394458D+02        -.60254490D+04
         20       -.24987116D-02        -.93396601D+02        -.60255873D+04
         21       -.97840054D-02        -.46076256D+03        -.29726617D+05
         22        .74319838D-02         .35111595D+03         .22652642D+05
         23       -.24987116D-02        -.93396601D+02        -.60255873D+04
         24       -.24986957D-02        -.93394458D+02        -.60254490D+04
Stop - Program terminated.
```

2.4.3 Example 2.3. A 72-Bar Double Layered Grid

A seventy two bar double layered square grid (see Figure 2.8) having members of equal length (54 inches) which is 1/8 symmetric is solved in this example. (See Levy and Hanaor, 1986 for a solution with optimal prestress.)

The grid is loaded vertically upwards at its central node with a load of 13,534 lbs. All circumferential nodes are restrained in the vertical direction only. The central node is restrained in both horizontal x and y directions and one of the corner nodes is restrained perpendicular to the main diagonal in a plan projection to prevent rigid body rotation.

Figure 2.9 shows 1/8 of the structure. It is plotted using PLOT3D.FOR with the appropriate analysis data file TR3D23.DAT. The plot shows 20 members instead of 12 and this is because an additional 8 "fictitious" members having areas of cross-section of 1,000,000 square inches were added to accommodate non fully fixed joints. The supports were established as follows: two fictitious bars are introduced at the circumferential nodes 4 and 5 in the vertical direction, two bars accommodate the original restraints of the central node, three fictitious bars perpendicular to the main diagonal in the plan projection restrain movement in that direction (because of symmetry) at nodes 2, 3, and 4, and finally one

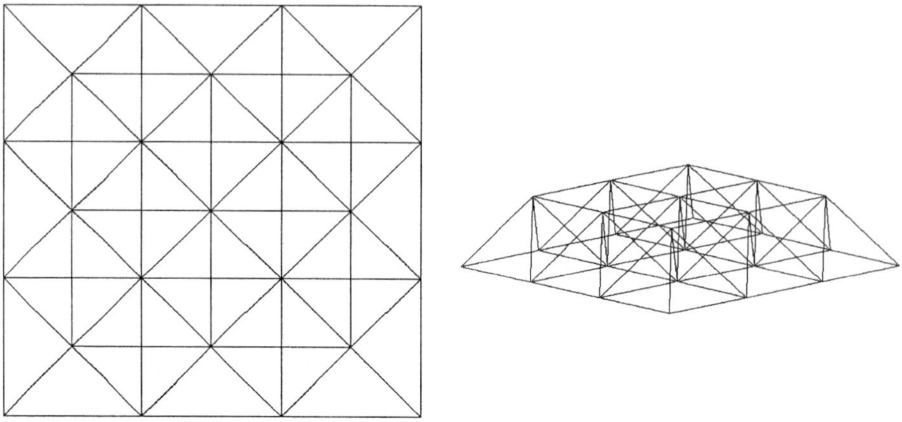

Figure 2.8 Plan and schematic views of the 72-bar double layered grid.

fictitious bar was added to node 6 in the x-direction (symmetry). Two fixed supports are added to the "cut" bars 3 and 10. Note that the areas of cross-section are halved for members 1,2,4 and 5 and the resulting forces should be doubled in these members when considering the full grid. The load is 1/8 that of the full load i.e. 1691.75 lbs.The output file TR3D23.OUT lists the analysis results in the now familiar format.

Data file TR3D23.DAT

```
20    16    10
      0.          0.          38.1834                        1691.75
      27.         27.
      54.         54.         38.1834
```

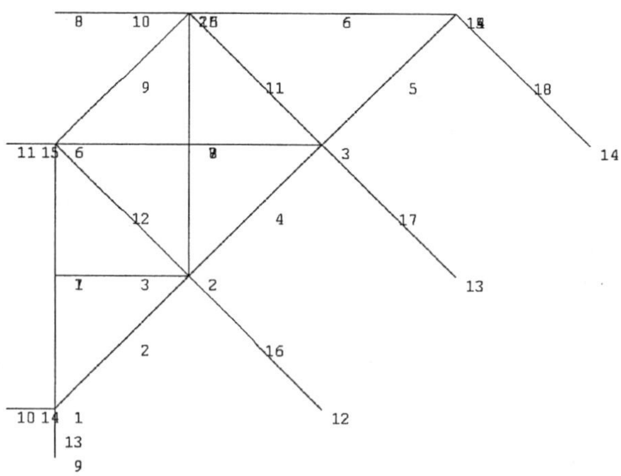

Figure 2.9 Node map of 1/8 of the double layered grid.

```
         81.         81.
         27.         81.
                     54.         38.1834
                     27.
                     81.
                     -10.        38.1834
         -10.                    38.1834
         -10.        54.         38.1834
         54.
         81.         27.         38.1834
         108.        54.
         81.         81.         -10.
         27.         81.         -10.
     1    6         .11065e+00
     1    2         .11065e+00
     7    2         .2213
     2    3         .11065e+00
     3    4         .11065e+00
     5    4         .2213
     6    3         .2213
     2    5         .2213
     6    5         .2213
     8    5         .2213
     3    5         .2213
     2    6         .2213
     1    9          10.e5
     1   10          10.e5
     6   11          10.e5
     2   12          10.e5
     3   13          10.e5
     4   14          10.e5
     4   15          10.e5
     5   16          10.e5
```

Output file TR3D23.OUT

```
  20  NO. MEMBERS
  16  NO. NODES
  10  NO. SUPPORTS

COORDINATES
                        X                    Y                    Z
  1         .00000000D+00        .00000000D+00        .38183400D+02
  2         .27000000D+02        .27000000D+02        .00000000D+00
  3         .54000000D+02        .54000000D+02        .38183400D+02
  4         .81000000D+02        .81000000D+02        .00000000D+00
  5         .27000000D+02        .81000000D+02        .00000000D+00
  6         .00000000D+00        .54000000D+02        .38183400D+02
  7         .00000000D+00        .27000000D+02        .00000000D+00
  8         .00000000D+00        .81000000D+02        .00000000D+00
  9         .00000000D+00       -.10000000D+02        .38183400D+02
 10        -.10000000D+02        .00000000D+00        .38183400D+02
 11        -.10000000D+02        .54000000D+02        .38183400D+02
 12         .54000000D+02        .00000000D+00        .00000000D+00
 13         .81000000D+02        .27000000D+02        .38183400D+02
 14         .10800000D+03        .54000000D+02        .00000000D+00
 15         .81000000D+02        .81000000D+02       -.10000000D+02
 16         .27000000D+02        .81000000D+02       -.10000000D+02

LOADS
                       PX                   PY                   PZ
  1         .00000000D+00        .00000000D+00        .16917500D+04
  2         .00000000D+00        .00000000D+00        .00000000D+00
  3         .00000000D+00        .00000000D+00        .00000000D+00
```

4	.00000000D+00	.00000000D+00	.00000000D+00
5	.00000000D+00	.00000000D+00	.00000000D+00
6	.00000000D+00	.00000000D+00	.00000000D+00
7	.00000000D+00	.00000000D+00	.00000000D+00
8	.00000000D+00	.00000000D+00	.00000000D+00
9	.00000000D+00	.00000000D+00	.00000000D+00
10	.00000000D+00	.00000000D+00	.00000000D+00
11	.00000000D+00	.00000000D+00	.00000000D+00
12	.00000000D+00	.00000000D+00	.00000000D+00
13	.00000000D+00	.00000000D+00	.00000000D+00
14	.00000000D+00	.00000000D+00	.00000000D+00
15	.00000000D+00	.00000000D+00	.00000000D+00
16	.00000000D+00	.00000000D+00	.00000000D+00

MEMBER	+ END	- END	AREA
1	1	6	.11065000E+00
2	1	2	.11065000E+00
3	7	2	.22130000E+00
4	2	3	.11065000E+00
5	3	4	.11065000E+00
6	5	4	.22130000E+00
7	6	3	.22130000E+00
8	2	5	.22130000E+00
9	6	5	.22130000E+00
10	8	5	.22130000E+00
11	3	5	.22130000E+00
12	2	6	.22130000E+00
13	1	9	.10000000E+07
14	1	10	.10000000E+07
15	6	11	.10000000E+07
16	2	12	.10000000E+07
17	3	13	.10000000E+07
18	4	14	.10000000E+07
19	4	15	.10000000E+07
20	5	16	.10000000E+07

DISPLACEMENTS

	X	Y	Z
1	.39875312D-09	.87725703D-09	.16718938D+00
2	-.18000648D-01	-.18000657D-01	.86690986D-01
3	.97300728D-02	.97300753D-02	.25456898D-01
4	-.43785307D-02	-.43785310D-02	.56391543D-10
5	-.24325188D-02	-.26757731D-01	.50752512D-09
6	.39875279D-09	.23352202D-01	.53665901D-01

MEMBER	DL	FORCE	STRESS
1	.23352201D-01	.14355117D+04	.12973445D+05
2	.38919949D-01	.23925073D+04	.21622297D+05
3	-.18000648D-01	-.44261595D+04	-.20000720D+05
4	-.15567972D-01	-.95700244D+03	-.86489147D+04
5	.38919864D-02	.23925020D+03	.21622250D+04
6	-.19460118D-02	-.23925135D+03	-.10811177D+04
7	.97300724D-02	.11962584D+04	.54055958D+04
8	-.87570745D-02	-.10766337D+04	-.48650414D+04
9	.11675988D-01	.14355048D+04	.64866913D+04
10	-.24325188D-02	-.59812937D+03	-.27027987D+04
11	.58379927D-02	.71775224D+03	.32433449D+04
12	-.11675988D-01	-.14355048D+04	-.64866913D+04
13	.87725703D-09	.26317711D+04	.26317711D-02
14	.39875312D-09	.11962594D+04	.11962594D-02
15	.39875279D-09	.11962584D+04	.11962584D-02
16	-.62444744D-08	-.49061224D+04	-.49061224D-02
17	.17226127D-08	.13534123D+04	.13534123D-02
18	-.21532622D-09	-.16917625D+03	-.16917625D-03
19	.56391543D-10	.16917463D+03	.16917463D-03

```
        20        .50752512D-09        .15225754D+04        .15225754D-02
Stop - Program terminated.
```

2.5 Problems

1. Show that Eq. 2.4 ($\mathbf{N}^{\mathrm{T}}\mathbf{F} = \mathbf{P}$) reproduces the node equilibrium in Eq. 2.3 for each node in the system.

2. Show that Eq. 2.7 ($\boldsymbol{\Delta} = \mathbf{N}\boldsymbol{\delta}$) reproduces Eq. 2.6 for each bar in the system.

3. Modify TR2D.FOR to accommodate prestress through "lack of fit". (Hint: use $F_i = K_i(\Delta_i - D_i)$ instead of $F_i = K_i \Delta_i$ for the constitutive relation. Here D_i is the "lack of fit" of member *i* in units of length.)

4. Comment on the use of the above modified program to accommodate thermal changes in truss members. (Hint: To what extent are "lack of fit" problems equivalent to thermal stress problems?)

5. Member 2 of Example 2.1 was manufactured 0.001667 inches longer than necessary. Use the modified program of problem 1 to analyze this truss. (For a discussion on use of "lack of fit" in the optimal design of trusses see Levy, 1989)

6. If the allowable stress is 10,000 psi, what percentage savings in material does prestress introduce if all members retain equal cross sectional areas?

CHAPTER 3

"EXACT" ANALYSIS OF TRUSSES

3.1 Introduction

A structure must satisfy the conditions of equilibrium (deformed equilibrium in the nonlinear case), the pieces of the structure must fit together in their loaded state and a constitutive equation must be satisfied. In this chapter all these conditions are met for the truss. The only restriction here is that the truss bars or elements be linear springs whose forces are proportional to their true length change.

This exact analysis of trusses is a nonlinear problem and therefore some iterative analysis procedure can be expected. (Actually, Newton's method is used). A typical step of this analysis can be described as follows. Given a fixed joint load matrix **P** and a starting configuration which is not in equilibrium with this joint load matrix (If it were there would be no analysis to perform.), the following sequence of actions must be taken:

- Compute the unbalanced load **P'**. Since the member forces **F** are not in equilibrium with the given load **P**, the unbalanced load can be computed as $\mathbf{P'} = \mathbf{P} - \mathbf{N}^{T}\mathbf{F}$.
- Solve for the incremental displacements. Under the unbalanced load **P'** the structure will displace. This computation involves solving the system $(\mathbf{K}_E + \mathbf{K}_G)\boldsymbol{\delta} = \mathbf{P'}$ for the node displacement matrix $\boldsymbol{\delta}$. (The geometric stiffness matrix $\mathbf{K}_G$ will be discussed below).
- Update the coordinates. The coordinate matrix **R** is updated to describe the deformed shape as $\mathbf{R} \rightarrow \mathbf{R} + \boldsymbol{\delta}$.
- Compute new member forces. The new coordinates of the structure imply new member lengths which in turn imply new member forces. The "exact" length change will be used here rather than the approximate expression $\boldsymbol{\Delta} = \mathbf{N}\boldsymbol{\delta}$.
- Repeat this sequence of calculations. Computation stops when the unbalanced load $\mathbf{P'} \rightarrow \mathbf{0}$. It will be argued below that these calculations comprise the application of Newton's method to the equations of node equilibrium.

3.2 Linearization of the Joint Equilibrium Equations

The above analysis is based on the node equilibrium equations which will be written here as $\mathbf{N}^{\mathrm{T}}\mathbf{F} = \mathbf{P}$. While this is the same as the form used for equilibrium in linear structural analysis, here it is assumed that $\mathbf{N}^{\mathrm{T}}\mathbf{F} = \mathbf{P}$. describes deformed equilibrium. Under a perturbation this equation takes the form,

$$\mathbf{N}^{\mathrm{T}}\mathbf{F} = \mathbf{P} \Rightarrow \mathrm{d}\mathbf{N}^{\mathrm{T}}\mathbf{F} + \mathbf{N}^{\mathrm{T}}\mathrm{d}\mathbf{F} = \mathrm{d}\mathbf{P} \tag{3.1}$$

It is assumed that all variables can be described in terms of the joint coordinate matrix **R** and its perturbation, the node displacement matrix $\boldsymbol{\delta}$. In Eq. 3.1 the term $\mathbf{N}^{\mathrm{T}}\mathrm{d}\mathbf{F}$ describes (under the chain rule of partial differentiation) the change in member forces with the matrix **N** fixed which is simply linear elastic analysis. That is,

$$\mathbf{N}^{\mathrm{T}}\mathrm{d}\mathbf{F} \rightarrow \mathbf{K}_{\mathrm{E}}\boldsymbol{\delta} \tag{3.2}$$

It is argued here that the variation of the member forces, d**F**, with the equilibrium "operator", **N**, fixed simply returns linear elastic theory. It is then the term $\mathrm{d}\mathbf{N}^{\mathrm{T}}\mathbf{F}$ of Eq. 3.1 which gives rise to the geometric stiffness matrix $\mathbf{K}_{\mathrm{G}}$, i.e.

$$\mathrm{d}(\mathbf{N}^{\mathrm{T}}\mathbf{F})_{\mathrm{F\,fixed}} \Rightarrow \mathbf{K}_{\mathrm{G}}\boldsymbol{\delta} \tag{3.3}$$

How the matrix $\mathbf{K}_{\mathrm{G}}$ can be computed as a gradient will now be explained. This argument is based on the fact that a small variation **df** of a function **f** can be represented as

$$\mathbf{df} \sim \nabla\mathbf{f} \cdot \mathbf{dx} \tag{3.4}$$

$\nabla\mathbf{f}$ is of course the gradient matrix of the function **f**. A vector, **f**, of n functions with m variables has an $n{\times}m$ gradient matrix as indicated in Eq. 3.5 below.

$$\mathbf{f} = \begin{Bmatrix} f_1 \\ f_2 \\ \vdots \\ \\ f_n \end{Bmatrix}; \quad \mathbf{x} = \begin{Bmatrix} x_1 \\ x_2 \\ \vdots \\ \\ x_m \end{Bmatrix}; \quad \nabla\mathbf{f} = \begin{bmatrix} \partial f_1/\partial x_1 & \partial f_1/\partial x_2 & \cdots & \partial f_1/\partial x_m \\ \partial f_2/\partial x_1 & \partial f_2/\partial x_2 & \cdots & \partial f_2/\partial x_m \\ \vdots & \vdots & & \vdots \\ \\ \partial f_n/\partial x_1 & \partial f_n/\partial x_2 & \cdots & \partial f_n/\partial x_m \end{bmatrix} \tag{3.5}$$

3.3 The Geometric Stiffness Matrix

In order to complete the details of the derivation of the geometric stiffness matrix as a gradient, it is convenient to work with $(\mathbf{N}^{\mathrm{T}}\mathbf{F})_i$, the contribution of bar i to the equilibrium equations. More explicitly, this term is shown below for the typical bar of Fig. 2.3 as

$$\left(\mathbf{N}^{\mathrm{T}}\mathbf{F}\right)_i = \begin{Bmatrix} \vdots \\ (\mathbf{n}_i)_x \\ (\mathbf{n}_i)_y \\ (\mathbf{n}_i)_z \\ \vdots \\ -(\mathbf{n}_i)_x \\ -(\mathbf{n}_i)_y \\ -(\mathbf{n}_i)_z \\ \vdots \end{Bmatrix} F_i \quad \begin{matrix} \\ \\ -\text{rowA} \\ \\ \\ \\ -\text{rowC} \\ \\ \\ \end{matrix} \tag{3.6}$$

It is the gradient of Eq. 3.6 which gives the contribution of bar i to the geometric stiffness matrix. Referring back to the original discussion of Newton's method, for the case of the equilibrium equations discussed here,

$$\begin{aligned} \mathbf{f} &\rightarrow (\mathbf{N}^{\mathrm{T}}\mathbf{F})_i = \mathbf{N}_i^{\mathrm{T}} F_i \\ \mathbf{x} &\rightarrow \mathbf{R} \\ \mathbf{dx} &\rightarrow \boldsymbol{\delta} \end{aligned} \tag{3.7}$$

and

$$\nabla \mathbf{f} \;\rightarrow\; \nabla\left(\mathbf{N}^{\mathrm{T}}\mathbf{F}\right)_i \;=\; \left(\nabla \mathbf{N}_i^{\mathrm{T}}\right)F_i \tag{3.8}$$

Now since $\mathbf{N}_i^{\mathrm{T}}$ is a function of the coordinates $x_{\mathrm{A}}, y_{\mathrm{A}}, z_{\mathrm{A}}$, and $x_{\mathrm{C}}, y_{\mathrm{C}}, z_{\mathrm{C}}$, chain rule of differentiation yields,

$$\mathrm{d}\left(\mathbf{N}_i^{\mathrm{T}}\mathbf{F}\right)_i = \left(\left(\nabla \mathbf{N}_i^{\mathrm{T}}\right)\cdot \mathrm{d}x\right)F_i =$$

$$\left(\frac{\partial \mathbf{N}_i^{\mathrm{T}}}{\partial x_{\mathrm{A}}}\mathrm{d}x_{\mathrm{A}} + \frac{\partial \mathbf{N}_i^{\mathrm{T}}}{\partial y_{\mathrm{A}}}\mathrm{d}y_{\mathrm{A}} + \frac{\partial \mathbf{N}_i^{\mathrm{T}}}{\partial z_{\mathrm{A}}}\mathrm{d}z_{\mathrm{A}} + \frac{\partial \mathbf{N}_i^{\mathrm{T}}}{\partial x_{\mathrm{C}}}\mathrm{d}x_{\mathrm{C}} + \frac{\partial \mathbf{N}_i^{\mathrm{T}}}{\partial y_{\mathrm{C}}}\mathrm{d}y_{\mathrm{C}} + \frac{\partial \mathbf{N}_i^{\mathrm{T}}}{\partial z_{\mathrm{C}}}\mathrm{d}z_{\mathrm{C}}\right)F_i \tag{3.9}$$

With $\mathrm{d}x_{\mathrm{A}} = (\boldsymbol{\delta}_{\mathrm{A}})_x$, $\mathrm{d}y_{\mathrm{A}} = (\boldsymbol{\delta}_{\mathrm{A}})_y$, $\mathrm{d}z_{\mathrm{A}} = (\boldsymbol{\delta}_{\mathrm{A}})_z$, $\mathrm{d}x_{\mathrm{C}} = (\boldsymbol{\delta}_{\mathrm{C}})_x$, $\mathrm{d}y_{\mathrm{C}} = (\boldsymbol{\delta}_{\mathrm{C}})_y$, and $\mathrm{d}z_{\mathrm{C}} = (\boldsymbol{\delta}_{\mathrm{C}})_z$, i.e. the changes in coordinates are interpreted as nodal displacements, the contribution of member i to the geometric stiffness matrix becomes

$$(\mathbf{K}_{\mathrm{G}})_i = \nabla(\mathbf{N}_i^{\mathrm{T}}\mathbf{F})_i = \begin{array}{c} \begin{array}{cc} \text{col A} & \text{col C} \end{array} \\ \begin{bmatrix} \nabla(\mathbf{N}_i^{\mathrm{T}})^{\mathrm{AA}} & \nabla(\mathbf{N}_i^{\mathrm{T}})^{\mathrm{AC}} \\ \nabla(\mathbf{N}_i^{\mathrm{T}})^{\mathrm{CA}} & \nabla(\mathbf{N}_i^{\mathrm{T}})^{\mathrm{CC}} \end{bmatrix} F_i \end{array} \begin{array}{c} \\ \text{row A} \\ \text{row C} \end{array} \tag{3.10}$$

where

$$\nabla\left(N_i^T\right)^{\mathrm{AA}} = \begin{bmatrix} \dfrac{\partial(\mathbf{n}_i)_x}{\partial x_{\mathrm{A}}} & \dfrac{\partial(\mathbf{n}_i)_x}{\partial y_{\mathrm{A}}} & \dfrac{\partial(\mathbf{n}_i)_x}{\partial z_{\mathrm{A}}} \\ \dfrac{\partial(\mathbf{n}_i)_y}{\partial x_{\mathrm{A}}} & \dfrac{\partial(\mathbf{n}_i)_y}{\partial y_{\mathrm{A}}} & \dfrac{\partial(\mathbf{n}_i)_y}{\partial z_{\mathrm{A}}} \\ \dfrac{\partial(\mathbf{n}_i)_z}{\partial x_{\mathrm{A}}} & \dfrac{\partial(\mathbf{n}_i)_z}{\partial y_{\mathrm{A}}} & \dfrac{\partial(\mathbf{n}_i)_z}{\partial z_{\mathrm{A}}} \end{bmatrix}$$

$$\nabla\left(\mathbf{N}_i^{\mathrm{T}}\right)^{\mathrm{AC}} = \begin{bmatrix} \dfrac{\partial(\mathbf{n}_i)_x}{\partial x_{\mathrm{C}}} & \dfrac{\partial(\mathbf{n}_i)_x}{\partial y_{\mathrm{C}}} & \dfrac{\partial(\mathbf{n}_i)_x}{\partial z_{\mathrm{C}}} \\ \dfrac{\partial(\mathbf{n}_i)_y}{\partial x_{\mathrm{C}}} & \dfrac{\partial(\mathbf{n}_i)_y}{\partial y_{\mathrm{C}}} & \dfrac{\partial(\mathbf{n}_i)_y}{\partial z_{\mathrm{C}}} \\ \dfrac{\partial(\mathbf{n}_i)_z}{\partial x_{\mathrm{C}}} & \dfrac{\partial(\mathbf{n}_i)_z}{\partial y_{\mathrm{C}}} & \dfrac{\partial(\mathbf{n}_i)_z}{\partial z_{\mathrm{C}}} \end{bmatrix}$$

$$\nabla\left(\mathbf{N}_i^{\mathrm{T}}\right)^{\mathrm{CC}} = \begin{bmatrix} -\dfrac{\partial(\mathbf{n}_i)_x}{\partial x_{\mathrm{C}}} & -\dfrac{\partial(\mathbf{n}_i)_x}{\partial y_{\mathrm{C}}} & -\dfrac{\partial(\mathbf{n}_i)_x}{\partial z_{\mathrm{C}}} \\ -\dfrac{\partial(\mathbf{n}_i)_y}{\partial x_{\mathrm{C}}} & -\dfrac{\partial(\mathbf{n}_i)_y}{\partial y_{\mathrm{C}}} & -\dfrac{\partial(\mathbf{n}_i)_y}{\partial z_{\mathrm{C}}} \\ -\dfrac{\partial(\mathbf{n}_i)_z}{\partial x_{\mathrm{C}}} & -\dfrac{\partial(\mathbf{n}_i)_z}{\partial y_{\mathrm{C}}} & -\dfrac{\partial(\mathbf{n}_i)_z}{\partial z_{\mathrm{C}}} \end{bmatrix}$$

and

$$\nabla\left(\mathbf{N}_i^{\mathrm{T}}\right)^{\mathrm{CA}} = \begin{bmatrix} -\dfrac{\partial(\mathbf{n}_i)_x}{\partial x_{\mathrm{A}}} & -\dfrac{\partial(\mathbf{n}_i)_x}{\partial y_{\mathrm{A}}} & -\dfrac{\partial(\mathbf{n}_i)_x}{\partial z_{\mathrm{A}}} \\ -\dfrac{\partial(\mathbf{n}_i)_y}{\partial x_{\mathrm{A}}} & -\dfrac{\partial(\mathbf{n}_i)_y}{\partial y_{\mathrm{A}}} & -\dfrac{\partial(\mathbf{n}_i)_y}{\partial z_{\mathrm{A}}} \\ -\dfrac{\partial(\mathbf{n}_i)_z}{\partial x_{\mathrm{A}}} & -\dfrac{\partial(\mathbf{n}_i)_z}{\partial y_{\mathrm{A}}} & -\dfrac{\partial(\mathbf{n}_i)_z}{\partial z_{\mathrm{A}}} \end{bmatrix}$$

The derivatives indicated in Eq. 3.10 can now be computed completing the description of the geometric stiffness matrix. Symbolically,

$$\begin{matrix} & \text{col A} & \text{col C} & \\ (\mathbf{K}_{\mathrm{G}})_i = & \left[\begin{matrix} (\mathbf{K}_{\mathrm{G}})_i^{\mathrm{AA}} \\ \\ -(\mathbf{K}_{\mathrm{G}})_i^{\mathrm{AA}} \end{matrix}\right. & \left.\begin{matrix} -(\mathbf{K}_{\mathrm{G}})_i^{\mathrm{AA}} \\ \\ (\mathbf{K}_{\mathrm{G}})_i^{\mathrm{AA}} \end{matrix}\right] & \begin{matrix} \text{row A} \\ \\ \text{row C} \end{matrix} \end{matrix} \tag{3.11}$$

with

$$(\mathbf{K}_G)_i^{AA} = \frac{F_i}{L_i}\begin{bmatrix} 1-(\mathbf{n}_i)_x^2 & -(\mathbf{n}_i)_x(\mathbf{n}_i)_y & -(\mathbf{n}_i)_x(\mathbf{n}_i)_z \\ -(\mathbf{n}_i)_y(\mathbf{n}_i)_x & 1-(\mathbf{n}_i)_y^{\,2} & -(\mathbf{n}_i)_y(\mathbf{n}_i)_z \\ -(\mathbf{n}_i)_z(\mathbf{n}_i)_x & -(\mathbf{n}_i)_z(\mathbf{n}_i)_y & 1-(\mathbf{n}_i)_z^{\,2} \end{bmatrix} \tag{3.12}$$

or

$$(\mathbf{K}_G)_i^{AA} = \frac{F_i}{L_i}\left(\mathbf{I} - \mathbf{n}_i\mathbf{n}_i^T\right) \tag{3.13}$$

where

$$\mathbf{n}_i = \left\{(\mathbf{n}_i)_x \quad (\mathbf{n}_i)_y \quad (\mathbf{n}_i)_z\right\}^T \tag{3.14}$$

Equation 3.12 follows directly from differentiation as will be shown below. It is appropriate to mention though that the geometric matrix of Eq 3.11 is identical to that of Eq 1.19 which was derived via a geometric observation concerning string effects. It is also identical to that of Eq. 1.44 which was derived via the fundamental theorem of prestress.

Consider the x-component of the unit vector $\boldsymbol{n}_i$ when written in terms of nodal coordinates,

$$(\mathbf{n}_i)_x = \frac{x_A - x_C}{\sqrt{(x_A - x_C)^2 + (y_A - y_C)^2 + (z_A - z_C)^2}} = \frac{x_A - x_C}{L_i} \tag{3.15}$$

Differentiation yields,

$$\frac{\partial(\mathbf{n}_i)_x}{\partial x_A} = \frac{1}{L_i} - \frac{1}{2}\frac{x_A - x_C}{L_i^3}2(x_A - x_C) = \frac{1}{L_i}\left[1-(\mathbf{n}_i)_x^2\right] \tag{3.16}$$

$$\frac{\partial(\mathbf{n}_i)_x}{\partial y_A} = -\frac{1}{2}\frac{x_A - x_C}{L_i^{\,3}}2(y_A - y_C) = -\frac{1}{L_i}(\mathbf{n}_i)_x(\mathbf{n}_i)_y \tag{3.17}$$

$$\frac{\partial(\mathbf{n}_i)_x}{\partial z_A} = -\frac{1}{2}\frac{x_A - x_C}{L_i^{\,3}}2(z_A - z_C) = -\frac{1}{L_i}(\mathbf{n}_i)_x(\mathbf{n}_i)_z \tag{3.18}$$

The derivatives of $(\mathbf{n}_i)_y$ and $(\mathbf{n}_i)_z$ with respect to ∂x_{A}, ∂y_{A} and ∂z_{A} follow in a similar fashion given the form of Eq. 3.15. Derivatives with respect to the coordinates at end C are the negative of derivatives with respect to the coordinates at end A.

3.4 Overall Buckling

The approach to buckling used in this book is one in which load is incremented and the determinant of the system matrix, $|\mathbf{K}_{\mathrm{E}} + \mathbf{K}_{\mathrm{G}}|$, is monitored. When this determinant goes to zero the structure is said to be unstable. (See Appendix 2.)

The relationship between the physical phenomenon of buckling (collapse) and the formal concept of the determinant becoming zero is non-trivial except perhaps in the scalar case shown in Fig. 3.1. In this case it seems clear that when the load displacement curve becomes nearly flat, a "small" increment of load will produce a "large" increment of displacement. But even in this simple case it is not common to follow the phenomenon to collapse which is what bucking implies. (Perhaps the structure will "snap-through" and recover its stability.)

There are several ways to argue the relationship of the determinant of the system matrix and buckling. It is most simple to argue that the determinant is the generalization of the slope of the curve in the scalar case above to the case of n dimensions. A more formal approach is through the idea of a "condition number" (Isaacson and Keller, 1966). Roughly a system **Ax**=**f** can be considered in its perturbed state as **(A**+d**A)(x**+d**x)**=**f**+d**f**. In a well-posed problem, a small perturbation of the system parameters produces a small perturbation of the state of the system. The condition number μ,

$$\mu = \left\|\mathbf{A}^{-1}\right\| \cdot \left\|\mathbf{A}\right\|$$

can be used to formalize the idea of an ill-conditioned system. It can be shown that when the condition number is not too large, bounded d**f** and d**A** imply bounded d**x**.

There are approaches to buckling other than incrementing the load on a structure and monitoring its behavior (the value of the determinant) but these other approaches are argued here to be less accurate. One such approach involves reducing buckling to an eigenvalue problem. This approach was briefly discussed in Chapter 1 and termed "linear buckling". For the case of the truss linear (eigenvalue) buckling is particularly simple to discuss.

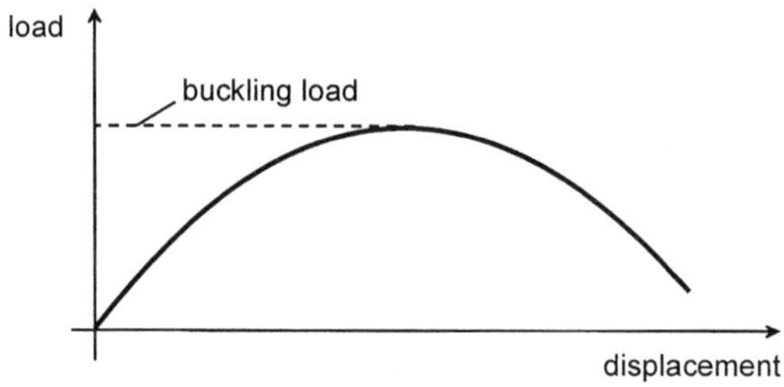

Figure 3.1 Load deflection curve showing buckling.

For trusses as has just been indicated, the geometric stiffness matrix is linear in the member forces. For example, if $\mathbf{K}_G(\mathbf{F}_1)$ is the geometric stiffness matrix corresponding to some load $\mathbf{P}_1$, then $\beta\mathbf{K}_G(\mathbf{F}_1)$ is the geometric stiffness matrix which corresponds to the load $\beta\mathbf{P}_1$. It is argued that, at buckling, for a given load $\overline{\mathbf{P}}$ and corresponding buckling load, $\mathbf{P}_{CR} = \beta\overline{\mathbf{P}}$ there are many deformed equilibrium positions. If that is true,

$$(\mathbf{K}_E + \mathbf{K}_G)\boldsymbol{\delta}_1 = \mathbf{P}_{CR} \tag{3.19}$$

and

$$(\mathbf{K}_E + \mathbf{K}_G)\boldsymbol{\delta}_2 = \mathbf{P}_{CR} \tag{3.20}$$

which implies that there exists a $\boldsymbol{\delta}$ such that

$$(\mathbf{K}_E + \mathbf{K}_G)\boldsymbol{\delta} = \mathbf{0} \tag{3.21}$$

or

$$(\mathbf{K}_E + \beta\overline{\mathbf{K}}_G)\boldsymbol{\delta} = \mathbf{0} \tag{3.22}$$

which implies a generalized eigenvalue problem. That is, the given load $\overline{\mathbf{P}}$ implies $\overline{\mathbf{K}}_G$ and buckling becomes an eigenvalue problem for the unknowns β and $\boldsymbol{\delta}$. Clearly, β itself can be computed from the determinantal equation

$$\left|\mathbf{K}_E + \beta\overline{\mathbf{K}}_G\right| = \mathbf{0} \tag{3.23}$$

This type of analysis depends heavily upon the initial configuration which is used to construct both $\mathbf{K}_E$ and $\mathbf{K}_G$. Since incrementing the loads to find the buckling load does not depend upon the initial configuration (this relates also to the question of "imperfection sensitivity") in this manner, it is argued that doing so is superior to the eigenvalue approach. There are other arguments against the eigenvalue approach which have been presented in section 1.5.6 where it was shown that the two approaches to buckling can lead to quite different results.

3.5 Computer Programs

This section presents four computer programs which are direct extensions of the linear analysis programs P1-TR3D.FOR and P2-TR2D.FOR that are discussed in Chapter 2. The first two programs perform nonlinear analysis and nonlinear buckling for two and three dimensional trusses. The other two programs determine linear buckling loads using the eigenvalue approach.

3.5.1 Programs P3-TR3DNL.FOR and P4-TR2DNL.FOR

These two powerful nonlinear analysis programs are iterative but converging rapidly. They signal buckling as well and indeed achieve critical loads if fine tuned to desired accuracy.

A brief comment on "exact" buckling loads. The procedures described below involve incrementing the load and monitoring the determinant of the system matrix. When this determinant becomes negative the buckling load has been exceeded. Clearly, smaller steps will give a closer bound on the buckling load.

There are two reasons for not attempting extremely small load steps. From an engineering point of view it is typically unnecessary to have an "exact" buckling load; from a computational point of view, as you get closer and closer to the buckling load the system becomes more and more singular and convergence becomes more and more difficult.

Following are some of the steps that have to be taken to go from linear analysis to nonlinear analysis:

• *Assembling* the geometric stiffness matrix, $\mathbf{K}_G$.

• *Iterating*. In order to solve a nonlinear system it is typically necessary to use some type of iterative method. (Here Newton's method is used.) In the programs this iteration is described by the parameter `ITER`. These are iterations performed with fixed load. Another parameter LDSTP describes the load incrementing. This is first of all required for buckling problems where the buckling load is not known and some load range which includes it is to be specified. Alternatively as a practical matter, nonlinear problems are sometimes sensitive to the manner in which load is applied. It is not uncommon for an analysis to diverge if the load is applied in one step but to converge when it is applied in increments. (This relates to the convergence of Newton's method which is most commonly shown to be local. That is, Newton's method can be shown to converge when you start near the answer.)

• *Saving data.* Since more than one analysis is to be performed, some data which would ordinarily be discarded must be kept. This is true for the joint loads PSAVE and the member forces FSAVE.

• *Computing the unbalanced load.* It is the unbalanced load rather than the total load which drives incremental analysis. The unbalanced load is computed here during the construction of the system matrix where the appropriate data is available.

• *Computing Error.* The error is measured here by computing the "length" of the unbalanced load $\mathbf{P}'$. That is the error is defined to be $\sqrt{\sum\left(P_i'\right)^2}$.

• *True length.* When member forces are updated their true length change is used rather than the approximate length change of linear structures.

• *Buckling.* The linear equation solver is set up so that the determinant can be monitored by simply looking at its diagonal terms during the process of Gaussian elimination. A zero or negative term along the diagonal generates a message in the output.

• *Computer output.* Full output for linear analysis is collected from each step of Newton's method and each load increment. This is generally too much output but the selection of output has been left to the reader who may wish to "scan" the output file. It is for example sometimes useful to scan the output file for the term ERROR to examine convergence of the algorithm. (It would have been better to plot this error and display it at as the program runs.) It can also be useful to scan the output file for the term NEG when concerned with buckling.

3.5.1.1 *Nonlinear Analysis and Newton's Method*

The iterative process for the nonlinear analysis of trusses is basically of the form

$$\left(\mathbf{K}_{\mathrm{E}}+\mathbf{K}_{\mathrm{G}}(\mathbf{F}^{(n)})\right)\boldsymbol{\delta}^{(n+1)}=\mathbf{P}-\mathbf{N}^{\mathrm{T}}\mathbf{F}^{(n)} \tag{3.24}$$

Letting $\mathbf{f}(\mathbf{x}^{(n)})=\mathbf{P}-\mathbf{N}^{\mathrm{T}}\mathbf{F}^{(n)}$ and $d\mathbf{x}=\boldsymbol{\delta}^{(n+1)}$ Newton's method of solving a nonlinear system implies

$$d\mathbf{x}=-\left[\nabla\mathbf{f}(\mathbf{x}^{(n)})\right]^{-1}\cdot\mathbf{f}(\mathbf{x})$$

which becomes

$$\boldsymbol{\delta}^{(n+1)}=-\left[\nabla(\mathbf{P}-\mathbf{N}^{\mathrm{T}}\mathbf{F}^{(n)})\right]^{-1}\cdot(\mathbf{P}-\mathbf{N}^{\mathrm{T}}\mathbf{F}^{(n)}) \tag{3.25}$$

when applied to Eq. 3.24. For a direct correspondence between Eqs. 3.24 and 3.25 it remains to show that the gradient of the unbalanced load is the tangential stiffness matrix. This gradient is now written as

$$\nabla(\mathbf{P}-\mathbf{N}^{\mathrm{T}}\mathbf{F}^{(n)})=\nabla\mathbf{P}-\nabla(\mathbf{N}^{\mathrm{T}}\mathbf{F}^{(n)})$$

But $\nabla\mathbf{P}$ is zero since $\mathbf{P}$ is constant for that particular load step and it has been shown in the previous sections that the gradient of the equilibrium equations i.e. $\nabla(\mathbf{N}^{\mathrm{T}}\mathbf{F}^{(n)})$ reduces to the tangential stiffness matrix, $\left(\mathbf{K}_{\mathrm{E}}+\mathbf{K}_{\mathrm{G}}(\mathbf{F}^{(n)})\right)$ making the recursive procedure of Eq. 3.24 identical to Newton's method with its quadratic convergence properties.

3.5.2 Programs P5-BUCK3D.FOR and P6-BUCK2D.FOR

These two programs demonstrate the eigenvalue approach to truss buckling. For eigenvalue truss buckling it is necessary to solve the "generalized " eigenvalue problem

$$\left(\mathbf{K}_{\mathrm{E}}+\beta\overline{\mathbf{K}}_{\mathrm{G}}\right)\boldsymbol{\delta}=\mathbf{0} \tag{3.26}$$

These programs have been assembled informally for specific eigenvalue buckling problems. This has been done by taking parts from P3 and P4 which are used to generate $\mathbf{K}_E$ and $\overline{\mathbf{K}}_G$ and then calling a "canned" eigenvalue solver `NROOT` taken from the old IBM Scientific Subroutine Package. Since `NROOT` requires that $\mathbf{K}_E$ and $\overline{\mathbf{K}}_G$ be dimensioned exactly, some re-coding must be done each time a problem of a different size is run. (See also problem 7 at the end of the chapter.)

As can be seem from the listing of the code, this program starts with an elastic analysis program which is first used (`ITER=1`) to generate the member forces. As a second step (`ITER=2`) the stiffness matrices are generated and the eigenvalue routine called. It should be noted that `NROOT`, because of its structure generates $-1/\beta$ as eigenvalues. This implies that the critical buckling load is obtained by multiplying the given load by a factor of (-1)/(smallest eigenvalue).

3.6 Examples

In this section six problems have been selected to give the reader an overview of geometrically nonlinear analysis for trusses.

3.6.1 Example 3.1. Biot's 2-Bar Prestressed Truss

Nonlinear analysis is now performed on Biot's 2-bar prestressed truss of Figure 1.5 which is "linearly unstable". (See Figure 3.2 generated by PLOT3D.FOR.) It is, of course, the geometric stiffness matrix which makes the analysis possible.

The analysis is performed using TR2DNL.FOR on the data file TR2DNL31.DAT. Both the data file and a partial listing of the output are given below. The first line in the data file includes the number of load steps (10 in this case) and the number of iterations in each load step (5 in this case) in addition to the usual number of bars, number of nodes and number of supports. Node information is supplied in the same manner as for linear analysis whereas member information includes the prestress (in this case 1,000 lbs) for each member. Thus the center node of this truss is loaded downwards with 70 lbs and each member carries a prestress of 1,000 lbs for equal areas of cross section of 0.0127 square inches. The given program was edited and Young's modulus was compiled to be E=10,000,000 psi.

As for the output, the input data is first echoed and then for each iteration of each load step nodal displacements, member deformations,

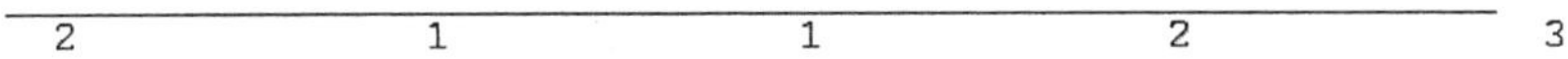

Figure 3.2 The linearly unstable 2-bar truss.

member forces and member stresses are provided in the same manner as in the linear analysis. Moreover, here the new deformed equilibrium position (UPDATED COORDINATES) and the resulting member forces at that deformed position (UPDATED FORCES) are also given. In this particular output set the last iteration (ITERATION 5)of the last load step (LOADSTEP 10) is listed. The equilibrium position was attained with an ERROR of .62413028D-11 which is basically zero error implying that the structure is in complete equilibrium with no unbalanced forces.

Data file TR2DNL31.DAT

```
2    3    2   10    5
        200.         00.        0.0          -70.
         00.          0.
        400.          0.
1    2              .0127           1000.0
1    3              .0127           1000.0
```

Output file TR2DNL31.OUT

```
  2  NO. MEMBERS
  3  NO. NODES
  2  NO. SUPPORTS
 10  NO. LOAD STEPS
  5  NO. ITERATIONS

                    COORDINATES                          LOADS
                      X                Y                  PX                PY
 1      .20000000D+03   .00000000D+00     .00000000D+00    -.70000000D+02
 2      .00000000D+00   .00000000D+00     .00000000D+00     .00000000D+00
 3      .40000000D+03   .00000000D+00     .00000000D+00     .00000000D+00

  MEMBER      + END      - END            AREA          PRESTRESS
       1          1          2     .12700000E-01    .10000000E+04
       2          1          3     .12700000E-01    .10000000E+04

****ITERATION NUMBER    5
    LOAD STEP          10

LOADSTEP  10;  ITERATION   5;  ERROR   .62413028D-11

DISPLACEMENTS
                               X                 Y
       1     -.42816502D-16     .51889630D-12

 MEMBER              DL             FORCE            STRESS  UPDATED FORCES
      1  -.17044482D-13   -.10817434D-10   -.85176650D-09    .10682100D+04
      2  -.16958895D-13   -.10763116D-10   -.84748945D-09    .10682100D+04
```

```
UPDATED COORDINATES
                              X                         Y
        1      .20000000D+03      -.65565399D+01
Stop - Program terminated.
```

3.6.2 Example 3.2. A Prestressed Cablenet

In this case nonlinear analysis is performed on a prestressed cable-net.

A 4×4 square grid with 6 ft. spans for a total of 24 ft is transversely loaded downwards with 10,000 lbs. Each cable has an area of cross section of 0.25 square inches and carries a prestressing force of 20,000 lbs. Figure 3.3 displays the net. It was plotted using PLOT3D.FOR and the analysis data file TR3DNL32.DAT.

The net is an extension of Example 3.1 to three dimensions and resembles a trampoline in its loading. The structure is linearly unstable and cannot be solved without the aid of the geometric stiffness matrix. The nonlinear analysis program TR3DNL.FOR is invoked with E = 30,000,000 psi to yield a converged solution at iteration 6 of load step 20. (These parameters are read in from the first line of the data file.) Listed below are the first few lines of output that echo the input as well as the results of the last iteration.

Output file TR3DNL32.OUT

```
  24  NO. MEMBERS
  25  NO. NODES
  16  NO. SUPPORTS
  20  NO. LOAD STEPS
   6  NO. ITERATIONS

COORDINATES
                          X                        Y                        Z
   1       .60000000D+02       .18000000D+03       .00000000D+00
   2       .12000000D+03       .18000000D+03       .00000000D+00
   3       .18000000D+03       .18000000D+03       .00000000D+00
   4       .60000000D+02       .12000000D+03       .00000000D+00
   5       .12000000D+03       .12000000D+03       .00000000D+00
   6       .18000000D+03       .12000000D+03       .00000000D+00
   7       .60000000D+02       .60000000D+02       .00000000D+00
   8       .12000000D+03       .60000000D+02       .00000000D+00
   9       .18000000D+03       .60000000D+02       .00000000D+00
  10       .00000000D+00       .24000000D+03       .00000000D+00
  11       .60000000D+02       .24000000D+03       .00000000D+00
  12       .12000000D+03       .24000000D+03       .00000000D+00
  13       .18000000D+03       .24000000D+03       .00000000D+00
  14       .24000000D+03       .24000000D+03       .00000000D+00
  15       .00000000D+00       .18000000D+03       .00000000D+00
  16       .00000000D+00       .12000000D+03       .00000000D+00
  17       .00000000D+00       .60000000D+02       .00000000D+00
  18       .00000000D+00       .00000000D+00       .00000000D+00
  19       .60000000D+02       .00000000D+00       .00000000D+00
  20       .12000000D+03       .00000000D+00       .00000000D+00
```

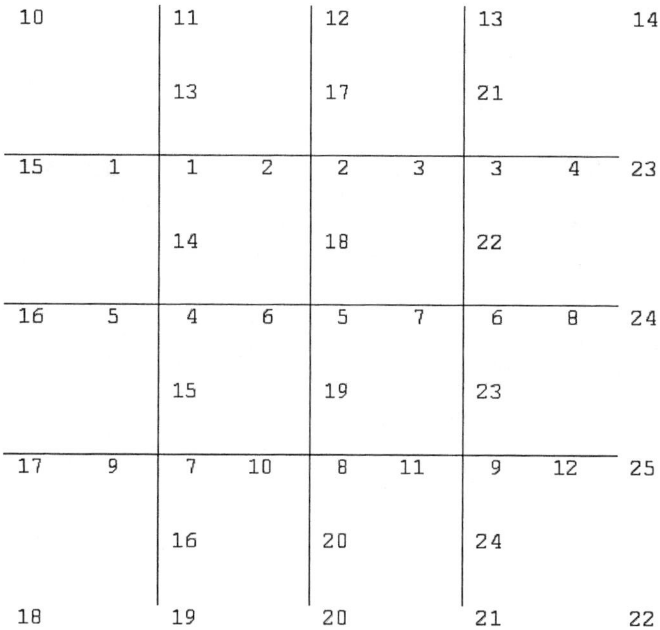

Figure 3.3 The transversely loaded flat cable-net.

21	.18000000D+03	.00000000D+00	.00000000D+00
22	.24000000D+03	.00000000D+00	.00000000D+00
23	.24000000D+03	.18000000D+03	.00000000D+00
24	.24000000D+03	.12000000D+03	.00000000D+00
25	.24000000D+03	.60000000D+02	.00000000D+00

LOADS

	PX	PY	PZ
1	.00000000D+00	.00000000D+00	.00000000D+00
2	.00000000D+00	.00000000D+00	.00000000D+00
3	.00000000D+00	.00000000D+00	.00000000D+00
4	.00000000D+00	.00000000D+00	.00000000D+00
5	.00000000D+00	.00000000D+00	-.10000000D+05
6	.00000000D+00	.00000000D+00	.00000000D+00
7	.00000000D+00	.00000000D+00	.00000000D+00
8	.00000000D+00	.00000000D+00	.00000000D+00
9	.00000000D+00	.00000000D+00	.00000000D+00
10	.00000000D+00	.00000000D+00	.00000000D+00
11	.00000000D+00	.00000000D+00	.00000000D+00
12	.00000000D+00	.00000000D+00	.00000000D+00
13	.00000000D+00	.00000000D+00	.00000000D+00
14	.00000000D+00	.00000000D+00	.00000000D+00
15	.00000000D+00	.00000000D+00	.00000000D+00
16	.00000000D+00	.00000000D+00	.00000000D+00
17	.00000000D+00	.00000000D+00	.00000000D+00
18	.00000000D+00	.00000000D+00	.00000000D+00
19	.00000000D+00	.00000000D+00	.00000000D+00
20	.00000000D+00	.00000000D+00	.00000000D+00
21	.00000000D+00	.00000000D+00	.00000000D+00
22	.00000000D+00	.00000000D+00	.00000000D+00

```
 23        .00000000D+00        .00000000D+00        .00000000D+00
 24        .00000000D+00        .00000000D+00        .00000000D+00
 25        .00000000D+00        .00000000D+00        .00000000D+00

   MEMBER     + END     - END            AREA       PRESTRESS
        1        15         1   .25000000E+00   .20000000E+05
        2         1         2   .25000000E+00   .20000000E+05
        3         2         3   .25000000E+00   .20000000E+05
        4         3        23   .25000000E+00   .20000000E+05
        5        16         4   .25000000E+00   .20000000E+05
        6         4         5   .25000000E+00   .20000000E+05
        7         5         6   .25000000E+00   .20000000E+05
        8         6        24   .25000000E+00   .20000000E+05
        9        17         7   .25000000E+00   .20000000E+05
       10         7         8   .25000000E+00   .20000000E+05
       11         8         9   .25000000E+00   .20000000E+05
       12         9        25   .25000000E+00   .20000000E+05
       13        11         1   .25000000E+00   .20000000E+05
       14         1         4   .25000000E+00   .20000000E+05
       15         4         7   .25000000E+00   .20000000E+05
       16         7        19   .25000000E+00   .20000000E+05
       17        12         2   .25000000E+00   .20000000E+05
       18         2         5   .25000000E+00   .20000000E+05
       19         5         8   .25000000E+00   .20000000E+05
       20         8        20   .25000000E+00   .20000000E+05
       21        13         3   .25000000E+00   .20000000E+05
       22         3         6   .25000000E+00   .20000000E+05
       23         6         9   .25000000E+00   .20000000E+05
       24         9        21   .25000000E+00   .20000000E+05

.
.
.

****ITERATION NUMBER   6
    LOAD STEP         20

ERROR =         .64700348D-02

DISPLACEMENTS
                           X                 Y                 Z
        1      .39541289D-08    -.39541284D-08    -.27070998D-09
        2      .24647582D-15    -.11617400D-07    -.44732138D-09
        3     -.39541284D-08    -.39541285D-08    -.27070975D-09
        4      .11617399D-07     .22924538D-15    -.44732168D-09
        5     -.21671971D-15    -.22742396D-15     .78463653D-09
        6     -.11617400D-07     .22923367D-15    -.44732142D-09
        7      .39541289D-08     .39541289D-08    -.27071047D-09
        8      .24649088D-15     .11617399D-07    -.44732183D-09
        9     -.39541284D-08     .39541289D-08    -.27071005D-09

 MEMBER              DL            FORCE            STRESS  UPDATED FORCES
      1   .39592112D-08    .49477617D-03    .19791047D-02    .21898010D+05
      2  -.39426830D-08   -.49271188D-03   -.19708475D-02    .21879540D+05
      3  -.39426830D-08   -.49271188D-03   -.19708475D-02    .21879540D+05
      4   .39592107D-08    .49477611D-03    .19791044D-02    .21898010D+05
      5   .11625776D-07    .14505332D-02    .58021330D-02    .33889930D+05
      6  -.11676549D-07   -.14568497D-02   -.58273989D-02    .33985020D+05
      7  -.11676549D-07   -.14568497D-02   -.58273989D-02    .33985020D+05
      8   .11625776D-07    .14505333D-02    .58021332D-02    .33889930D+05
      9   .39592112D-08    .49477617D-03    .19791047D-02    .21898010D+05
     10  -.39426830D-08   -.49271188D-03   -.19708475D-02    .21879540D+05
     11  -.39426830D-08   -.49271188D-03   -.19708475D-02    .21879540D+05
     12   .39592107D-08    .49477611D-03    .19791044D-02    .21898010D+05
     13   .39592107D-08    .49477611D-03    .19791044D-02    .21898010D+05
     14  -.39426830D-08   -.49271188D-03   -.19708475D-02    .21879540D+05
     15  -.39426830D-08   -.49271188D-03   -.19708475D-02    .21879540D+05
```

```
      16    .39592112D-08    .49477617D-03    .19791047D-02    .21898010D+05
      17    .11625776D-07    .14505333D-02    .58021332D-02    .33889930D+05
      18   -.11676549D-07   -.14568497D-02   -.58273989D-02    .33985020D+05
      19   -.11676549D-07   -.14568497D-02   -.58273989D-02    .33985020D+05
      20    .11625776D-07    .14505332D-02    .58021330D-02    .33889930D+05
      21    .39592107D-08    .49477611D-03    .19791044D-02    .21898010D+05
      22   -.39426830D-08   -.49271188D-03   -.19708475D-02    .21879540D+05
      23   -.39426830D-08   -.49271188D-03   -.19708475D-02    .21879540D+05
      24    .39592112D-08    .49477617D-03    .19791047D-02    .21898010D+05

UPDATED COORDINATES
                               X                Y                 Z
          1      .60000097D+02    .17999990D+03   -.13456794D+01
          2      .12000000D+03    .17994911D+03   -.26924913D+01
          3      .17999990D+03    .17999990D+03   -.13456794D+01
          4      .60050886D+02    .12000000D+03   -.26924913D+01
          5      .12000000D+03    .12000000D+03   -.71144376D+01
          6      .17994911D+03    .12000000D+03   -.26924913D+01
          7      .60000097D+02    .60000097D+02   -.13456794D+01
          8      .12000000D+03    .60050886D+02   -.26924913D+01
          9      .17999990D+03    .60000097D+02   -.13456794D+01
Stop - Program terminated.
```

3.6.3 Example 3.3. Plane Truss Buckling

A seven bar plane truss is solved for buckling (Figure 3.4). This problem is taken form Timoshenko (1951) who solves it for

$$P_{CR} = \frac{AE}{\cot^2\theta\left(3 + \dfrac{2L_1 A}{L A_1 \cos^2\theta}\right)}$$

Here A and L refer to the horizontal bars and L_1 and A_1 refer to the inclined bars. Two solutions are presented to this problem. The first is obtained by iterative nonlinear analysis using TR2DNL.FOR whereas the second is obtained using BUCK2D33.FOR to solve a generalized eigenvalue problem ("linear buckling"). For A_1=2 square inches, A=200 square inches, E=30,000,000 psi, L_1=120 inches and θ=45° Timoshenko's solution yields $P_{CR} = 21\times10^6$ lbs.

A maximum load of 30,000,000 lbs is chosen as input for the **nonlinear buckling** analysis. The output file TR2DNL33.OUT indicates a negative term in the diagonal at load step 10. The last load for which the analysis converged is then load step 9. This implies that the buckling load is bounded by the loads at load steps 9 and 10. Thus:

Load step 9: $$30{,}000{,}000 \times \frac{10}{15} = 20{,}000{,}000 \text{ lb}$$

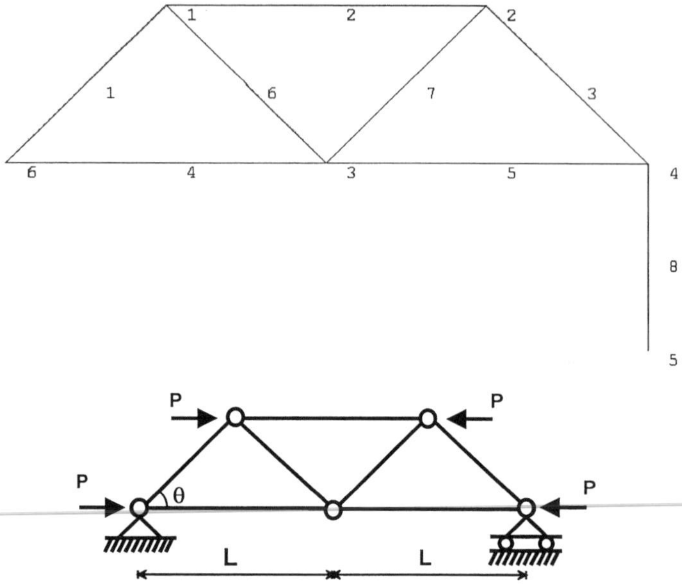

Figure 3.4 Schematic and node map of the 7-bar truss.

Timoshenko:⇒ P_{CR} =21,000,000 lbs

Load step 10: $$30{,}000{,}000 \times \frac{11}{15} = 22{,}000{,}000 \text{ lbs}$$

If a more accurate value for the buckling load is needed a different scheme can be used to generate load. But it should be noted again that as you get closer to the buckling load, computation becomes more and more difficult. In this sense a very accurate determination of the buckling load can be counter productive.

The same load of 30,000,000 lbs is selected as input for the **linear buckling** analysis. The data from BUCK2D33.DAT is first echoed in the output file BUCK2D33.OUT, then the linear analysis results are printed and finally the column of $-1/\beta$ is listed. The last entry in this column provides the smallest eigenvalue which is to be multiplied by the selected initial load, i.e. $P_{CR} = 30{,}000{,}000 \times (-1/-.14292360\text{E}+01) = 21 \times 10^6$ lbs.

It is interesting to note that both the linear and nonlinear buckling loads are the same for this truss. This is rather an exceptional case. In the examples that follow (single layered reticulated type) the linear buckling load is much higher than the nonlinear buckling load.

Output file TR2DNL33.OUT

```
   8  NO. MEMBERS
   6  NO. NODES
   2  NO. SUPPORTS
  15  NO. LOAD STEPS
   5  NO. ITERATIONS

                      COORDINATES                         LOADS
                       X                Y                 PX                PY
  1     .84850000D+02    .84850000D+02    .30000000D+08    .00000000D+00
  2     .25456000D+03    .84850000D+02   -.30000000D+08    .00000000D+00
  3     .16970000D+03    .00000000D+00    .00000000D+00    .00000000D+00
  4     .33941000D+03    .00000000D+00   -.30000000D+08    .00000000D+00
  5     .33941000D+03   -.10000000D+03    .00000000D+00    .00000000D+00
  6     .00000000D+00    .00000000D+00    .00000000D+00    .00000000D+00

   MEMBER      + END     - END            AREA         PRESTRESS
      1          1         6      .20000000E+01    .00000000E+00
      2          2         1      .20000000E+03    .00000000E+00
      3          2         4      .20000000E+01    .00000000E+00
      4          3         6      .20000000E+03    .00000000E+00
      5          3         4      .20000000E+03    .00000000E+00
      6          1         3      .20000000E+01    .00000000E+00
      7          2         3      .20000000E+01    .00000000E+00
      8          4         5      .20000000E+08    .00000000E+00

:

****ITERATION NUMBER    5
    LOAD STEP           9


 LOADSTEP    9;  ITERATION    5;  ERROR    .82974001D+00

DISPLACEMENTS
                           X                 Y
        1      .14560879D-07    -.75826807D-07
        2      .25488995D-07    -.69831642D-07
        3      .21600028D-07    -.12735170D-06
        4      .37381528D-07     .42232683D-09

 MEMBER              DL             FORCE            STRESS   UPDATED FORCES
      1   -.43532891D-07   -.21767026D-01   -.10883513D-01     .38932170D+03
      2    .10927904D-07    .38764006D+00    .19382003D-02    -.20000150D+08
      3   -.41461073D-07   -.20731433D-01   -.10365717D-01    -.60647490D+03
      4    .21602921D-07    .76635474D+00    .38317737D-02    -.20000280D+08
      5    .15779581D-07    .55974110D+00    .27987055D-02    -.19999580D+08
      6    .41516244D-07    .20758601D-01    .10379300D-01     .60644710D+03
      7    .43545338D-07    .21772249D-01    .10886125D-01    -.38933380D+03
      8   -.10561904D-14   -.63371423D-02   -.31685712D-09    -.73047570D+03

 UPDATED COORDINATES
                           X                 Y
        1      .84569203D+02     .85130968D+02
        2      .25371433D+03     .85124994D+02
        3      .16913515D+03    -.38420104D-02
        4      .33828030D+03    -.63813696D-02

:

****ITERATION NUMBER    2
    LOAD STEP          10
```

```
LOADSTEP  10;  ITERATION   2;  ERROR   .38143457D+09

**NEG TERM ON THE DIAGONAL AT ROW    6
```

Output file BUCK2D33.OUT

```
   8  NO. MEMBERS
   6  NO. NODES
   2  NO. SUPPORTS
                    COORDINATES                          LOADS
                      X                Y                  PX                 PY
  1    .84850000D+02    .84850000D+02    .30000000D+08     .00000000D+00
  2    .25456000D+03    .84850000D+02   -.30000000D+08     .00000000D+00
  3    .16970000D+03    .00000000D+00    .00000000D+00     .00000000D+00
  4    .33941000D+03    .00000000D+00   -.30000000D+08     .00000000D+00
  5    .33941000D+03   -.10000000D+03    .00000000D+00     .00000000D+00
  6    .00000000D+00    .00000000D+00    .00000000D+00     .00000000D+00

   MEMBER       + END      - END              AREA
        1           1          6       .20000000E+01
        2           2          1       .20000000E+03
        3           2          4       .20000000E+01
        4           3          6       .20000000E+03
        5           3          4       .20000000E+03
        6           1          3       .20000000E+01
        7           2          3       .20000000E+01
        8           4          5       .20000000E+08

DISPLACEMENTS
                           X                    Y
        1       -.42422500D+00       .42422500D+00
        2       -.12727750D+01       .42427500D+00
        3       -.84850000D+00      -.50001473D-04
        4       -.16970500D+01      -.15926995D-21
        5        .00000000D+00       .00000000D+00
        6        .00000000D+00       .00000000D+00

   MEMBER                 DL                FORCE               STRESS
        1        .27869146D-14       .13935035D-08        .69675175D-09
        2       -.84855000D+00      -.30000000D+08       -.15000000D+06
        3       -.28313940D-14      -.14157439D-08       -.70787197D-09
        4       -.84850000D+00      -.30000000D+08       -.15000000D+06
        5       -.84855000D+00      -.30000000D+08       -.15000000D+06
        6       -.27277442D-14      -.13639174D-08       -.68195868D-09
        7        .27469888D-14       .13734590D-08        .68672951D-09
        8       -.15926995D-21      -.95561972D-09       -.47780986D-16

             -1/β
     -.68074820E-12
     -.17469420E-11
     -.24195970E-11
     -.20062000E-08
     -.75163970E-08
     -.14731450E-07
     -.70959640E+00
     -.14292360E+01
Stop - Program terminated.
```

3.6.4 Example 3.4. Buckling of a Symmetric 2-Bar Truss

The symmetric 2-bar truss of Figure 3.5 with E=10,000,000 psi and A=0.0127 square inches was already discussed in Chapter 1 and is included here in order to demonstrate the distinct features of the nonlinear analysis program with respect to the manner in which length change is represented.

Nonlinear buckling analysis for an initial load of 30,000 lbs reveals that buckling occurs somewhere between load step 43 and load step 44. Printed below in TR2DNL34.OUT is the last converged iteration number 6 of load step 43. This implies that the buckling load is bound between 44/50× 30,000=26,400 lbs and 45/50×30,000=27,000 lbs.

Linear buckling analysis is performed using BUCK2D34.FOR on BUCK2D34.DAT and the results are listed below in BUCK2D34.OUT. The critical buckling load is P_{CR} =30,000×(-1/-.16703310E+00)=179,605 lbs.

The details of the computational model used can have some surprising effects. Equation 1.22 models nonlinear behavior by considering equilibrium in the deformed configuration and a strain model which defines strain as the total length change divided by the initial length. It is followed by Eq. 1.24 which provides the critical buckling load. For the 2-bar truss of this example the critical load is calculated as, P_{CR} =23,800 lbs using Eq. 1.24. Now if strain is modeled as the total length change divided by the final length, the following relationships replace Eqs. 1.23 and 1.24 respectively:

$$\frac{\mathrm{d}P}{\mathrm{d}u} = 0 \Rightarrow \left(\frac{H_1}{L}\right)^2 = \frac{L_0 - L}{2L_0 - L} \tag{3.27}$$

and

$$P = 2\frac{AE}{L}\left(L_0 - L\right)\sqrt{\frac{L_0 - L}{2L_0 - L}} \tag{3.28}$$

Equation 3.27 yields u =106.355 (making H_1=93.645) which is substituted into Eq. 3.28 to yield P_{CR} =30,240 lbs. The result of P_{CR} =26,400 lbs which is obtained using iterative incremental analysis is approximately the mean of the above two models. It comes from the fact that this program is coded with strain modeled as the length change divided by the actual length but this is done for each increment of load rather than the total load. Thus,

$$F_i^{n+1} = F_i^n + \Delta F_i^{n+1} \quad \text{where} \quad \Delta F_i^{n+1} = \frac{AE}{L_i^{n+1}}\left(L_i^{n+1} - L_i^n\right)$$

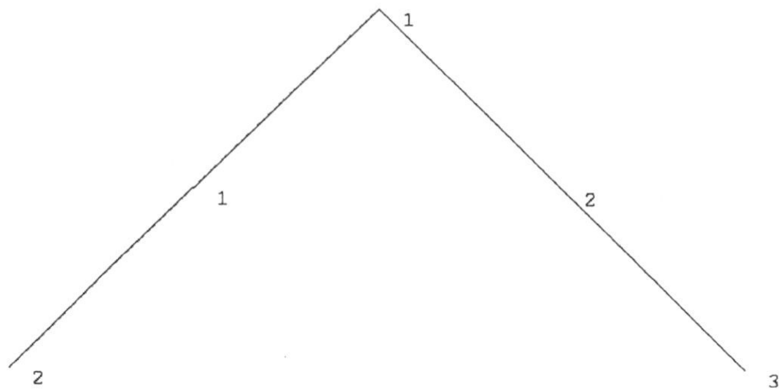

Figure 3.5 The symmetric 2-bar truss.

compares to $F_i = AE\left(L_{0i} - L_i\right)/L_i$ in Eq. 3.28. The question of whether the sum of the increments is always bounded by the two linear models that consider total length changes remains open. Irrespective of the model chosen the crucial buckling load factor λ=179,605/26,400=6.80 remains.

Output file TR2DNL34.OUT

```
  2   NO. MEMBERS
  3   NO. NODES
  2   NO. SUPPORTS
 50   NO. LOAD STEPS
  6   NO. ITERATIONS

                 COORDINATES                              LOADS
                     X                  Y                  PX                PY
 1      .20000000D+03     .20000000D+03     .00000000D+00     -.30000000D+05
 2      .00000000D+00     .00000000D+00     .00000000D+00      .00000000D+00
 3      .40000000D+03     .00000000D+00     .00000000D+00      .00000000D+00

  MEMBER     + END     - END           AREA          PRESTRESS
       1         1         2     .12700000E-01     .00000000E+00
       2         1         3     .12700000E-01     .00000000E+00
```

:

```
****ITERATION NUMBER    6
    LOAD STEP          43

LOADSTEP  43;   ITERATION   6;   ERROR   .63713672D-03

DISPLACEMENTS
                           X                  Y
        1     -.77705446D-15     .11648657D-04

 MEMBER              DL             FORCE            STRESS   UPDATED FORCES
      1    .54763203D-05    .30692093D-02    .24167003D+00   -.28077660D+05
      2    .54763203D-05    .30692093D-02    .24167003D+00   -.28077660D+05
```

```
UPDATED COORDINATES
                                X                  Y
        1      .20000000D+03     .10653176D+03
```

Output file BUCK2D34.OUT

```
   2  NO. MEMBERS
   3  NO. NODES
   2  NO. SUPPORTS

                COORDINATES                              LOADS
                 X                 Y                 PX                 PY
  1      .20000000D+03     .20000000D+03     .00000000D+00     -.30000000D+05
  2      .00000000D+00     .00000000D+00     .00000000D+00      .00000000D+00
  3      .40000000D+03     .00000000D+00     .00000000D+00      .00000000D+00

   MEMBER       + END      - END              AREA
        1           1          2      .12700000E-01
        2           1          3      .12700000E-01

DISPLACEMENTS
                              X                  Y
        1      -.20753441D-14     -.66813240D+02
        2       .00000000D+00      .00000000D+00
        3       .00000000D+00      .00000000D+00

   MEMBER                    DL                FORCE              STRESS
        1      -.47244095D+02     -.21213203D+05     -.16703310D+07
        2      -.47244095D+02     -.21213203D+05     -.16703310D+07


                -1/β
     -.74505810E-08
     -.16703310E+00
Stop - Program terminated.
```

3.6.5 Example 3.5. Buckling of a 4-Bar Shallow Space Truss

The nonlinear and linear buckling loads of the 4-bar space truss (E = 10,000,000 psi) of Figure 3.6 are determined using TR3DNL.FOR and BUCK3D35.FOR on TR3DNL35.DAT and BUCK3D35.DAT respectively. Areas of cross section and loads may be read from the output files TR3DNL35.OUT and BUCK3D35.OUT which are listed below.

This shallow truss is rather "soft" in terms of overall buckling. Once again loss of stability is encountered at loads much lower than anticipated by the eigenvalue formulation. In fact, it is established that the nonlinear buckling load occurs at a load bounded by 17/18×180=170 lbs and 18/18× 180=180 lbs in the downward direction. It will be taken as 170 lbs since load step 16 is the last converged load step. For the linear buckling load it is calculated from the output as -1/-.19274340E+00×180=934 lbs. The buckling load factor is λ=934/170=5.5 which is very high and warrants added caution in design.

TR3DNL35.OUT

```
     4  NO. MEMBERS
     5  NO. NODES
     4  NO. SUPPORTS
    18  NO. LOAD STEPS
    10  NO.ITERATIONS

COORDINATES
                          X                   Y                   Z
  1        .14400000D+03       .12000000D+03       .30000000D+01
  2        .20400000D+03       .00000000D+00       .00000000D+00
  3        .00000000D+00       .00000000D+00       .00000000D+00
  4        .00000000D+00       .19200000D+03       .00000000D+00
  5        .20400000D+03       .19200000D+03       .00000000D+00

LOADS
                         PX                  PY                  PZ
  1        .00000000D+00       .00000000D+00      -.18000000D+03
  2        .00000000D+00       .00000000D+00       .00000000D+00
  3        .00000000D+00       .00000000D+00       .00000000D+00
  4        .00000000D+00       .00000000D+00       .00000000D+00
  5        .00000000D+00       .00000000D+00       .00000000D+00

    MEMBER     + END     - END             AREA          PRESTRESS
         1         1         2     .19421000E+01     .00000000E+00
         2         1         5     .25994000E+01     .00000000E+00
         3         1         4     .15521000E+01     .00000000E+00
         4         1         3     .14900000E+01     .00000000E+00

:

****ITERATION NUMBER  10

    LOAD STEP           16

ERROR =          .23648477D-03
```

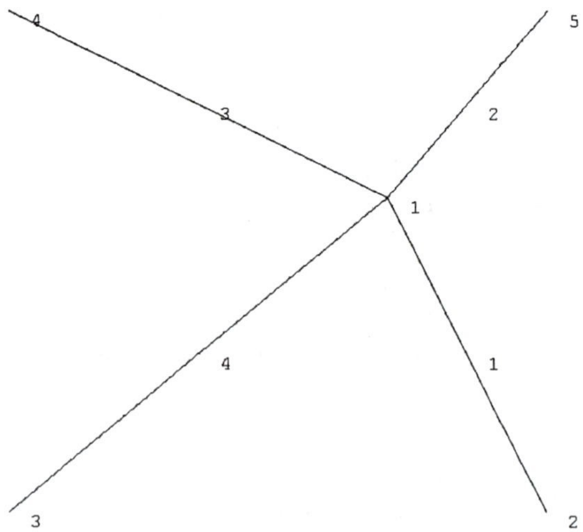

Figure 3.6 The 4-bar space truss.

```
DISPLACEMENTS
                         X                   Y                   Z
       1     -.10418559D-08    -.75207808D-09    -.57874319D-07

MEMBER              DL              FORCE             STRESS     UPDATED FORCES
     1  -.10913058D-08  -.15795503D-03  -.81332076D-04  -.26922100D+04
     2  -.21665792D-10  -.60067474D-05  -.23108207D-05  -.36035370D+04
     3  -.13327117D-08  -.12847635D-03  -.82775826D-04  -.21514200D+04
     4  -.19150728D-08  -.15222991D-03  -.10216772D-03  -.20653590D+04

UPDATED COORDINATES
                         X                   Y                   Z
       1      .14398836D+03     .11999335D+03     .20511593D+01

:

****ITERATION NUMBER    5
    LOAD STEP          17

 ERROR =          .11937231D+02

**NEG TERM ON THE DIAGONAL AT ROW     3
```

Output file BUCK3D35.OUT

```
   4  NO. MEMBERS
   5  NO. NODES
   4  NO.SUPPORTS

COORDINATES
                    X                    Y                    Z
  1        .14400000D+03        .12000000D+03        .30000000D+01
  2        .20400000D+03        .00000000D+00        .00000000D+00
  3        .00000000D+00        .00000000D+00        .00000000D+00
  4        .00000000D+00        .19200000D+03        .00000000D+00
  5        .20400000D+03        .19200000D+03        .00000000D+00

LOADS
                   PX                   PY                   PZ
  1        .00000000D+00        .00000000D+00       -.18000000D+03
  2        .00000000D+00        .00000000D+00        .00000000D+00
  3        .00000000D+00        .00000000D+00        .00000000D+00
  4        .00000000D+00        .00000000D+00        .00000000D+00
  5        .00000000D+00        .00000000D+00        .00000000D+00

   MEMBER       + END       - END                AREA
        1          1           2        .19421000E+01
        2          1           5        .25994000E+01
        3          1           4        .15521000E+01
        4          1           3        .14900000E+01

DISPLACEMENTS
                             X                    Y                    Z
       1         -.84268028D-02       -.48165264D-02       -.57836826D+00

   MEMBER                   DL                FORCE               STRESS
        1        -.13468792D-01       -.19491957D+04       -.10036536D+04
        2        -.94134299D-02       -.26094710D+04       -.10038744D+04
        3        -.16157597D-01       -.15574122D+04       -.10034226D+04
        4        -.18811264D-01       -.14951077D+04       -.10034280D+04

             -1/β
  -.55872000E-04
```

```
     -.18070120E-03
     -.19274340E+00
Stop - Program terminated.
```

3.6.6 Example 3.6. Buckling of the 24-Bar Dome

The 24-bar dome was encountered in Chapter 2 where it was linearly analyzed for a vertically downward load of 220.46 lbs. As is observed from the output below it actually buckles non linearly at a load 14/15×150=140 lbs. Linear analysis cannot anticipate buckling where the phenomenon is simply unaccounted for. The linear buckling load is calculated from the output below as -1/-0.21208881E+00×150=707 lbs. Note that the listing from NROOT changes sign in the calculation of the 14-th eigenvalue. This does not imply negative eigenvalues which of course are not possible! Rather, NROOT loses accuracy in single precision arithmetic in the very high frequency range. Finally the linear buckling load factor becomes λ =707/140=5.1, a very high factor indeed.

Output file TR3DNL36.OUT

```
  24   NO. MEMBERS
  13   NO. NODES
   6   NO. SUPPORTS
  15   NO. LOAD STEPS
   6   NO.ITERATIONS

 COORDINATES
                     X                    Y                    Z
   1         .00000000D+00        .00000000D+00        .32346000D+01
   2         .49212500D+01        .85239000D+01        .24472000D+01
   3        -.49212500D+01        .85239000D+01        .24472000D+01
   4        -.98425000D+01        .00000000D+00        .24472000D+01
   5        -.49212500D+01       -.85239000D+01        .24472000D+01
   6         .49212500D+01       -.85239000D+01        .24472000D+01
   7         .98425000D+01        .00000000D+00        .24472000D+01
   8         .00000000D+00        .19685000D+02        .00000000D+00
   9        -.17047200D+02        .98425000D+01        .00000000D+00
  10        -.17047200D+02       -.98425000D+01        .00000000D+00
  11         .00000000D+00       -.19685000D+02        .00000000D+00
  12         .17047200D+02       -.98425000D+01        .00000000D+00
  13         .17047200D+02        .98425000D+01        .00000000D+00

 LOADS
                    PX                   PY                   PZ
   1         .00000000D+00        .00000000D+00       -.15000000D+03
   2         .00000000D+00        .00000000D+00        .00000000D+00
   3         .00000000D+00        .00000000D+00        .00000000D+00
   4         .00000000D+00        .00000000D+00        .00000000D+00
   5         .00000000D+00        .00000000D+00        .00000000D+00
   6         .00000000D+00        .00000000D+00        .00000000D+00
   7         .00000000D+00        .00000000D+00        .00000000D+00
   8         .00000000D+00        .00000000D+00        .00000000D+00
   9         .00000000D+00        .00000000D+00        .00000000D+00
  10         .00000000D+00        .00000000D+00        .00000000D+00
```

```
11        .00000000D+00        .00000000D+00        .00000000D+00
12        .00000000D+00        .00000000D+00        .00000000D+00
13        .00000000D+00        .00000000D+00        .00000000D+00

  MEMBER     + END     - END          AREA          PRESTRESS
       1         1         3     .15500000E-01    .00000000E+00
       2         2         3     .15500000E-01    .00000000E+00
       3         3         8     .15500000E-01    .00000000E+00
       4         2         8     .15500000E-01    .00000000E+00
       5         1         4     .15500000E-01    .00000000E+00
       6         3         4     .15500000E-01    .00000000E+00
       7         4         9     .15500000E-01    .00000000E+00
       8         3         9     .15500000E-01    .00000000E+00
       9         1         5     .15500000E-01    .00000000E+00
      10         4         5     .15500000E-01    .00000000E+00
      11         5        10     .15500000E-01    .00000000E+00
      12         4        10     .15500000E-01    .00000000E+00
      13         1         6     .15500000E-01    .00000000E+00
      14         5         6     .15500000E-01    .00000000E+00
      15         6        11     .15500000E-01    .00000000E+00
      16         5        11     .15500000E-01    .00000000E+00
      17         1         7     .15500000E-01    .00000000E+00
      18         6         7     .15500000E-01    .00000000E+00
      19         7        12     .15500000E-01    .00000000E+00
      20         6        12     .15500000E-01    .00000000E+00
      21         1         2     .15500000E-01    .00000000E+00
      22         7         2     .15500000E-01    .00000000E+00
      23         2        13     .15500000E-01    .00000000E+00
      24         7        13     .15500000E-01    .00000000E+00

:

****ITERATION NUMBER    6
   LOAD STEP           13

 ERROR =          .40512682D-04

DISPLACEMENTS
                      X                  Y                  Z
       1    -.75034262D-18    -.18483798D-18     .36610039D-08
       2    -.15101719D-09    -.26843517D-09    -.92847311D-09
       3     .15101718D-09    -.26843517D-09    -.92847310D-09
       4     .28897376D-09     .22457035D-18    -.82729047D-09
       5     .15101718D-09     .26843517D-09    -.92847310D-09
       6    -.15101718D-09     .26843517D-09    -.92847310D-09
       7    -.28897376D-09    -.53907912D-18    -.82729048D-09

 MEMBER              DL            FORCE             STRESS     UPDATED FORCES
      1  -.55429356D-10  -.26127454D-05  -.16856423D-03  -.42480780D+03
      2  -.30203437D-09  -.14258422D-04  -.91989822D-03   .35593260D+03
      3  -.27935858D-11  -.10442775D-06  -.67372741D-05  -.58962080D+02
      4  -.27935859D-11  -.10442775D-06  -.67372744D-05  -.58962080D+02
      5  -.42008062D-10  -.19801205D-05  -.12774971D-03  -.42480230D+03
      6  -.30145001D-09  -.14230779D-04  -.91811480D-03   .35593370D+03
      7   .35023673D-11   .13092558D-06   .84468116D-05  -.58961200D+02
      8  -.82325055D-11  -.30775282D-06  -.19855021D-04  -.58961710D+02
      9  -.55429356D-10  -.26127455D-05  -.16856423D-03  -.42480780D+03
     10  -.30145001D-09  -.14230779D-04  -.91811480D-03   .35593370D+03
     11  -.82325059D-11  -.30775284D-06  -.19855022D-04  -.58961710D+02
     12   .35023677D-11   .13092559D-06   .84468124D-05  -.58961200D+02
     13  -.55429355D-10  -.26127454D-05  -.16856422D-03  -.42480780D+03
     14  -.30203437D-09  -.14258422D-04  -.91989822D-03   .35593260D+03
     15  -.27935862D-11  -.10442776D-06  -.67372751D-05  -.58962080D+02
     16  -.27935857D-11  -.10442775D-06  -.67372740D-05  -.58962080D+02
     17  -.42008061D-10  -.19801205D-05  -.12774971D-03  -.42480230D+03
```

```
18  -.30145001D-09  -.14230779D-04  -.91811480D-03   .35593370D+03
19   .35023672D-11   .13092557D-06   .84468112D-05  -.58961200D+02
20  -.82325051D-11  -.30775281D-06  -.19855020D-04  -.58961710D+02
21  -.55429356D-10  -.26127455D-05  -.16856423D-03  -.42480780D+03
22  -.30145001D-09  -.14230779D-04  -.91811480D-03   .35593370D+03
23  -.82325059D-11  -.30775284D-06  -.19855022D-04  -.58961710D+02
24   .35023680D-11   .13092560D-06   .84468133D-05  -.58961200D+02

UPDATED COORDINATES
                   X                Y                Z
     1   -.63623184D-17    .18422895D-17    .30031710D+01
     2    .49250183D+01    .85304268D+01    .24613183D+01
     3   -.49250183D+01    .85304268D+01    .24613183D+01
     4   -.98500367D+01    .30818523D-17    .24613180D+01
     5   -.49250183D+01   -.85304268D+01    .24613183D+01
     6    .49250183D+01   -.85304268D+01    .24613183D+01
     7    .98500367D+01   -.18711817D-17    .24613180D+01
```

Output file BUCK3D36.OUT

```
          -1/β
      0.51235702E-01
      0.26026258E-01
      0.26026156E-01
      0.90162409E-03
      0.56519511E-03
      0.56519359E-03
      0.17111958E-03
      0.17111933E-03
     -0.54412139E-04
     -0.14497516E-03
     -0.14497591E-03
     -0.18869033E-03
     -0.24152090E-03
     -0.26560493E-03
     -0.26560537E-03
     -0.11734122E-02
     -0.11734152E-02
     -0.43631340E-02
     -0.13986786E-01
     -0.13986799E-01
     -0.21208881E+00
Stop - Program terminated.
```

3.7 Problems

1. Find the linear buckling load of the 3-bar truss of Example 2.1 using BUCK2D.FOR given that only one load acts vertically downwards, the areas of cross section of members 1 and 2 are 0.1 square inches and that member 3 is rigid. (For computational purposes an area of 1,000,000 square inches has been used.)

2. Calculate the buckling load of problem 1 using principles of mechanics.

3. Using TR2DNL.FOR find the nonlinear buckling load of the 2-bar truss of Figure 1.14 given that E=10,000,000 psi, H=10 inches, B_1=200 inches, B_2=50 inches, and $A_1 = A_2 = 0.0127\,\text{in}^2$. (Ans: P=60 lbs).

4. Write a few lines of code that will verify your results in problem 3. (Increment the displacement and obtain the load via equilibrium in the deformed configuration. Buckling occurs when the slope of the P-δ curve changes sign.)

5. a) Derive Eq. 3.20.

b) Use $\varepsilon = \dfrac{L^2 - L_0^2}{2L_0^2}$ and re derive Eqs. 3.20 and 3.21.

c) Use these expressions to obtain a buckling load for Example 3.4.

6. Figure 3.7 shows a 12-member cable net which consists of a two-directional network of cables attached to a rigid frame. The cross sectional area of the cable is 0.0003038 square inches, L=96.85 inches and f=22.24 inches. The horizontal component of initial prestress is 26.455 lbs and E=30,000,000 psi. If one of the free nodes is loaded downwards with 18.298 lbs find the forces in the members and the displacements of the nodes using TR3DNL.FOR. (For a discussion on the type of analysis required for cable networks (linear vs. nonlinear) the reader is referred to Levy *et al.*, 1991.)

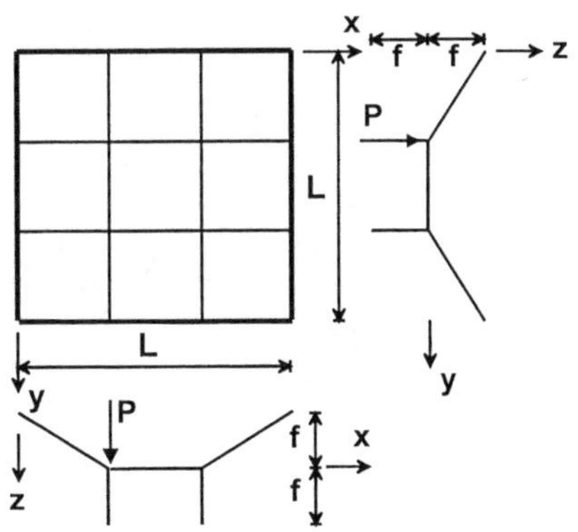

Figure 3.7 The 12-member cable network.

7. Modify programs P5 and P6 for more general usage when computing eigenvalue buckling loads. (Hint: If NROOT is placed within a subroutine, arrays with variable DIMENSION statements can be used avoiding re-coding when problems of different size are to be run.)

CHAPTER 4

NONLINEAR ANALYSIS OF PLANE FRAMES

4.1 Linear Analysis

The discussion of plane frames adds little to the earlier discussion of trusses. Plane frames are, of course, more complex and include trusses as a special case as the moment of inertia of frame members is set to zero.

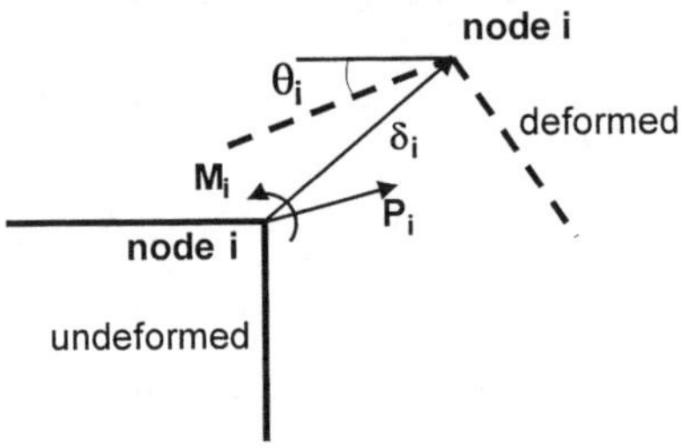

Figure 4.1 Node *i*.

Figure 4.1 shows the typical variables associated with the joint of a plane frame. In terms of kinematics each joint is described by a two-dimensional displacement vector $\boldsymbol{\delta}_i$ and a scalar rotation θ_i. The conjugate variables are $\mathbf{P}_i$ and $\mathbf{M}_i$, the joint vector and the scalar moment. These node quantities give rise to the joint load matrix **P** and the joint displacement matrix δ

$$\mathbf{P} = \begin{Bmatrix} (\mathbf{P}_1)_x \\ (\mathbf{P}_1)_y \\ M_1 \\ (\mathbf{P}_2)_x \\ (\mathbf{P}_2)_y \\ M_2 \\ \vdots \end{Bmatrix}; \quad \boldsymbol{\delta} = \begin{pmatrix} (\boldsymbol{\delta}_1)_x \\ (\boldsymbol{\delta}_1)_y \\ \theta_1 \\ (\boldsymbol{\delta}_2)_x \\ (\boldsymbol{\delta}_2)_y \\ \theta_2 \\ \vdots \end{pmatrix} \tag{4.1}$$

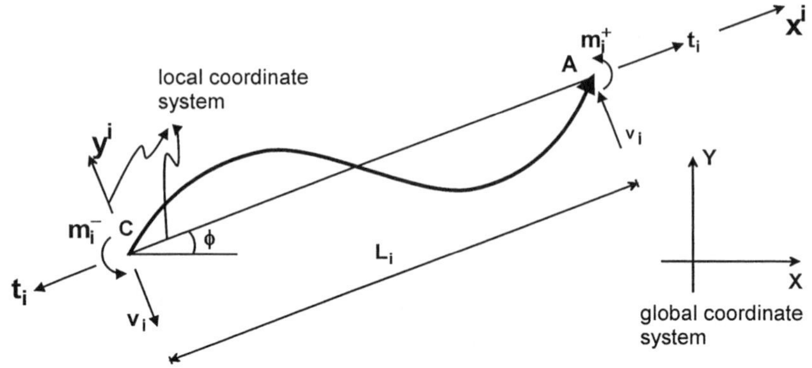

Figure 4.2 Typical member.

Clearly, the size of the system matrix is three times the number of movable joints.

Figure 4.2 describes a typical member. At each end there is a shear v_i, a moment $m_i^{\pm}$ and thrust t_i. Obviously, $m_i^+ \neq m_i^-$. The member force matrix, $\mathbf{F}_i$, which contained only bar forces in the case of the truss, now contains three terms per member. It can be shown that the number of these terms is invariant but that the specific choice of terms is somewhat arbitrary. Here the three force components that are chosen per member are the two end moments and the thrust of the "+ end". The other three quantities i.e. the two end shear forces and the thrust in the "- end" are related to the first three quantities through the equations of equilibrium. The member force matrix, thus, becomes

$$\mathbf{F} = \begin{Bmatrix} t_1 \\ m_1^+ \\ m_1^- \\ t_2 \\ m_2^+ \\ m_2^- \\ \vdots \end{Bmatrix} \tag{4.2}$$

Let the member forces at the positive and negative ends be represented by $\mathbf{f}_i^+$ and $\mathbf{f}_i^-$ as

$$\mathbf{f}_i^+ = \begin{Bmatrix} t_i \\ v_i \\ m_i^+ \end{Bmatrix}; \ \mathbf{f}_i^- = \begin{Bmatrix} -t_i \\ -v_i \\ m_i^- \end{Bmatrix} \tag{4.3}$$

Member equilibrium relates $\mathbf{f}_i^+$ and $\mathbf{f}_i^-$ to $\mathbf{F}_i$ as

$$\mathbf{f}_i^+ = \mathbf{N}_i^{+\mathrm{T}} \mathbf{F}_i; \ \mathbf{f}_i^- = \mathbf{N}_i^{-\mathrm{T}} \mathbf{F}_i \tag{4.4}$$

where

$$\mathbf{N}_i^+ = \begin{bmatrix} 1 & 0 & 0 \\ 0 & -L_i^{-1} & 1 \\ 0 & -L_i^{-1} & 0 \end{bmatrix}; \ \mathbf{N}_i^- = \begin{bmatrix} -1 & 0 & 0 \\ 0 & L_i^{-1} & 0 \\ 0 & L_i^{-1} & 1 \end{bmatrix} \tag{4.5}$$

Since the equilibrium equations at each node must be written in the global coordinate system, these end of member forces must be transformed using the usual rotation matrix $\mathbf{R}_i$ which is described in Appendix 3 as

$$\mathbf{R}_i = \begin{bmatrix} \cos\phi_i & \sin\phi_i & 0 \\ -\sin\phi_i & \cos\phi_i & 0 \\ 0 & 0 & 1 \end{bmatrix} \tag{4.6}$$

It is the transpose of the rotation matrix that is directly applied to the member forces because member end forces are considered "local". The transformation here goes from a "local" to a "global" system whereas the rotation matrix, $\mathbf{R}_i$, (Appendix 3) takes vectors from the "global" to a "local" system. This inverse rotation is performed using the transpose of $\mathbf{R}_i$ to yield $\mathbf{R}_i^{\mathrm{T}} \mathbf{N}_i^{+\mathrm{T}} \mathbf{F}_i$ and $\mathbf{R}_i^{\mathrm{T}} \mathbf{N}_i^{-\mathrm{T}} \mathbf{F}_i$ as the end of member forces.

Consistent with the notation of the node method that was introduced in Chapter 2 it is now possible to define the generalized incidence matrix $\mathbf{N}$ (in $\mathbf{N}^{\mathrm{T}}\mathbf{F} = \mathbf{P}$) as

$$N_{ij} = \begin{cases} \mathbf{N}_i^+ \mathbf{R}_i & \text{if node } j \text{ is the + end of member } i \\ \mathbf{N}_i^- \mathbf{R}_i & \text{if node } j \text{ is the -- end of member } i \\ 0 & \text{otherwise} \end{cases} \tag{4.7}$$

Proceeding to the constitutive relations, it is obvious that the member displacements Δ_i entering the equations must be the member elongation and the two member end rotations. The node method relates member displacements to nodal displacements through $\mathbf{N\delta} = \mathbf{\Delta}$. With $\mathbf{\Delta}$ described as

$$\mathbf{\Delta} = \begin{Bmatrix} \Delta L_1 \\ \alpha_1^+ \\ \alpha_1^- \\ \Delta L_2 \\ \alpha_2^+ \\ \alpha_2^- \\ \vdots \end{Bmatrix} \tag{4.8}$$

and $\mathbf{N}$ and δ already defined, $\mathbf{N\delta} = \mathbf{\Delta}$ reduces for a single member to

$$\mathbf{\Delta}_i = \mathbf{N}_i^+ \mathbf{R}_i \boldsymbol{\delta}_\mathrm{A} + \mathbf{N}_i^- \mathbf{R}_i \boldsymbol{\delta}_\mathrm{C} \tag{4.9}$$

If $\mathbf{R}_i\boldsymbol{\delta}_\mathrm{A}$ and $\mathbf{R}_i\boldsymbol{\delta}_\mathrm{C}$ are regarded as the nodal displacements of member i in the local coordinate system indicated by the subscript x^i, then the above expression, when expanded, becomes

$$\mathbf{\Delta}_i = \begin{Bmatrix} \Delta L_i \\ \alpha_i^+ \\ \alpha_i^- \end{Bmatrix} = \begin{Bmatrix} (\delta_\mathrm{A})_{x^i} - (\delta_\mathrm{C})_{x^i} \\ \theta_\mathrm{A} - \dfrac{1}{L_i}\left[(\delta_\mathrm{A})_{y^i} - (\delta_\mathrm{C})_{y^i}\right] \\ \theta_\mathrm{C} - \dfrac{1}{L_i}\left[(\delta_\mathrm{A})_{y^i} - (\delta_\mathrm{C})_{y^i}\right] \end{Bmatrix} \tag{4.10}$$

with

ΔL_i = member length change of member i

α_i^+ = rotation of the "+ end" minus the rigid body rotation of the member

α_i^- = rotation of the "- end" minus the rigid body rotation of the member

Finally the member "primitive" or "local" stiffness matrix is presented in its usual form as

$$\mathbf{K}_i = \begin{bmatrix} \frac{EA_i}{L_i} & 0 & 0 \\ 0 & \frac{4EI_i}{L_i} & \frac{2EI_i}{L_i} \\ 0 & \frac{2EI_i}{L_i} & \frac{4EI_i}{L_i} \end{bmatrix} \tag{4.11}$$

where A_i, E, L_i are the member area, Young's modulus, and the member length respectively.

In terms of global assemblage for use in the node method ($\mathbf{N}^{\mathrm{T}}\mathbf{KN\delta} = \mathbf{P}$), the contribution of each member to the global stiffness matrix becomes

$$\mathbf{N}_i{}^{\mathrm{T}}\mathbf{K}_i\mathbf{N}_i = \begin{Bmatrix} \vdots \\ \mathbf{R}_i{}^{\mathrm{T}}\mathbf{N}_i^{+\mathrm{T}} \\ \vdots \\ \mathbf{R}_i{}^{\mathrm{T}}\mathbf{N}_i^{-\mathrm{T}} \\ \vdots \end{Bmatrix} \mathbf{K}_i \left\{ \cdots \quad \mathbf{N}_i^{+}\mathbf{R}_i \quad \cdots \quad \mathbf{N}_i^{+}\mathbf{R}_i \quad \cdots \right\} \tag{4.12}$$

or

$$(\mathbf{K}_{\mathrm{E}})_i = \begin{array}{c} \\ \\ \end{array} \begin{array}{cc} \text{col A} & \text{col C} \end{array}$$

$$(\mathbf{K}_{\mathrm{E}})_i = \begin{bmatrix} \mathbf{R}_i{}^{\mathrm{T}}\mathbf{N}_i^{+\mathrm{T}}\mathbf{K}_i\mathbf{N}_i^{+}\mathbf{R}_i & \mathbf{R}_i{}^{\mathrm{T}}\mathbf{N}_i^{+\mathrm{T}}\mathbf{K}_i\mathbf{N}_i^{-}\mathbf{R}_i \\ \mathbf{R}_i{}^{\mathrm{T}}\mathbf{N}_i^{-\mathrm{T}}\mathbf{K}_i\mathbf{N}_i^{+}\mathbf{R}_i & \mathbf{R}_i{}^{\mathrm{T}}\mathbf{N}_i^{-\mathrm{T}}\mathbf{K}_i\mathbf{N}_i^{-}\mathbf{R}_i \end{bmatrix} \begin{array}{l} \text{row A} \\ \text{row C} \end{array} \tag{4.13}$$

4.2 Computer Program P7-FR2D.FOR

This program performs linear analysis of plane frames. The coding follows the development of the node method for plane trusses. Appendix 5 contains a full listing of this program.

The program begins with some initialization of variables and some input. The system matrix $\mathbf{K}_{\mathrm{E}}$ is then generated as C by summing the contribution of each member to it. Joint displacements are computed and finally member forces are determined.

The subroutines in the program are RNK, INSERT and PROD. Subroutine RNK defines the matrices **R**, $\mathbf{N}^+$, $\mathbf{N}^-$ and **K** and then forms the products $\mathbf{N}^+\mathbf{R}$ and $\mathbf{N}^-\mathbf{R}$ as NRP and NRM. Subroutine INSERT is used to place elements in the system matrix C. Subroutine PROD forms the product of three matrices $\mathtt{D}=\mathtt{A}^{\mathtt{T}}\mathtt{BC}$.

The variables in the program are the rotation matrix R=**R**, the member stiffness matrix SK = **K**, SNP = $\mathbf{N}^+$, SNM = $\mathbf{N}^-$, the number of nodes NN, the number of bars NB, the joint load matrix P, the system matrix C, the member area A, the member length AL, the member moment of inertia SI, the member orientation θ, as TH, and the member end nodes NP and MI.

Here again as in the truss programs all supports are regarded to be fully fixed and only nodal loads are considered.

4.2.1 Example 4.1. A Simple Plane Frame

This example is presented to familiarize the reader with some of the details of FR2D.FOR in terms of both input and output. Figure 4.3 shows a plane frame which is loaded with a horizontal force P = 10,000 lbs, a downward vertical force of 5,000 lbs and a moment M = 12,000 in-lbs. The span is 24 feet (L=12 ft.) and H=8 ft. One support is fully fixed whereas the other is a roller. The roller is handled by modeling it as a fictitious fixed bar of zero moment of inertia and an infinite area of cross section.

Program FR2D.FOR is invoked with the input file FR2D41.DAT and the results are listed in the output file FR2D41.OUT. The moment diagram of Figure 4.4 is drawn from the output columns "M+" and "M-". The moments are drawn on the tension side of the beam.

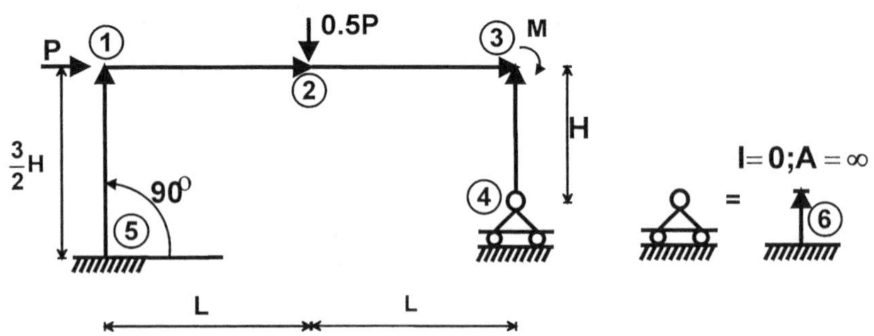

Figure 4.3 A three member plane frame.

The first line of the input file supplies NB, NN and NS. The next four lines provide nodal load information and the last five lines provide member information in terms of cross sectional area, length, moment of inertia, angle of orientation and "+ end" node number and "- end" node number respectively. Note that the last line in the member information which relates to the fictitious bar of length of one inch, zero moment of inertia and infinite area (actually 10,000,000 square inches was used).

Input file FR2D41.DAT

```
    5    6    2                                            (input format: 3I5)
10000.                                                     (input format: 3D10.2)
          -5000.                                           (input format: 3D10.2)
                    -12000.                                (input format: 3D10.2)
                                                           (input format: 3D10.2)
100.      144.      800.      90.        1    5    (input format: 4E10.2,2I5)
100.      144.      800.       0.        2    1    (input format: 4E10.2,2I5)
100.      144.      800.       0.        3    2    (input format: 4E10.2,2I5)
100.       96.      800.      90.        3    4    (input format: 4E10.2,2I5)
10000000. 1.        0.0000000 90.        4    6    (input format: 4E10.2,2I5)
```

Output file FR2D41.OUT

```
    5  NO. MEMBERS
    6  NO. NODES
    2  NO. SUPPORTS

JOINT LOADS
                    PX                    PY                    M
  1        .10000000D+05        .00000000D+00        .00000000D+00
  2        .00000000D+00       -.50000000D+04        .00000000D+00
  3        .00000000D+00        .00000000D+00       -.12000000D+05
  4        .00000000D+00        .00000000D+00        .00000000D+00

MEMBER PROPERTIES
             AREA           LENGTH                I        ANGLE+END-END
  1  .10000000E+03   .14400000E+03   .80000000E+03   .90000000E+02    1    5
  2  .10000000E+03   .14400000E+03   .80000000E+03   .00000000E+00    2    1
  3  .10000000E+03   .14400000E+03   .80000000E+03   .00000000E+00    3    2
  4  .10000000E+03   .96000000E+02   .80000000E+03   .90000000E+02    3    4
  5  .10000000E+08   .10000000E+01   .00000000E+00   .90000000E+02    4    6
```

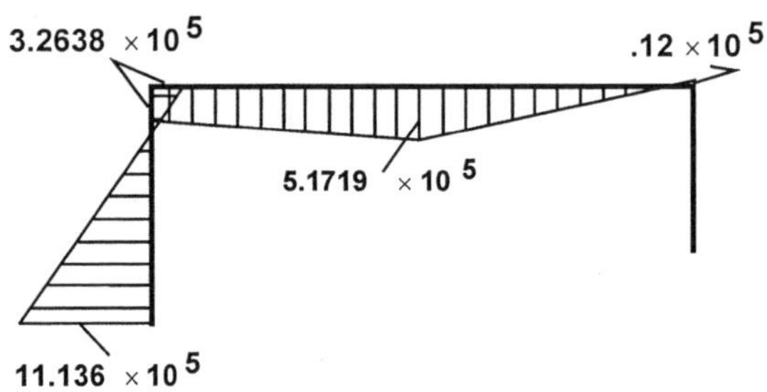

Figure 4.4 Moment diagram.

```
JOINT DISPLACEMENTS
                 DX                  DY                  TH
 1      .28316217D+00     -.66131430D-04     -.24431519D-02
 2      .28316217D+00     -.17759724D+00      .17483097D-03
 3      .28316217D+00     -.12198879D-03      .17426643D-02
 4      .45045795D+00     -.53409112D-06      .17426643D-02

MEMBER DISPLACEMENTS-FORCES
          DL          AL+          AL-          T           M+          M-
1 -.65796D-04-.47675D-03  .19664D-02-.13251D+04  .32638D+06  .11136D+07
2  .21635D-09 .14077D-02 -.12103D-02 .43571D-02  .51719D+06 -.32638D+06
3  .21635D-09 .51020D-03 -.10576D-02 .43571D-02 -.12000D+05 -.51719D+06
4 -.12165D-03-.16079D-17  .86736D-18-.36749D+04 -.11350D-08  .61321D-10
5 -.12672D-10 .45220D+00  .45046D+00-.36749D+04  .00000D+00  .00000D+00
Stop - Program terminated.
```

4.3 The Geometric Stiffness Matrix

The geometric stiffness matrix for plane frames will be derived in an identical manner to that used for trusses and similar arguments will be made to those of Section 3.2. First the contribution of member i to the equilibrium equations is formulated and then the gradient of this contribution is taken in order to generate the geometric stiffness matrix. Following the node method for plane frames member i will contribute the following terms to the node equilibrium equations

$$\left(\mathbf{N}^{\mathrm{T}}\mathbf{F}\right)_i = \begin{Bmatrix} \vdots \\ \mathbf{R}_i^{\mathrm{T}}\mathbf{N}_i^{+\mathrm{T}}\mathbf{F}_i \\ \vdots \\ \mathbf{R}_i^{\mathrm{T}}\mathbf{N}_i^{-\mathrm{T}}\mathbf{F}_i \\ \vdots \end{Bmatrix} \begin{matrix} \\ \text{joint A} \\ \\ \text{joint C} \\ \\ \end{matrix} \tag{4.14}$$

which, after multiplication of the matrices yields

$$\left(\mathbf{N}^{\mathrm{T}}\mathbf{F}\right)_i = \begin{Bmatrix} \vdots \\ t_i\cos\varphi_i - v_i\sin\varphi_i \\ t_i\sin\varphi_i + v_i\cos\varphi_i \\ m_i^+ \\ \vdots \\ -t_i\cos\varphi_i + v_i\sin\varphi_i \\ -t_i\sin\varphi_i - v_i\cos\varphi_i \\ m_i^- \\ \vdots \end{Bmatrix} = \begin{Bmatrix} \vdots \\ A_{i1} \\ A_{i2} \\ m_i^+ \\ \vdots \\ -A_{i1} \\ -A_{i2} \\ m_i^- \\ \vdots \end{Bmatrix} \begin{matrix} \\ \\ \text{jointA} \\ \\ \\ \\ \text{jointC} \\ \\ \\ \end{matrix} \tag{4.15}$$

The geometric stiffness matrix can be computed directly as

$$(\mathbf{K}_G)_i = \nabla\left(\mathbf{N}^T\mathbf{F}\right)_i = \begin{bmatrix} \nabla\left(\mathbf{N}^T\mathbf{F}\right)_i^{AA} & \nabla\left(\mathbf{N}^T\mathbf{F}\right)_i^{AC} \\ \nabla\left(\mathbf{N}^T\mathbf{F}\right)_i^{CA} & \nabla\left(\mathbf{N}^T\mathbf{F}\right)_i^{CC} \end{bmatrix} \begin{matrix} \text{row A} \\ \text{row C} \end{matrix} \qquad (4.16)$$

(columns: col A, col C)

where the sub matrices are produced in the same manner as Eq. 3.10 to become

$$\nabla\left(\mathbf{N}_i^T\mathbf{F}\right)^{AA} = \begin{bmatrix} \frac{\partial A_{i1}}{\partial x_A} & \frac{\partial A_{i1}}{\partial y_A} & 0 \\ \frac{\partial A_{i2}}{\partial x_A} & \frac{\partial A_{i2}}{\partial y_A} & 0 \\ 0 & 0 & 0 \end{bmatrix}; \nabla\left(\mathbf{N}_i^T\mathbf{F}\right)^{AC} = \begin{bmatrix} \frac{\partial A_{i1}}{\partial x_C} & \frac{\partial A_{i1}}{\partial y_C} & 0 \\ \frac{\partial A_{i2}}{\partial x_C} & \frac{\partial A_{i2}}{\partial y_C} & 0 \\ 0 & 0 & 0 \end{bmatrix}$$

and

$$\nabla\left(\mathbf{N}_i^T\mathbf{F}\right)^{CA} = \begin{bmatrix} -\frac{\partial A_{i1}}{\partial x_A} & -\frac{\partial A_{i1}}{\partial y_A} & 0 \\ -\frac{\partial A_{i2}}{\partial x_A} & -\frac{\partial A_{i2}}{\partial y_A} & 0 \\ 0 & 0 & 0 \end{bmatrix}; \nabla\left(\mathbf{N}_i^T\mathbf{F}\right)^{CC} = \begin{bmatrix} -\frac{\partial A_{i1}}{\partial x_C} & -\frac{\partial A_{i1}}{\partial y_C} & 0 \\ -\frac{\partial A_{i2}}{\partial x_C} & -\frac{\partial A_{i2}}{\partial y_C} & 0 \\ 0 & 0 & 0 \end{bmatrix}$$

The derivatives are now computed to produce the terms of the geometric stiffness matrix. Symbolically

$$(\mathbf{K}_G)_i = \nabla\left(\mathbf{N}^T\mathbf{F}\right)_i = \begin{bmatrix} (\mathbf{K}_G)_i^{AA} & -(\mathbf{K}_G)_i^{AA} \\ -(\mathbf{K}_G)_i^{AA} & (\mathbf{K}_G)_i^{AA} \end{bmatrix} \begin{matrix} \text{row A} \\ \text{row C} \end{matrix} \qquad (4.17)$$

(columns: col A, col C)

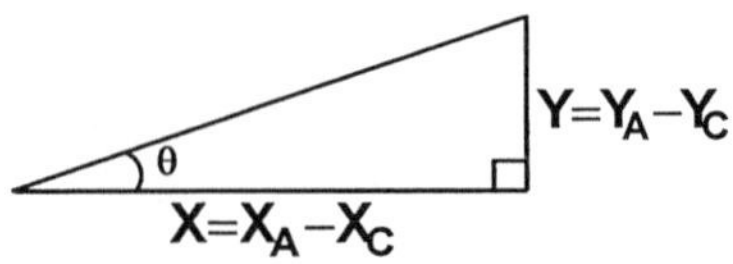

Figure 4.5 Member projections.

where

$$(\mathbf{K}_{\mathrm{G}})_i^{\mathrm{AA}} = \begin{bmatrix} a & b & 0 \\ b & c & 0 \\ 0 & 0 & 0 \end{bmatrix} \tag{4.18}$$

and

$$a = t_i \frac{1-(\mathbf{n}_i)_x^2}{L_i} - \left(m_i^+ + m_i^-\right)\frac{2(\mathbf{n}_i)_x(\mathbf{n}_i)_y}{L_i^2} \tag{4.19}$$

$$b = -t_i \frac{(\mathbf{n}_i)_x(\mathbf{n}_i)_y}{L_i} + \left(m_i^+ + m_i^-\right)\frac{1-2(\mathbf{n}_i)_y^2}{L_i^2} \tag{4.20}$$

$$c = t_i \frac{1-(\mathbf{n}_i)_y^2}{L_i} + \left(m_i^+ + m_i^-\right)\frac{2(\mathbf{n}_i)_x(\mathbf{n}_i)_y}{L_i^2} \tag{4.21}$$

These terms can be derived with the help of the following (Fig 4.5):

$$\cos\theta = \frac{x}{L} = \frac{x}{\sqrt{x^2+y^2}} = n_x$$

$$\frac{\partial}{\partial x}\cos\theta = \frac{1}{L}\left(1-n_x^2\right)$$

$$\frac{\partial}{\partial y}\cos\theta = -x\left(x^2+y^2\right)^{-3/2} y = -\frac{n_x n_y}{L}$$

$$\frac{\partial}{\partial x}\frac{\cos\theta}{L} = \frac{1-2n_x^2}{L^2}$$

$$\frac{\partial}{\partial y}\frac{\cos\theta}{L} = -2\frac{n_x n_y}{L^2}$$

$$\sin\theta = \frac{y}{L} = \frac{y}{\sqrt{x^2 + y^2}} = n_y$$

$$\frac{\partial}{\partial x}\sin\theta = -\frac{n_x n_y}{L}$$

$$\frac{\partial}{\partial y}\sin\theta = \frac{1 - n_y^2}{L}$$

$$\frac{\partial}{\partial x}\frac{\sin\theta}{L} = -2\frac{n_x n_y}{L^2}$$

$$\frac{\partial}{\partial y}\frac{\sin\theta}{L} = \frac{1 - 2n_y^2}{L^2}$$

4.4 Computer Program P8-FR2DNL.FOR

This program implements the material described above and is quite similar to the truss problem with one exception. The truss problem uses linear springs for members and does not include the possibility of member buckling. (This could of course have been built in but it would be necessary in this case to introduce the member moment of inertia. There is also the problem for cable nets of members having no stiffness in compression.) As was pointed out in Chapter 1, beams (beam-columns) have stress stiffening and softening in the presence of axial load. For that reason the member stiffness matrix for plane frames is based upon beam-column theory. The careful reader will see that complex exponentials are used in computing the member stiffness which allow a single form of solution to include both compression and tension.

New subroutines encountered in this program include `CORLD` which computes the unbalanced load, `GSTIFF` which computes the geometric stiffness matrix, `RESET` which computes the new member deformations and `BMCOL` which is used to construct the member stiffness matrix including the effect of axial load. Subroutine `RNK` has been expanded to accommodate and change the member stiffness matrix by calling `BMCOL`.

New variables include `AI` which is used for the triple product that comes out of subroutine `PROD3`. `AJ` is the transpose of `AI` and the matrix `AK` is the member geometric stiffness matrix. `AI` is later destroyed by inserting member deformations in its first row and member forces in its second row.

4.4.1 Examples

Three examples are presented in this section. The first two involve frame buckling whereas the third involves an analysis for large deflections (small strain) on a circular cantilever beam.

4.4.1.1 *Example 4.2. Buckling of a Portal Frame*

This is a simple buckling example taken from Timoshenko (1936). Again, no attempt is made to obtain an "exact" value of the buckling load. (See earlier comments.) Timoshenko gives the buckling load in this ($E = 29\times10^6$ psi $I = 800$ in^4 $L = 144$ in) case to be

$$P_{\mathrm{CR}} = \frac{1.82EI}{L^2} = 2.04\times10^6 \,\mathrm{lbs}$$

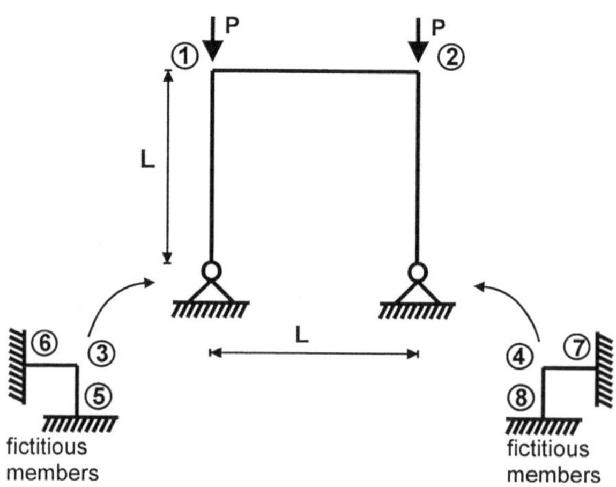

Figure 4.6 Plane frame buckling.

Figure 4.6 shows Timoshenko's buckling problem which is described in detail by the input file FR2DNL42.DAT which in turn is read by FR2DNL.FOR. Only partial results from the lengthy output file are listed below as the file FR2DNL42.OUT. The full load step is in this case specified to be 2,200,000 lbs. Note that fictitious bars with I=0 and 1.0 inch length have been added to model the pinned supports. Also large areas have been used in some members to simulate Timoshenko's neglect of member length change. By scanning the output file it is found that a negative term first appears on the diagonal of the system matrix during the 9th load step. The buckling load lies, therefore, between the 8th and 9th load steps. The 8th load step is of $\frac{9}{10} \times 2{,}200{,}000 = 1{,}980{,}000$ lbs. and the 9th load step is of $\frac{10}{10} \times 2{,}200{,}000 = 2{,}200{,}000$ lbs. This implies that the buckling load occurs between 1,980,000 lbs and 2,200,000 lbs which agrees with Timoshenko's result.

Input file FR2DNL42.DAT

```
     7      8     4    10      4
0.          -2200000.
            -2200000.
0.
0.
100.        144.        800.        90.          1    3
100.        144.        800.        90.          2    4
1.          144.        800.         0.          2    1
100.        1.          0.          90.          3    6
1.          1.          0.           0.          3    5
1.          1.          0.           0.          4    7
100.        1.          0.          90.          4    8
0.
0.
0.
0.
0.
0.
0.
```

Output file FR2DNL42.OUT

```
     7   NO. MEMBERS
     8   NO. NODES
     4   NO. SUPPORTS
    10   NO. LOAD STEPS
     4   NO. ITERATIONS/STEP

JOINT  LOADS
                           PX                       PY                      M
    1         .00000000D+00         -.22000000D+07          .00000000D+00
    2         .00000000D+00         -.22000000D+07          .00000000D+00
    3         .00000000D+00          .00000000D+00          .00000000D+00
    4         .00000000D+00          .00000000D+00          .00000000D+00
```

```
MEMBER PROPERTIES
          AREA            LENGTH              I              ANGLE  +END  -END

1  .10000000E+03  .14400000E+03  .80000000E+03  .90000000E+02    1     3
2  .10000000E+03  .14400000E+03  .80000000E+03  .90000000E+02    2     4
3  .10000000E+01  .14400000E+03  .80000000E+03  .00000000E+00    2     1
4  .10000000E+03  .10000000E+01  .00000000E+00  .90000000E+02    3     6
5  .10000000E+01  .10000000E+01  .00000000E+00  .00000000E+00    3     5
6  .10000000E+01  .10000000E+01  .00000000E+00  .00000000E+00    4     7
7  .10000000E+03  .10000000E+01  .00000000E+00  .90000000E+02    4     8

CURRENT MEMBER FORCES
   1       .00000000E+00        .00000000E+00        .00000000E+00
   2       .00000000E+00        .00000000E+00        .00000000E+00
   3       .00000000E+00        .00000000E+00        .00000000E+00
   4       .00000000E+00        .00000000E+00        .00000000E+00
   5       .00000000E+00        .00000000E+00        .00000000E+00
   6       .00000000E+00        .00000000E+00        .00000000E+00
   7       .00000000E+00        .00000000E+00        .00000000E+00

:

****ITERATION NUMBER    4
    LOAD STEP           8

JOINT DISPLACEMENTS
                  DX                  DY                TH
 1     -.64286104D-04     -.85467514D-07     .13680851D-06
 2     -.64287406D-04      .96456597D-07     .13680924D-06
 3     -.19722560D-10     -.60868362D-09     .61633452D-06
 4      .14104490D-10      .64596191D-09     .61635144D-06
MEMBER DISPLACEMENTS-FORCES
        DL          AL+         AL-           T           M+           M-
1-.87626D-07 -.30962D-06 .16990D-06 -.17647D+01 -.13103D+03 -.19380D-06
2 .93044D-07 -.30963D-06 .16991D-06  .18738D+01 -.13103D+03  .29267D-06
3-.13017D-08  .13555D-06 .13555D-06 -.26216D-03  .13103D+03  .13103D+03
4-.60868D-09 .61631D-06 -.19722D-10 -.17652D+01  .00000D+00  .00000D+00
5-.19307D-10 .61694D-06  .60870D-09 -.55990D-03  .00000D+00  .00000D+00
6 .13664D-10 .61571D-06 -.64597D-09  .39624D-03  .00000D+00  .00000D+00
7 .64596D-09 .61637D-06  .14104D-10  .18733D+01  .00000D+00  .00000D+00

MEMBER PROPERTIES
          AREA            LENGTH              I              ANGLE  +END  -END
1  .10000000E+03  .14400000E+03  .80000000E+03  .89997610E+02    1     3
2  .10000000E+03  .14400000E+03  .80000000E+03  .89997610E+02    2     4
3  .10000000E+01  .14400000E+03  .80000000E+03 -.67403680E-05    2     1
4  .10000000E+03  .10000000E+01  .00000000E+00  .90000000E+02    3     6
5  .10000000E+01  .10000000E+01  .00000000E+00 -.39115870E-01    3     5
6  .10000000E+01  .10000000E+01  .00000000E+00 -.39122570E-01    4     7
7  .10000000E+03  .10000000E+01  .00000000E+00  .90000000E+02    4     8

CURRENT MEMBER FORCES
   1      -.19798300E+07        .12208010E+05       -.25477250E-06
   2      -.19801700E+07        .12208020E+05        .27050080E-06
   3       .71968580E-02       -.12208020E+05       -.12208010E+05
   4      -.19798300E+07        .00000000E+00        .00000000E+00
   5       .23545330E+01        .00000000E+00        .00000000E+00
   6       .23405690E+01        .00000000E+00        .00000000E+00
   7      -.19801700E+07        .00000000E+00        .00000000E+00

*** ERROR =          .18657140E+00

:

****ITERATION NUMBER    2
```

```
    LOAD STEP          9

:

**NEG TERM ON THE DIAGONAL AT ROW    12

:

Stop - Program terminated.
```

4.4.1.2 *Example 4.3. Buckling of a Plane Frame*

This example which is described in Figure 4.7 is taken from the test problems that appear in the ANSYS (1987) manual. Ten load steps with 4 iterations per load step are used to solve the problem using FR2DNL.FOR and the input file FR2DNL43.DAT with a starting load of 30,000 lbs. Some of the output is listed in FR2DNL43.OUT. It includes the echo of the data file and the results of the last iteration of the last converged load step (load step 8, iteration 4). The buckling load is bounded between 9/10 × 30,000 = 27,000 lbs and 10/10 × 30,000 = 30,000 lbs. This result agrees with the ANSYS result of 27,789 lbs. (ANSYS, 1987, example no. 12.)

Output file FR2DNL43.OUT

```
    6  NO. MEMBERS
    7  NO. NODES
    4  NO. SUPPORTS
   10  NO. LOAD STEPS
    4  NO. ITERATIONS/STEP

 JOINT LOADS
                        PX                   PY                   M
   1        .00000000D+00        .00000000D+00        .00000000D+00
   2        .00000000D+00        .00000000D+00        .00000000D+00
   3       -.30000000D+05        .00000000D+00        .00000000D+00

 MEMBER PROPERTIES
           AREA           LENGTH               I             ANGLE  +END  -END
 1  .38730000E+01   .14400000E+03   .12500000E+01   .00000000E+00     2     1
 2  .33540000E+01   .96000000E+02   .93750000E+00   .90000000E+02     2     6
```

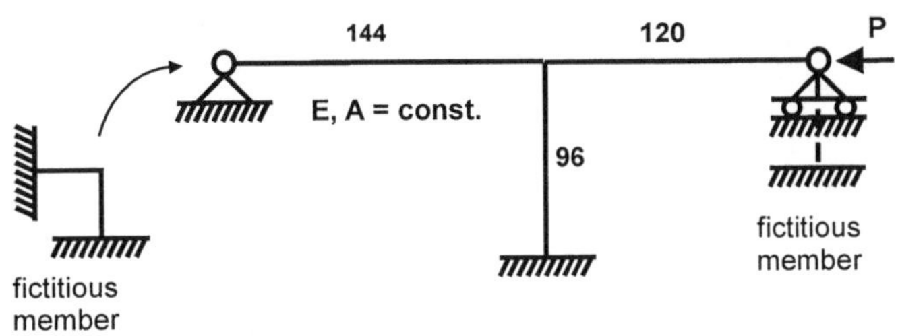

Figure 4.7 Plane frame of example 4.3.

```
3  .42430000E+01  .12000000E+03  .15000000E+01  .00000000E+00    3    2
4  .10000000E+05  .10000000E+01  .00000000E+00  .00000000E+00    1    4
5  .10000000E+05  .10000000E+01  .00000000E+00  .90000000E+02    1    5
6  .10000000E+05  .10000000E+03  .00000000E+00  .90000000E+02    3    7

CURRENT MEMBER FORCES
   1        .00000000E+00        .00000000E+00        .00000000E+00
   2        .00000000E+00        .00000000E+00        .00000000E+00
   3        .00000000E+00        .00000000E+00        .00000000E+00
   4        .00000000E+00        .00000000E+00        .00000000E+00
   5        .00000000E+00        .00000000E+00        .00000000E+00
   6        .00000000E+00        .00000000E+00        .00000000E+00

:

****ITERATION NUMBER    4
    LOAD STEP           8

JOINT DISPLACEMENTS
                     DX                  DY                  TH
 1     -.23518923D-14     -.10649286D-16     -.15353358D-09
 2     -.87463440D-09     -.10712609D-11      .75974923D-10
 3     -.10371566D-08     -.60452218D-12     -.67537903D-10


MEMBER DISPLACEMENTS-FORCES
          DL          AL+          AL-            T           M+           M-
1-.87462D-09 .75982D-10 -.15353D-09 -.70571D-03 -.87157D-04  .66174D-22
2-.76739D-12 .66864D-10 -.91108D-11 -.80432D-06  .73017D-04  .28502D-04
3-.16252D-09-.67542D-10  .75971D-10 -.17239D-03 -.16272D-19  .16048D-04
4-.23315D-14-.15353D-09  .10649D-16 -.69944D-03  .00000D+00  .00000D+00
5 .00000D+00-.15354D-09 -.23519D-14  .00000D+00  .00000D+00  .00000D+00
6 .00000D+00-.77909D-10 -.10372D-10  .00000D+00  .00000D+00  .00000D+00

MEMBER PROPERTIES
           AREA          LENGTH              I            ANGLE +END -END
1  .38730000E+01  .14400000E+03  .12500000E+01 -.29413460E-05    2    1
2  .33540000E+01  .96000000E+02  .93750000E+00  .90019980E+02    2    6
3  .42430000E+01  .12000000E+03  .15000000E+01 -.47256770E-05    3    2
4  .10000000E+05  .10000000E+01  .00000000E+00 -.19312930E-10    1    4
5  .10000000E+05  .10000000E+01  .00000000E+00  .90000000E+02    1    5
6  .10000000E+05  .10000000E+03  .00000000E+00  .90033780E+02    3    7

CURRENT MEMBER FORCES
   1      -.26993110E+05        .19375060E+02        .49922000E-19
   2      -.16752790E+01       -.23694280E+03       -.42474940E+03
   3      -.27000000E+05       -.22042710E-19        .21756780E+03
   4      -.26993110E+05        .00000000E+00        .00000000E+00
   5      -.13316010E+00        .00000000E+00        .00000000E+00
   6       .18108210E+01        .00000000E+00        .00000000E+00

*** ERROR =          .56081750E-03

****ITERATION NUMBER    2
    LOAD STEP           9

:

**NEG TERM ON THE DIAGONAL AT ROW
:

Stop - Program terminated.
```

4.4.1.3 *Example 4.4. Large Rotations of a Circular Cantilever Beam*

This example is intriguing because an initially circular cantilever beam is taken for an excursion of large rotations going through a straightened geometry and finally curving in the opposite direction. The beam is an aluminum (E = 10,000,000 psi) 1.0 inch square bar with a 45° circular bend and a radius of 100 in. It is subjected to incremental loading of up to 600 lbs in ten load steps. It is analyzed here, of course, as a plane beam with no thought given to lateral buckling effects which are discussed in Chapter 5.
The plane frame nonlinear analysis program uses the input file FR2DNL44.DAT with results of the input echoed at 300 lbs and 600 lbs loads listed below in the file FR2DNL44.OUT. For the 300 lbs load convergence is achieved at load step 4 and iteration 5 whereas at 600 lbs convergence is achieved at load step 10 and iteration 5. Figure 4.8 shows the beam unloaded, at its 300 lbs load position, and at the 600 lbs load position. The coordinates for these positions are calculated using its output data to be (64.40 ,44.54) and (49.76, 59.3) respectively.

Output file FR2DNL44.OUT

```
    8  NO. MEMBERS
    9  NO. NODES
    1  NO. SUPPORTS
   10  NO. LOAD STEPS
    5  NO. ITERATIONS/STEP

JOINT LOADS
                   PX                  PY                  M

   1        .00000000D+00       .00000000D+00       .00000000D+00
   2        .00000000D+00       .00000000D+00       .00000000D+00
   3        .00000000D+00       .00000000D+00       .00000000D+00
   4        .00000000D+00       .00000000D+00       .00000000D+00
   5        .00000000D+00       .00000000D+00       .00000000D+00
   6        .00000000D+00       .00000000D+00       .00000000D+00
   7        .00000000D+00       .00000000D+00       .00000000D+00
   8        .00000000D+00       .60000000D+03       .00000000D+00

MEMBER PROPERTIES
            AREA          LENGTH              I              ANGLE +END -END

1  .10000000E+01  .98140000E+01  .83300000E-01  .19690000E+02    1    9
2  .10000000E+01  .98140000E+01  .83300000E-01  .14060000E+02    2    1
3  .10000000E+01  .98140000E+01  .83300000E-01  .84400000E+01    3    2
4  .10000000E+01  .98140000E+01  .83300000E-01  .28100000E+01    4    3
5  .10000000E+01  .98140000E+01  .83300000E-01 -.28100000E+01    5    4
6  .10000000E+01  .98140000E+01  .83300000E-01 -.84400000E+01    6    5
7  .10000000E+01  .98140000E+01  .83300000E-01 -.14100000E+02    7    6
8  .10000000E+01  .98140000E+01  .83300000E-01 -.19700000E+02    8    7

CURRENT MEMBER FORCES
   1        .00000000E+00       .00000000E+00       .00000000E+00
   2        .00000000E+00       .00000000E+00       .00000000E+00
   3        .00000000E+00       .00000000E+00       .00000000E+00
   4        .00000000E+00       .00000000E+00       .00000000E+00
```

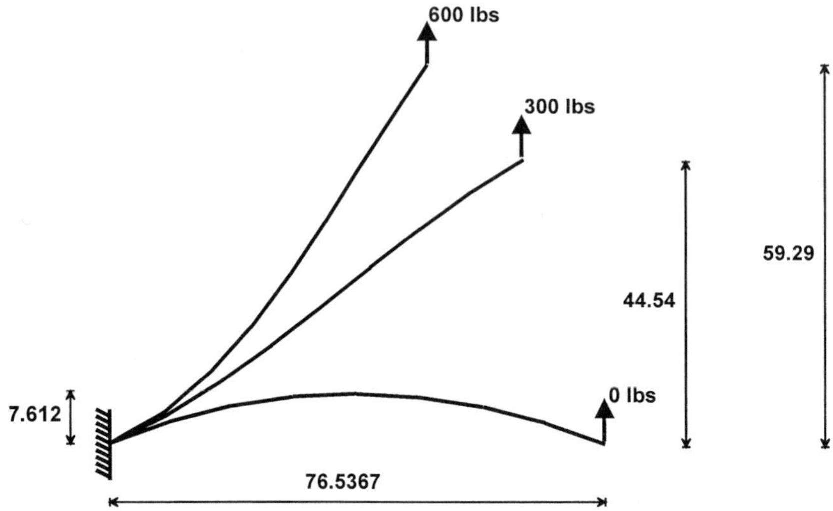

Figure 4.8 The Circular Cantilever Beam

```
   5        .00000000E+00        .00000000E+00        .00000000E+00
   6        .00000000E+00        .00000000E+00        .00000000E+00
   7        .00000000E+00        .00000000E+00        .00000000E+00
   8        .00000000E+00        .00000000E+00        .00000000E+00

:

****ITERATION NUMBER    5
    LOAD STEP           4
JOINT DISPLACEMENTS
                     DX                  DY                 TH
 1     -.28441240D-06      .58532742D-06      .12888902D-06
 2     -.11486064D-05      .19909044D-05      .18772723D-06
 3     -.20153363D-05      .31999202D-05      .79209549D-07
 4     -.16571571D-05      .27402412D-05     -.23386931D-06
 5      .11470688D-05     -.72113709D-06     -.68948487D-06
 6      .67864300D-05     -.78277321D-05     -.11489570D-05
 7      .14466492D-04     -.18315821D-04     -.14701809D-05
 8      .22566532D-04     -.31093912D-04     -.15801411D-05

MEMBER DISPLACEMENTS-FORCES
           DL          AL+          AL-            T           M+           M-

1-.14474D-10 .62579D-07 -.66310D-07 -.14748D-04  .10003D-01 -.11904D-01
2-.26657D-09 .19601D-07 -.39237D-07 -.27163D-03 -.62086D-07 -.10003D-01
3-.81091D-09-.72369D-07  .36148D-07 -.82628D-03 -.18453D-01  .62078D-07
4-.15612D-08-.17449D-06  .13859D-06 -.15908D-02 -.35766D-01  .17476D-01
5-.24384D-08-.23557D-06  .22005D-06 -.24846D-02 -.42695D-01  .34790D-01
6-.32851D-08-.22454D-06  .23494D-06 -.33473D-02 -.36420D-01  .41719D-01
7-.39654D-08-.14561D-06  .17562D-06 -.40405D-02 -.19668D-01  .34955D-01
8-.43699D-08-.38556D-07  .71404D-07 -.44528D-02 -.98066D-03  .17715D-01

MEMBER PROPERTIES
            AREA         LENGTH              I            ANGLE +END -END
1  .10000000E+01  .98140000E+01  .83300000E-01  .25914010E+02    1    9
2  .10000000E+01  .98140000E+01  .83300000E-01  .31575310E+02    2    1
```

```
3  .10000000E+01  .98140000E+01  .83300000E-01  .35605110E+02   3    2
4  .10000000E+01  .98140000E+01  .83300000E-01  .38079040E+02   4    3
5  .10000000E+01  .98140000E+01  .83300000E-01  .39043940E+02   5    4
6  .10000000E+01  .98140000E+01  .83300000E-01  .38454120E+02   6    5
7  .10000000E+01  .98140000E+01  .83300000E-01  .36231480E+02   7    6
8  .10000000E+01  .98140000E+01  .83300000E-01  .32387170E+02   8    7

CURRENT MEMBER FORCES
   1       .13110640E+03        .16673050E+05        -.19321220E+05
   2       .15708560E+03        .14164740E+05        -.16673050E+05
   3       .17465850E+03        .11770960E+05        -.14164740E+05
   4       .18502430E+03        .94533960E+04        -.11770960E+05
   5       .18897470E+03        .71667450E+04        -.94533960E+04
   6       .18656620E+03        .48611220E+04        -.71667450E+04
   7       .17731460E+03        .24862240E+04        -.48611220E+04
   8       .16069120E+03        .34999940E-09        -.24862240E+04

*** ERROR =          .15305700E-03

:

****ITERATION NUMBER   5
   LOAD STEP          10

JOINT DISPLACEMENTS
                   DX                  DY                  TH
 1     -.75269063D-08      .13462135D-07      .44011095D-08
 2     -.42210138D-07      .54311544D-07      .34856049D-08
 3     -.37527155D-07      .50153758D-07     -.38889369D-08
 4      .11075002D-07      .14945884D-07     -.81852663D-08
 5      .82273536D-07     -.30675756D-07     -.75555522D-08
 6      .13502084D-06     -.63018006D-07     -.63542219D-08
 7      .19694203D-06     -.10254573D-06     -.84039721D-08
 8      .26760207D-06     -.15348658D-06     -.91111953D-08

MEMBER DISPLACEMENTS-FORCES
          DL          AL+         AL-           T          M+          M-
1-.72085D-11 .28295D-08 -.15716D-08 -.73451D-05  .69511D-03 -.54112D-04
2-.11642D-10-.19747D-08 -.10592D-08 -.11863D-04 -.85111D-03 -.69511D-03
3-.93934D-11-.32508D-08  .41237D-08 -.95714D-05 -.40619D-03  .85111D-03
4 .14129D-10-.20701D-08  .22263D-08  .14397D-04 -.32656D-03  .40619D-03
5-.12925D-10 .10608D-08  .43110D-09 -.13170D-04  .43397D-03  .32656D-03
6 .34799D-11-.49632D-10 -.12510D-08  .35458D-05 -.22904D-03 -.43397D-03
7 .13495D-10-.91854D-09  .11312D-08  .13751D-04 -.12062D-03  .22904D-03
8-.12758D-10-.23530D-09  .47192D-09 -.13000D-04 -.75057D-18  .12062D-03

MEMBER PROPERTIES
           AREA          LENGTH               I           ANGLE +END -END
1  .10000000E+01  .98140000E+01  .83300000E-01  .29183540E+02   1    9
2  .10000000E+01  .98140000E+01  .83300000E-01  .40320540E+02   2    1
3  .10000000E+01  .98140000E+01  .83300000E-01  .48485710E+02   3    2
4  .10000000E+01  .98140000E+01  .83300000E-01  .54066620E+02   4    3
5  .10000000E+01  .98140000E+01  .83300000E-01  .57358410E+02   5    4
6  .10000000E+01  .98140000E+01  .83300000E-01  .58482090E+02   6    5
7  .10000000E+01  .98140000E+01  .83300000E-01  .57437180E+02   7    6
8  .10000000E+01  .98140000E+01  .83300000E-01  .54219520E+02   8    7

CURRENT MEMBER FORCES
   1       .29256510E+03        .24714460E+05        -.29855400E+05
   2       .38823760E+03        .20224930E+05        -.24714460E+05
   3       .44927400E+03        .16322060E+05        -.20224930E+05
   4       .48581960E+03        .12866480E+05        -.16322060E+05
   5       .50523640E+03        .96903770E+04        -.12866480E+05
   6       .51148580E+03        .66121240E+04        -.96903770E+04
   7       .50568080E+03        .34428410E+04        -.66121240E+04
```

```
   8          .48675760E+03          .58803160E-14          -.34428410E+04

*** ERROR =            .35856650E-03
:

Stop - Program terminated.
```

4.5 Problems

1.a. Show that the member stiffness matrix for a member with a hinge at the "+ end" is

$$(\mathbf{K}_{\mathrm{E}})_i = \begin{bmatrix} \left(\frac{AE}{L}\right)_i & 0 & 0 \\ 0 & 0 & 0 \\ 0 & 0 & \left(\frac{3EI}{L}\right)_i \end{bmatrix}$$

b. What is the stiffness matrix of a member with a hinge at its "-end".
c. Show that the two matrices can be used to perform an incremental plastic collapse analysis for a structure in which the formulation of plastic hinges can lead to collapse.

2. It is common to assume that the beams in a structural frame are rigid in order to simplify buckling calculations. Construct a real frame example and perform parametric studies to determine the extent to which this is true.

3. The AISC manual provides charts which can be used to determine the effective length of frame members appropriately. Construct examples and run parametric studies which will verify the accuracy of these charts.

4. Discuss the "complex" form of the solution to the beam column equations. Verify the code which is included in subroutine BMCOL of program FR2DNL.FOR. (Hint: Refer to section 1.5.3 and Appendix 1 .)

5. The geometric stiffness matrix for the plane frame is symmetric (as is the case of the truss). Find the potential from which the geometric stiffness matrix can be computed as a derivative. (Hint: For the truss let the potential $\phi = \sum_{members} F_i L_i$. When the F_i's are held constant $\nabla\phi = \mathbf{N}^{\mathrm{T}}\mathbf{F}$ and $\mathbf{K}_{\mathrm{G}} = \nabla\nabla\phi$.)

6. Compare the examples of this chapter with the results of some commercial program such as ANSYS. Discuss the effect of subdividing a member into many finite elements. How many element should be used? What is the effect of using a single finite element to model each member?

7. If a beam such as the one used in Example 4.4 is made with sufficient curvature it is possible to have such a beam buckle before reverse curvature develops. Construct such a case.

CHAPTER 5

NONLINEAR ANALYSIS OF SPACE FRAMES

5.1 Introduction

This chapter completes the sequence which began with trusses some pages back. As you move up in terms of complexity from trusses, space frames add new issues which revolve about the rotation matrix (see Appendix 3). Both trusses and plane frames can be described geometrically in terms of node coordinates. That is, certainly for the case of straight members, the geometry of trusses and plane frames can be described by giving the coordinates of the ends of the members. That is not the case for space frames where members can still be rotated about their axes after their ends have been located. From another side, joint rotations in plane frames can still be treated as scalars while joint rotations in space frames require the concept of a rotation matrix.

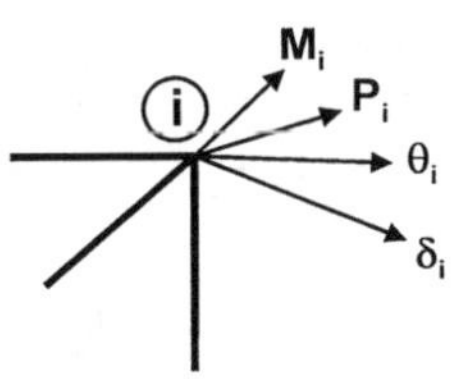

Figure 5.1 Typical joint *i*.

5.2 Linear Analysis

A typical joint i of a linear elastic space frame (Fig. 5.1) has four physical vectors associated with it,

$\boldsymbol{\delta}_i$	=	displacement vector
$\boldsymbol{\theta}_i$	=	(small) rotation vector
$\mathbf{P}_i$	=	applied load vector
$\mathbf{M}_i$	=	applied moment vector

These vectors are used to form the joint load matrix $\mathbf{P}$ and the joint displacement matrix $\boldsymbol{\delta}$ as

$$\mathbf{P} = \begin{Bmatrix} \mathbf{P}_1 \\ \mathbf{P}_2 \\ \vdots \\ \mathbf{P}_{NN} \end{Bmatrix}; \quad \boldsymbol{\delta} = \begin{Bmatrix} \boldsymbol{\delta}_1 \\ \boldsymbol{\delta}_2 \\ \vdots \\ \boldsymbol{\delta}_{NN} \end{Bmatrix} \tag{5.1}$$

with

$$\mathbf{P}_i = \begin{Bmatrix} (\mathbf{P}_i)_x \\ (\mathbf{P}_i)_y \\ (\mathbf{P}_i)_z \\ (\mathbf{M}_i)_x \\ (\mathbf{M}_i)_y \\ (\mathbf{M}_i)_z \end{Bmatrix}; \quad \boldsymbol{\delta}_i = \begin{Bmatrix} (\boldsymbol{\delta}_i)_x \\ (\boldsymbol{\delta}_i)_y \\ (\boldsymbol{\delta}_i)_z \\ (\boldsymbol{\theta}_i)_x \\ (\boldsymbol{\theta}_i)_y \\ (\boldsymbol{\theta}_i)_z \end{Bmatrix} \tag{5.2}$$

Here NN is the number of moveable nodes in the structure.

The members of the structure are described by the member force matrix $\mathbf{F}$ and the member displacement matrix $\boldsymbol{\Delta}$

$$\mathbf{F} = \begin{Bmatrix} \mathbf{F}_1 \\ \mathbf{F}_2 \\ \vdots \\ \mathbf{F}_{NB} \end{Bmatrix}; \quad \boldsymbol{\Delta} = \begin{Bmatrix} \boldsymbol{\Delta}_1 \\ \boldsymbol{\Delta}_2 \\ \vdots \\ \boldsymbol{\Delta}_{NB} \end{Bmatrix} \tag{5.3}$$

with

$$\mathbf{F}_i = \begin{Bmatrix} (\mathbf{F}_i^+)_{x^i} \\ (\mathbf{M}_i^+)_{x^i} \\ (\mathbf{M}_i^+)_{y^i} \\ (\mathbf{M}_i^+)_{z^i} \\ (\mathbf{M}_i^-)_{y^i} \\ (\mathbf{M}_i^-)_{z^i} \end{Bmatrix}; \quad \boldsymbol{\Delta}_i = \begin{Bmatrix} (\boldsymbol{\delta}_\mathrm{A})_{x^i} - (\boldsymbol{\delta}_\mathrm{C})_{x^i} \\ (\boldsymbol{\theta}_\mathrm{A})_{x^i} - (\boldsymbol{\theta}_\mathrm{C})_{x^i} \\ (\boldsymbol{\theta}_\mathrm{A})_{y^i} + [(\boldsymbol{\delta}_\mathrm{A})_{z^i} - (\boldsymbol{\delta}_\mathrm{C})_{z^i}]/L_i \\ (\boldsymbol{\theta}_\mathrm{A})_{z^i} - [(\boldsymbol{\delta}_\mathrm{A})_{y^i} - (\boldsymbol{\delta}_\mathrm{C})_{y^i}]/L_i \\ (\boldsymbol{\theta}_\mathrm{C})_{y^i} + [(\boldsymbol{\delta}_\mathrm{A})_{z^i} - (\boldsymbol{\delta}_\mathrm{C})_{z^i}]/L_i \\ (\boldsymbol{\theta}_\mathrm{C})_{z^i} - [(\boldsymbol{\delta}_\mathrm{A})_{y^i} - (\boldsymbol{\delta}_\mathrm{C})_{y^i}]/L_i \end{Bmatrix} \tag{5.4}$$

Here NB is the number of bars in the structure; $\mathbf{F}_i^+$ and $\mathbf{M}_i^+$ indicate the member forces and moments at the "+" end of the member; $\mathbf{F}_i^-$ and $\mathbf{M}_i^-$ the member forces and moments at the"-" end of the member and the subscripts x^i, y^i, z^i indicate vector components in the local coordinate system of member i. The generalized incidence matrix **N** has elements

$$N_{ij} = \begin{cases} \mathbf{N}_i^+ \mathcal{R}_i & \text{if node } j \text{ is the + end of member } i \\ \mathbf{N}_i^- \mathcal{R}_i & \text{if node } j \text{ is the -- end of member } i \\ \mathbf{0} & \text{otherwise} \end{cases} \tag{5.5}$$

where

$$\boldsymbol{N}_i^+ = \begin{bmatrix} 1 & 0 & 0 & 0 & 0 & 0 \\ 0 & 0 & 0 & 1 & 0 & 0 \\ 0 & 0 & 1/L_i & 0 & 1 & 0 \\ 0 & -1/L_i & 0 & 0 & 0 & 1 \\ 0 & 0 & 1/L_i & 0 & 0 & 0 \\ 0 & -1/L_i & 0 & 0 & 0 & 0 \end{bmatrix} \tag{5.6}$$

and

$$\boldsymbol{N}_i^- = \begin{bmatrix} -1 & 0 & 0 & 0 & 0 & 0 \\ 0 & 0 & 0 & -1 & 0 & 0 \\ 0 & 0 & -1/L_i & 0 & 0 & 0 \\ 0 & 1/L_i & 0 & 0 & 0 & 0 \\ 0 & 0 & -1/L_i & 0 & 1 & 0 \\ 0 & 1/L_i & 0 & 0 & 0 & 1 \end{bmatrix} \tag{5.7}$$

and the matrix $\mathcal{R}_i$ is formed from the three-dimensional rotation matrix $\mathbf{R}_i$ which describes the transformation from global to local coordinates for member i as

$$\mathcal{R}_i = \begin{bmatrix} \mathbf{R}_i & 0 \\ 0 & \mathbf{R}_i \end{bmatrix}$$

Equations 5.4-5.7 were obtained using identical steps to those of Chapter 4 for plane frames. If the member forces at the positive and negative ends are represented by f_i^+ and f_i^- then

$$f_i^+ = \begin{Bmatrix} \left(\mathbf{F}_i^+\right)_{x^i} \\ \left(\mathbf{F}_i^+\right)_{y^i} \\ \left(\mathbf{F}_i^+\right)_{z^i} \\ \left(\mathbf{M}_i^+\right)_{x^i} \\ \left(\mathbf{M}_i^+\right)_{y^i} \\ \left(\mathbf{M}_i^+\right)_{z^i} \end{Bmatrix}; \quad f_i^- = \begin{Bmatrix} \left(\mathbf{F}_i^-\right)_{x^i} \\ \left(\mathbf{F}_i^-\right)_{y^i} \\ \left(\mathbf{F}_i^-\right)_{z^i} \\ \left(\mathbf{M}_i^-\right)_{x^i} \\ \left(\mathbf{M}_i^-\right)_{y^i} \\ \left(\mathbf{M}_i^-\right)_{z^i} \end{Bmatrix}$$

Member equilibrium relates f_i^+ and f_i^- to $\mathbf{F}_i$ as

$$f_i^+ = \mathbf{N}_i^{+^1} \mathbf{F}_i\,; \quad f_i^- = \mathbf{N}_i^{-^\mathrm{T}} \mathbf{F}_i$$

where $\mathbf{N}_i^+$ and $\mathbf{N}_i^-$ have already been presented in Eqs 5.6 and 5.7. In order to develop the equations of equilibrium ($\mathbf{N}^\mathrm{T}\mathbf{F} = \mathbf{P}$) which must be assembled in a global coordinate system, it is necessary to transform the local member end forces using the transpose of the full rotation matrix $\mathbf{R}_i$ that is described in Appendix 3. Since 6 quantities (two vectors) are being transformed and not 3 as in the case of plane frames it is essential to stack two rotation matrices diagonally. This transformation yields $\mathcal{R}_i^\mathrm{T} \mathbf{N}_i^{+^\mathrm{T}} \mathbf{F}_i$ and $\mathcal{R}_i^\mathrm{T} \mathbf{N}_i^{-^\mathrm{T}} \mathbf{F}_i$. Thus for example the x-global component of a moment vector in the local coordinate system becomes

$$\left(\mathbf{M}_i^+\right)_x = R_{11}\left(\mathbf{M}_i^+\right)_{x^i} + R_{21}\left(\mathbf{M}_i^+\right)_{y^i} + R_{31}\left(\mathbf{M}_i^+\right)_{z^i}$$

Member displacements are related to nodal displacements in the local coordinate system of the member i as

$$\boldsymbol{\Delta}_i = \mathbf{N}_i^+ \mathcal{R}_i \boldsymbol{\delta}_\mathrm{A} + \mathbf{N}_i^- \mathcal{R}_i \boldsymbol{\delta}_\mathrm{C}$$

and the primitive stiffness matrix is given by

$$
(\mathbf{K})_i = \begin{bmatrix}
\frac{EA_i}{L_i} & 0 & 0 & 0 & 0 & 0 \\
0 & \frac{GJ_i}{L_i} & 0 & 0 & 0 & 0 \\
0 & 0 & \frac{4EI_i^{y^i}}{L_i} & 0 & \frac{2EI_i^{y^i}}{L_i} & 0 \\
0 & 0 & 0 & \frac{4EI_i^{z^i}}{L_i} & 0 & \frac{2EI_i^{z^i}}{L_i} \\
0 & 0 & \frac{2EI_i^{y^i}}{L_i} & 0 & \frac{4EI_i^{y^i}}{L_i} & 0 \\
0 & 0 & 0 & \frac{2EI_i^{z^i}}{L_i} & 0 & \frac{4EI_i^{z^i}}{L_i}
\end{bmatrix}
$$

where

L_i	=	length of member i
A_i	=	area of member i
J_i	=	torsional stiffness of member i
$I_i^{y^i}$	=	bending stiffness about the local y-axis of member i
$I_i^{z^i}$	=	bending stiffness about the local z-axis of member i
E	=	Young's modulus
G	=	modulus of rigidity

Global assemblage is identical to that presented for plane frames (Eq 4.11) to yield:

$$
\begin{array}{ccc}
 & \text{col A} & \text{col C} \\
(\mathbf{K}_\mathrm{E})_i = & \left[\begin{array}{c} \mathcal{R}_i^{\mathrm{T}}\mathbf{N}_i^{+\mathrm{T}}\mathbf{K}_i\mathbf{N}_i^{+}\mathcal{R}_i \\ \mathcal{R}_i^{\mathrm{T}}\mathbf{N}_i^{-\mathrm{T}}\mathbf{K}_i\mathbf{N}_i^{+}\mathcal{R}_i \end{array}\right. & \left.\begin{array}{c} \mathcal{R}_i^{\mathrm{T}}\mathbf{N}_i^{+\mathrm{T}}\mathbf{K}_i\mathbf{N}_i^{-}\mathcal{R}_i \\ \mathcal{R}_i^{\mathrm{T}}\mathbf{N}_i^{-\mathrm{T}}\mathbf{K}_i\mathbf{N}_i^{-}\mathcal{R}_i \end{array}\right] \begin{array}{l} \text{row A} \\ \text{row C} \end{array}
\end{array}
$$

5.3 Computer Program P9-FR3D.FOR

This program performs linear analysis of space frames. It is identical in form to FR2D.FOR and in fact is its direct extension into 3 dimensions. The only complication is the rotation matrix which will be discussed in detail below.

The orientation of members is described using three angles as input (instead of one for plane frames) thus paving the way to define the rotation matrix as a sequence of three rotations about the global coordinate axes that take the global coordinate system into the local system. Appendix 3 provides the three individual rotation matrices about the different axes and discusses compound rotations. One generic angle, θ, has been used in the appendix for illustrative purposes and ease of reference to the same 2-dimensional Figure A3.2. In general each member will posses three different angles of orientation. Figure 5.2 shows a member in space with the angles of orientation θ (anticlockwise), ϕ, (clockwise) and γ (clockwise) that the Z-axis, Y-axis and X-axis undergo respectively to rotate the global coordinate system (XYZ) into the local coordinate system (xyz). Anticlockwise directions are positive.

It is now a matter of direct matrix multiplication to obtain the rotation matrix for member i from the individual rotations about the X, Y, and Z-axes. This rotation matrix performs the following transformation

$$\begin{Bmatrix} x_i \\ y_i \\ z_i \end{Bmatrix} = \mathbf{R}_i \begin{Bmatrix} X \\ Y \\ Z \end{Bmatrix}$$

or

$$\begin{Bmatrix} x_i \\ y_i \\ z_i \end{Bmatrix} = \mathbf{R}_{iX} \mathbf{R}_{iY} \mathbf{R}_{iZ} \begin{Bmatrix} X \\ Y \\ Z \end{Bmatrix}$$

The product of the three rotation matrices forms $\mathbf{R}_i$ as

$$\mathbf{R}_i = \begin{bmatrix} 1 & 0 & 0 \\ 0 & \cos\gamma & \sin\gamma \\ 0 & -\sin\gamma & \cos\gamma \end{bmatrix} \begin{bmatrix} \cos\varphi & 0 & -\sin\varphi \\ 0 & 1 & 0 \\ \sin\varphi & 0 & \cos\varphi \end{bmatrix} \begin{bmatrix} \cos\theta & \sin\theta & 0 \\ -\sin\theta & \cos\theta & 0 \\ 0 & 0 & 1 \end{bmatrix}$$

to result in

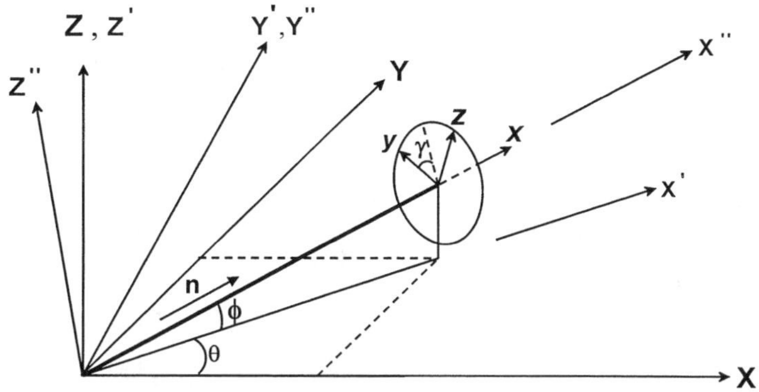

Figure 5.2 Rotations in 3D.

$$\mathbf{R}_i = \begin{bmatrix} \cos\varphi\cos\theta & \cos\varphi\sin\theta & -\sin\varphi \\ -\sin\theta\cos\gamma + \sin\gamma\sin\varphi\cos\theta & \cos\gamma\cos\theta + \sin\gamma\sin\varphi\sin\theta & \sin\gamma\cos\varphi \\ \sin\gamma\sin\theta + \cos\gamma\sin\varphi\cos\theta & -\sin\gamma\cos\theta + \cos\gamma\sin\varphi\sin\theta & \cos\gamma\cos\varphi \end{bmatrix}$$

Finally it is worth noting that the terms of each row are the projections of the local base vectors $(\mathbf{n}_x, \mathbf{n}_y, \mathbf{n}_z)$ of member i with respect to the global coordinate system.

5.3.1 Example 5.1 A 2-Storey Simple Space Frame

A two storey frame is loaded horizontally by two point loads as shown in Figure 5.3 L = 100 inches and P = 1.0 lb. Also shown are the global coordinate system and the joint numbering. Young's modulus is taken as E=29,000,000 psi and the shear modulus is taken as G=12,000,000 psi.

This problem is solved using FR3D.FOR and the data file FR3D51.DAT. The results are listed in the output file FR3D51.OUT. The reader is particularly alerted to the manner of supplying the angles of member orientation. The *X,Y,Z* global coordinates are assigned numbers 1,2,3 respectively and if member *i* rotates by, say, -90°, about the *Y*-axis only (clockwise) then in the `I` th line of input for member information use `NT(1,I)=2` and `TH(1,I)=-90;` `NT(2,I)=1` and `TH(2,I)=0;` and `NT(3,I)=1` and `TH(3,I)=0.` Any order of inserting a pair of axis and angle is actually valid as long as the zero value for that member does not appear first since NT(J,I) is used for specifying the angle of rotation for the J-th rotation of member I and TH(J,I) is the value of the J-th rotation about the axis NT(J,I) of member I (See input file below).

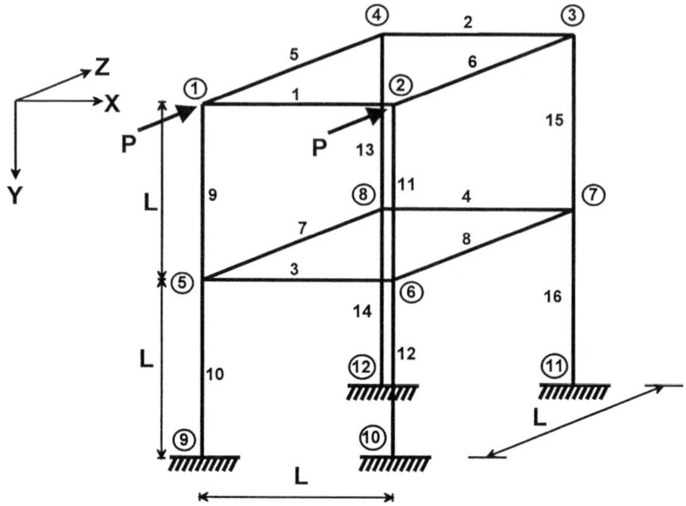

Figure 5.3 Simple space frame linear analysis.

Input file FR3D51.DAT

```
16      12      4
    1         0.0E0      0.0E0      1.0E0      0.0E0      0.0E0      0.0E0
    1         0.0E0      0.0E0      1.0E0      0.0E0      0.0E0      0.0E0
    1         0.0E0      0.0E0      0.0E0      0.0E0      0.0E0      0.0E0
    1         0.0E0      0.0E0      0.0E0      0.0E0      0.0E0      0.0E0
    1         0.0E0      0.0E0      0.0E0      0.0E0      0.0E0      0.0E0
    1         0.0E0      0.0E0      0.0E0      0.0E0      0.0E0      0.0E0
    1         0.0E0      0.0E0      0.0E0      0.0E0      0.0E0      0.0E0
    1         0.0E0      0.0E0      0.0E0      0.0E0      0.0E0      0.0E0
     1.0E0      1.0E2      2.0E0      1.0E0      1.0E0    2   1
     1.0E0      1.0E2      2.0E0      1.0E0      1.0E0    3   4
     1.0E0      1.0E2      2.0E0      1.0E0      1.0E0    6   5
     1.0E0      1.0E2      2.0E0      1.0E0      1.0E0    7   8
     1.0E0      1.0E2      2.0E0      1.0E0      1.0E0    4   1
     1.0E0      1.0E2      2.0E0      1.0E0      1.0E0    3   2
     1.0E0      1.0E2      2.0E0      1.0E0      1.0+0    8   5
     1.0E0      1.0E2      2.0E0      1.0E0      1.0E0    7   6
     1.0E0      1.0E2      2.0E0      1.0E0      1.0E0    1   5
     1.0E0      1.0E2      2.0E0      1.0E0      1.0E0    5   9
     1.0E0      1.0E2      2.0E0      1.0E0      1.0E0    2   6
     1.0E0      1.0E2      2.0E0      1.0E0      1.0E0    6  10
     1.0E0      1.0E2      2.0E0      1.0E0      1.0E0    4   8
     1.0E0      1.0E2      2.0E0      1.0E0      1.0E0    8  12
     1.0E0      1.0E2      2.0E0      1.0E0      1.0E0    3   7
     1.0E0      1.0E2      2.0E0      1.0E0      1.0E0    7  11
   1    0.0000   1    0.0000   1    0.0000
   1    0.00     1     .00     1    0.00
   1    0.00     1    0.00     1    0.00
   1    0.00     1    0.00     1    0.00
   2  -90.00     1    0.00     1    0.00
   2  -90.00     1    0.00     1    0.00
   2  -90.00     1    0.00     1    0.00
   2  -90.00     1    0.00     1    0.00
   3  -90.00     1    0.00     1    0.00
```

```
   3  -90.00      1    0.00     1    0.00
   3  -90.00      1    0.00     1    0.00
   3  -90.00      1    0.00     1    0.00
   3  -90.00      1    0.00     1    0.00
   3  -90.00      1    0.00     1    0.00
   3  -90.00      1    0.00     1    0.00
   3  -90.00      1    0.00     1    0.00
```

Output file FR3D51.OUT

```
   16  NO. MEMBERS
   12  NO. NODES
    4  NO. SUPPORTS

APPLIED LOADS-MOMENTS
              PX         PY         PZ         MX         MY         MZ
    1       0.00       0.00       1.00       0.00       0.00       0.00
    2       0.00       0.00       1.00       0.00       0.00       0.00
    3       0.00       0.00       0.00       0.00       0.00       0.00
    4       0.00       0.00       0.00       0.00       0.00       0.00
    5       0.00       0.00       0.00       0.00       0.00       0.00
    6       0.00       0.00       0.00       0.00       0.00       0.00
    7       0.00       0.00       0.00       0.00       0.00       0.00
    8       0.00       0.00       0.00       0.00       0.00       0.00

MEMBER PROPERTIES
          AREA     LENGTH           IX         IY          IZ +END -END
    1   1.0000  100.0000       2.0000     1.0000      1.0000    2    1
    2   1.0000  100.0000       2.0000     1.0000      1.0000    3    4
    3   1.0000  100.0000       2.0000     1.0000      1.0000    6    5
    4   1.0000  100.0000       2.0000     1.0000      1.0000    7    8
    5   1.0000  100.0000       2.0000     1.0000      1.0000    4    1
    6   1.0000  100.0000       2.0000     1.0000      1.0000    3    2
    7   1.0000  100.0000       2.0000     1.0000      1.0000    8    5
    8   1.0000  100.0000       2.0000     1.0000      1.0000    7    6
    9   1.0000  100.0000       2.0000     1.0000      1.0000    1    5
   10   1.0000  100.0000       2.0000     1.0000      1.0000    5    9
   11   1.0000  100.0000       2.0000     1.0000      1.0000    2    6
   12   1.0000  100.0000       2.0000     1.0000      1.0000    6   10
   13   1.0000  100.0000       2.0000     1.0000      1.0000    4    8
   14   1.0000  100.0000       2.0000     1.0000      1.0000    8   12
   15   1.0000  100.0000       2.0000     1.0000      1.0000    3    7
   16   1.0000  100.0000       2.0000     1.0000      1.0000    7   11

MEMBER PROPERTIES
              AXIS/ANG       AXIS/ANG       AXIS/ANG
    1    1      0.0000 1       0.0000 1       0.0000
    2    1      0.0000 1       0.0000 1       0.0000
    3    1      0.0000 1       0.0000 1       0.0000
    4    1      0.0000 1       0.0000 1       0.0000
    5    2    -90.0000 1       0.0000 1       0.0000
    6    2    -90.0000 1       0.0000 1       0.0000
    7    2    -90.0000 1       0.0000 1       0.0000
    8    2    -90.0000 1       0.0000 1       0.0000
    9    3    -90.0000 1       0.0000 1       0.0000
   10    3    -90.0000 1       0.0000 1       0.0000
   11    3    -90.0000 1       0.0000 1       0.0000
   12    3    -90.0000 1       0.0000 1       0.0000
   13    3    -90.0000 1       0.0000 1       0.0000
   14    3    -90.0000 1       0.0000 1       0.0000
   15    3    -90.0000 1       0.0000 1       0.0000
   16    3    -90.0000 1       0.0000 1       0.0000
```

```
JOINT DISPLACEMENTS
              X            Y            Z          TH X         TH Y          TH Z
 1-.66223D-08 -.65806D-05 .60241D-02 -.15806D-04 .38511D-10 -.26220D-10
 2-.66223D-08 -.65806D-05 .60241D-02 -.15806D-04 .38511D-10 -.26220D-10
 3 .30631D-11  .65806D-05 .60224D-02 -.15792D-04 .38516D-10 -.12565D-10
 4 .30632D-11  .65806D-05 .60224D-02 -.15792D-04 .38516D-10 -.12565D-10
 5-.27658D-08 -.47005D-05 .26166D-02 -.23601D-04 .62792D-11 -.35262D-10
 6-.27658D-08 -.47005D-05 .26166D-02 -.23601D-04 .62792D-11 -.35262D-10
 7 .42579D-09  .47005D-05 .26166D-02 -.23593D-04 .62814D-11 -.11539D-10
 8 .42579D-09  .47005D-05 .26166D-02 -.23593D-04 .62814D-11 -.11539D-10

MEMBER FORCES
        THRUST      TORQUE       MY(+)        MZ(+)        MY(-)        MZ(-)
 1-.93113E-12 .23039E-05 .10460E-04 -.45321E-04  .10460E-04 -.45321E-04
 2-.38750E-12 .23039E-05 .10468E-04 -.21647E-04  .10468E-04 -.21647E-04
 3-.15192E-11 .26915E-05-.65876E-05 -.61143E-04 -.65876E-05 -.61143E-04
 4-.56272E-12 .26915E-05-.65837E-05 -.19934E-04 -.65837E-05 -.19934E-04
 5-.49989E+00 .32813E-05-.48302E-04  .27257E+02 -.48305E-04  .27266E+02
 6-.49989E+00 .32813E-05-.48302E-04  .27257E+02 -.48305E-04  .27266E+02
 7-.17124E-03 .56956E-05-.44606E-04  .40893E+02 -.44607E-04  .40898E+02
 8-.17124E-03 .56956E-05-.44605E-04  .40893E+02 -.44607E-04  .40898E+02
 9 .54523E+00-.55178E-05 .27266E+02  .16276E-04  .22745E+02  .11031E-04
10 .13631E+01-.82227E-05 .18153E+02  .73185E-05  .31841E+02  .27770E-04
11 .54523E+00-.55178E-05 .27266E+02  .16276E-04  .22745E+02  .11031E-04
12 .13631E+01-.82227E-05 .18153E+02  .73185E-05  .31841E+02  .27770E-04
13-.54523E+00-.55164E-05 .27257E+02 -.13951E-04  .22732E+02 -.13356E-04
14-.13631E+01-.82211E-05 .18161E+02 -.20891E-04  .31845E+02 -.14198E-04
15-.54523E+00-.55164E-05 .27257E+02 -.13951E-04  .22732E+02 -.13356E-04
16-.13631E+01-.82211E-05 .18161E+02 -.20891E-04  .31845E+02 -.14198E-04

Stop - Program terminated.
```

5.3.1 Example 5.2 A More Complex Space Frame

A 16 member space frame having four members rigidly held at the corners of a square layout of 504 inches is loaded at nodes 1-6 (Fig. 5.4). Three slanted members reduce the base of the second floor to a rectangle of 324×396 inches. Young's modulus is taken as E=29,000,000 psi, the shear modulus is taken as G=12,000,000 psi and *P*=10,000 lbs. The problem is solved using FR3D.FOR and the data file FR3D52.DAT which is listed below together with the results FR3D52.OUT.

Input file FR3D52.DAT

```
16     12      4
     0.0E0      1.0E4      -1.0E4     0.0E0      0.0E0      0.0E0
     0.0E0      1.0E4      -1.5E4     0.0E0      0.0E0      0.0E0
     0.0E0      0.0E0      -0.75E4    0.0E0      0.0E0      0.0E0
     0.0E0      0.0E0      -0.5E4     0.0E0      0.0E0      0.0E0
     0.0E0      1.25E4     0.0E0      0.0E0      0.0E0      0.0E0
     0.0E0      1.25E4     0.0E0      0.0E0      0.0E0      0.0E0
     0.0E0      0.0E0      0.0E0      0.0E0      0.0E0      0.0E0
     0.0E0      0.0E0      0.0E0      0.0E0      0.0E0      0.0E0
    1.5E0    324.0E0    1.5E0     1.0E0     0.5E0    2    1
    1.5E0    324.0E0    1.5E0     1.0E0     0.5E0    3    4
    1.5E0    324.0E0    1.5E0     1.0E0     0.5E0    6    5
    1.5E0    324.0E0    1.5E0     1.0E0     0.5E0    7    8
```

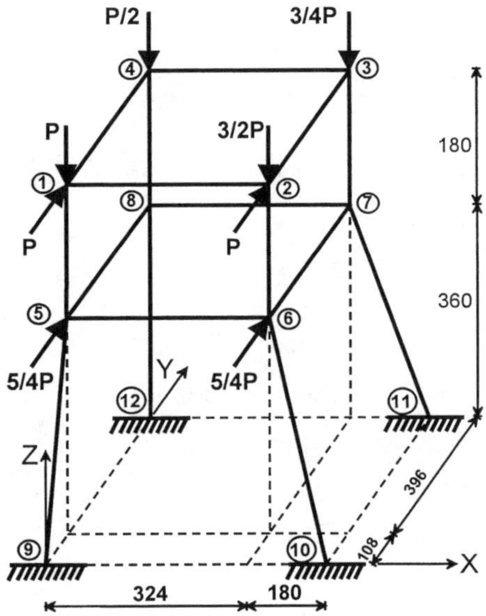

Figure 5.4 A more complex space frame linear analysis.

```
      1.5E0   396.0E0       1.5E0     1.0E0      0.5E0   4   1
      1.5E0   396.0E0       1.5E0     1.0E0      0.5E0   3   2
      1.5E0   396.0E0       1.5E0     1.0E0      0.5+0   8   5
      1.5E0   396.0E0       1.5E0     1.0E0      0.5E0   7   6
      1.5E0   180.0E0       1.5E0     1.0E0      0.5E0   1   5
      1.5E0375.8510E0       1.5E0     1.0E0      0.5E0   5   9
      1.5E0   180.0E0       1.5E0     1.0E0      0.5E0   2   6
      1.5E0416.7301E0       1.5E0     1.0E0      0.5E0   6  10
      1.5E0   180.0E0       1.5E0     1.0E0      0.5E0   4   8
      1.5E0   360.0E0       1.5E0     1.0E0      0.5E0   8  12
      1.5E0   180.0E0       1.5E0     1.0E0      0.5E0   3   7
      1.5E0402.4922E0       1.5E0     1.0E0      0.5E0   7  11
    1     0.00      1    0.00     1    0.00
    1     0.00      1    0.00     1    0.00
    1     0.00      1    0.00     1    0.00
    1     0.00      1    0.00     1    0.00
    3    90.00      1    0.00     1    0.00
    3    90.00      1    0.00     1    0.00
    3    90.00      1    0.00     1    0.00
    3    90.00      1    0.00     1    0.00
    2   -90.00      1    0.00     1    0.00
    3    90.00      2  -73.3008   1    0.00
    2   -90.00      1    0.00     1    0.00
    3   149.0362    2  -59.7537   1    0.00
    2   -90.00      1    0.00     1    0.00
    2   -90.00      1    0.00     1    0.00
    2   -90.00      1    0.00     1    0.00
    3   180.00      2  -63.4349   1    0.00
```

Output file FR3D52.OUT

```
    16  NO. MEMBERS
    12  NO. NODES
```

4 NO. SUPPORTS

APPLIED LOADS-MOMENTS

	PX	PY	PZ	MX	MY	MZ
1	0.00	10000.00	-10000.00	0.00	0.00	0.00
2	0.00	10000.00	-15000.00	0.00	0.00	0.00
3	0.00	0.00	-7500.00	0.00	0.00	0.00
4	0.00	0.00	-5000.00	0.00	0.00	0.00
5	0.00	12500.00	0.00	0.00	0.00	0.00
6	0.00	12500.00	0.00	0.00	0.00	0.00
7	0.00	0.00	0.00	0.00	0.00	0.00
8	0.00	0.00	0.00	0.00	0.00	0.00

MEMBER PROPERTIES

	AREA	LENGTH	IX	IY	IZ	+END	-END
1	1.5000	324.0000	1.5000	1.0000	0.5000	2	1
2	1.5000	324.0000	1.5000	1.0000	0.5000	3	4
3	1.5000	324.0000	1.5000	1.0000	0.5000	6	5
4	1.5000	324.0000	1.5000	1.0000	0.5000	7	8
5	1.5000	396.0000	1.5000	1.0000	0.5000	4	1
6	1.5000	396.0000	1.5000	1.0000	0.5000	3	2
7	1.5000	396.0000	1.5000	1.0000	0.5000	8	5
8	1.5000	396.0000	1.5000	1.0000	0.5000	7	6
9	1.5000	180.0000	1.5000	1.0000	0.5000	1	5
10	1.5000	375.8510	1.5000	1.0000	0.5000	5	9
11	1.5000	180.0000	1.5000	1.0000	0.5000	2	6
12	1.5000	416.7301	1.5000	1.0000	0.5000	6	10
13	1.5000	180.0000	1.5000	1.0000	0.5000	4	8
14	1.5000	360.0000	1.5000	1.0000	0.5000	8	12
15	1.5000	180.0000	1.5000	1.0000	0.5000	3	7
16	1.5000	402.4922	1.5000	1.0000	0.5000	7	11

MEMBER PROPERTIES

		AXIS/ANG		AXIS/ANG		AXIS/ANG
1	1	0.0000	1	0.0000	1	0.0000
2	1	0.0000	1	0.0000	1	0.0000
3	1	0.0000	1	0.0000	1	0.0000
4	1	0.0000	1	0.0000	1	0.0000
5	3	90.0000	1	0.0000	1	0.0000
6	3	90.0000	1	0.0000	1	0.0000
7	3	90.0000	1	0.0000	1	0.0000
8	3	90.0000	1	0.0000	1	0.0000
9	2	-90.0000	1	0.0000	1	0.0000
10	3	90.0000	2	-73.3008	1	0.0000
11	2	-90.0000	1	0.0000	1	0.0000
12	3	149.0362	2	-59.7537	1	0.0000
13	2	-90.0000	1	0.0000	1	0.0000
14	2	-90.0000	1	0.0000	1	0.0000
15	2	-90.0000	1	0.0000	1	0.0000
16	3	180.0000	2	-63.4349	1	0.0000

JOINT DISPLACEMENTS

	X	Y	Z	TH X	TH Y	TH Z
1	.31121D+03	.32938D+04	-.86769D+03	.77391D+00	.15139D-02	.20835D+01
2	.31120D+03	.39788D+04	-.93953D+03	.45530D+00	.35516D+00	.17371D+01
3	-.45338D+03	.39788D+04	-.24543D+03	.17178D+00	.53939D+00	.16669D+01
4	-.45339D+03	.32938D+04	-.23672D+00	.14223D+00	.52633D+00	.21566D+01
5	.30038D+03	.28925D+04	-.86766D+03	-.39089D+01	.42206D+00	.20801D+01
6	.30040D+03	.36323D+04	-.93948D+03	-.27790D+01	-.53028D+00	.55792D+00
7	-.49041D+03	.36323D+04	-.24539D+03	-.20742D+01	-.11810D+00	.37585D+00
8	-.49039D+03	.28925D+04	-.19700D+00	-.19504D+01	-.13323D+00	.22746D+01

	THRUST	TORQUE	MY(+)	MZ(+)	MY(-)	MZ(-)
1	-.18385E+04	-.17700E+05	.83465E+04	-.70244E+05	-.54960E+05	-.39243E+05
2	.30677E+03	.16418E+04	-.11907E+06	-.76283E+05	-.12141E+06	-.32451E+05

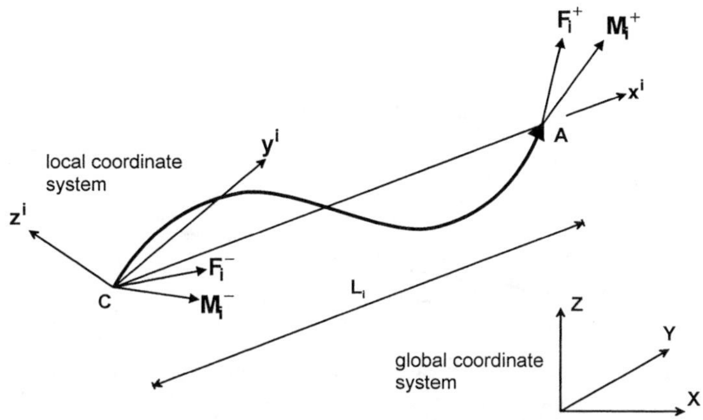

Figure 5.5 Typical Member *i*

```
 3 .25319E+04 .62773E+05-.23334E+06 -.32712E+06 -.62864E+05 -.19087E+06
 4-.33754E+04-.68816E+04-.47254E+06 -.34226E+06 -.47525E+06 -.17231E+06
 5-.67818E+04 .23855E+05 .80750E+06  .44256E+05  .71498E+06  .38902E+05
 6-.55644E+04 .83744E+04 .65316E+06 -.52823E+05  .61163E+06 -.47679E+05
 7-.77236E+03-.25240E+05 .21064E+07  .46774E+05  .23932E+07  .32525E+05
 8-.51463E+04 .18736E+05 .17848E+07 -.34283E+06  .18880E+07 -.32950E+06
 9-.62992E+04 .34248E+03 .78815E+05 -.69728E+06  .21432E+06  .57182E+05
10 .88253E+04 .10122E+06-.25132E+07  .21485E+06 -.31164E+07  .19992E+06
11-.11662E+05 .11792E+06 .27161E+02 -.62933E+06 -.28528E+06 -.10825E+06
12 .24232E+04 .66729E+05-.13442E+07  .14232E+07 -.16065E+07  .15305E+07
13-.95869E+04-.11803E+05-.97552E+05  .80914E+06 -.11497E+06 -.47200E+06
14-.23875E+05 .11373E+06 .61546E+06 -.16275E+07  .63692E+06 -.17846E+07
15-.99517E+04 .12911E+06 .11070E+06 -.65152E+06 -.10116E+06 -.28966E+06
16-.16825E+05 .56519E+05-.55497E+06  .17076E+07 -.57199E+06  .18291E+07
```

5.4 Nonlinear Effects

While the approach to geometric nonlinearities is consistent throughout the text, it is not the only approach possible. It is of some interest to note that there is an on-going discussion of three-dimensional frames in the literature. This discussion includes the work of Chan and Zhou (1994), Kuo, Yang and Chou (1993), Izzudin and Elnashai (1993), Krysl (1993).

Trusses and plane frames have moved forward with few surprises when developing the geometric stiffness matrix. That is not the case with three-dimensional frames where the use of the three-dimensional rotation matrix (see Appendix 3) leads to a geometric stiffness matrix which is unsymmetric. There is another piece to the description of the space frame. Earlier in the case of the plane frame it was argued that the member stiffness matrix should come from beam-column theory. For the space frame there is an analogous three-dimensional beam column which should be used for the member stiffness. That is discussed below in Chapter 8.

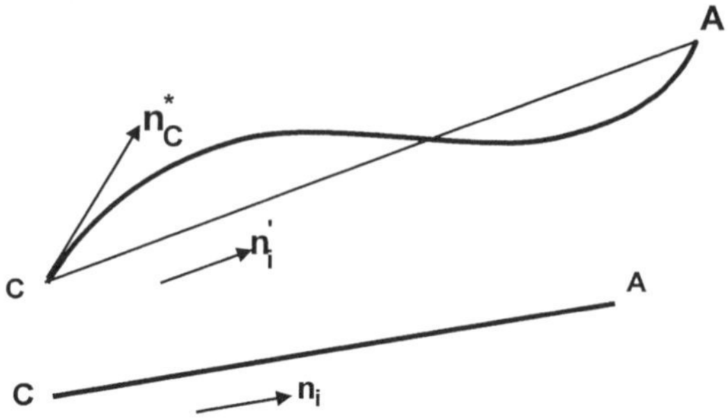

Figure 5.6 Unit vector notation.

If the discussion of space frames is to include finite rotations (large displacements but small strains) then it is not possible to use a rotation vector to describe either joint rotations or member rotations. What must be used in both of these cases is a rotation matrix. Euler's Theorem (see Appendix 3) shows the equivalence between a rotation matrix and a finite rotation about a single axis. This allows the small rotation vector of linear analysis to be described as a perturbation of this rotation about a single axis.

The assumption of small rotational strain comes in when member forces are to be computed given a deformed structure. Their computation is essential if you want to perform incremental analysis since these internal forces (the equivalent of prestress in trusses) must be saved within the members for the next increment of load to be applied. It is argued that small strain implies that the rotation of one end of a member with respect to the other end can be described using a small rotation vector. This is done in the following manner. Figure 5.5 represents a typical case of deformed equilibrium. Associated with joints A and C are the rotation matrices $\mathbf{R}_\mathrm{A}$ and $\mathbf{R}_\mathrm{C}$ which describe the finite rotation of these joints (similar to the rotation matrix, **R**, of Eq. A3.3 in Appendix 3); the rigid body motion of member i is described by two rotation matrices $\mathbf{R}_i^{\mathrm{old}}$ and $\mathbf{R}_i^{\mathrm{new}}$ which correspond to the undeformed and deformed rigid body positions.

In order to apply linear structural theory to compute member forces, it is necessary to have the rotation of joints A and C relative to the rigid body motion of member i. Let the unit vector $\mathbf{n}_i$ (Figure 5.6) represent the undeformed member axis (i.e. $\mathbf{n}_i$ describes the direction of the initial line from joint C to joint A). In general, the local coordinate system of member i defines the rigid body motion of this member. Note that the rigid body motion of a member is not fully determined by the unit vector $\mathbf{n}_i$ since rotation about the member axis is still possible given $\mathbf{n}_i$. In order to complete the description of the rigid body motion it will be assumed that each member is fixed to its negative joint with regard to rigid body rotation (twist) about the member axis.

Let the unit vector $\mathbf{n}_i^{'}$ describe the deformed member axis. ($\mathbf{n}_i^{'}$ is determined from the deformed coordinates of joints A and C.) Furthermore, let the unit vector $\mathbf{n}_\mathrm{C}^{*} = \mathbf{R}_\mathrm{C}^\mathrm{T}\ \mathbf{n}_i$ describe the deformed member axis at joint C. The end of member rotation of member i at joint C is then the vector $\boldsymbol{\omega}_C$ (see Appendix 3, Figure A3.4) with the properties

$$\begin{gathered}\boldsymbol{\omega}_C \sim \mathbf{n}_i^{'} \times \mathbf{n}_\mathrm{C}^{*} \\ \left|\boldsymbol{\omega}_\mathrm{C}\right| = \cos^{-1}\left(\mathbf{n}_i^{'} \cdot \mathbf{n}_\mathrm{C}^{*}\right)\end{gathered} \tag{5.8}$$

Equation 5.8 implies that the direction of $\boldsymbol{\omega}_\mathrm{C}$ is perpendicular to the plane containing the vectors $\mathbf{n}_i^{'}$ and $\mathbf{n}_\mathrm{C}^{*}$ and has a magnitude in radians equal to the angle between the two vectors. It remains to compute the end of member rotation vector $\boldsymbol{\omega}_\mathrm{A}$ for joint A. Let $\boldsymbol{\omega}_\mathrm{AC}$ be the relative rotation of joint A with respect to joint C. It follows that

$$\boldsymbol{\omega}_\mathrm{A} = \boldsymbol{\omega}_\mathrm{C} + \boldsymbol{\omega}_\mathrm{AC} \tag{5.9}$$

However, $\boldsymbol{\omega}_\mathrm{AC}$ is simply the vector form (Eqs A3.4, A3.5 of Appendix 3) of the matrix product $\mathbf{R}_\mathrm{A}^\mathrm{T}\mathbf{R}_\mathrm{C}$. The matrix product itself is the compound rotation describing the relative rotation of joint A with respect to joint C. Adding the two rotations in Eq. 5.9 is valid since we are actually dealing with pure rotations that are small. It follows that the kinematic description of a structure is now complete.

5.5 The Geometric Stiffness Matrix

Again, the starting point for constructing the geometric stiffness matrix is the contribution of bar i to the equilibrium equations

$$\left(\mathbf{N}^\mathrm{T}\mathbf{F}\right)_i = \begin{Bmatrix} \vdots \\ \mathcal{R}_i^\mathrm{T}\mathbf{N}_i^{+\mathrm{T}}\mathbf{F}_i \\ \vdots \\ \mathcal{R}_i^\mathrm{T}\mathbf{N}_i^{-\mathrm{T}}\mathbf{F}_i \\ \vdots \end{Bmatrix} \begin{matrix} \vdots \\ \text{row A} \\ \vdots \\ \text{row C} \\ \vdots \end{matrix} \tag{5.10}$$

The gradient of Eq. 5.10 is then the contribution of member i to the geometric stiffness matrix. This contribution (Eq. 5.11) has the usual form of four sub matrices that have to be properly inserted into the global geometric stiffness matrix according to the member's incidence. Here the $(\mathbf{K}_G)_{iIJ}$ terms are 6×6 matrices.

$$
(\mathbf{K}_G)_i = \begin{bmatrix} & \vdots & & \vdots & \\ \cdots & (\mathbf{K}_G)_{i11} & \cdots & (\mathbf{K}_G)_{i12} & \cdots \\ & \vdots & & \vdots & \\ \cdots & (\mathbf{K}_G)_{i21} & \cdots & (\mathbf{K}_G)_{i22} & \cdots \\ & \vdots & & \vdots & \end{bmatrix} \begin{matrix} \vdots \\ \text{row A} \\ \vdots \\ \text{row C} \\ \vdots \end{matrix} \qquad (5.11)
$$

(columns: col A, col C)

There are two points to be made with regard to the member equilibrium equation. First, the end-of-member forces constitute two vectors, a force and a moment at each end of a member. Second, because of the choice which has been made in defining the member force matrix, the shears are written in terms of the moments. This will lead to two kinds of terms when a gradient is used to form the geometric stiffness matrix. In one type of terms ('stretch' terms) the moments are kept fixed and the gradients of the L_i^{-1} terms are computed; in the other type the vectors are held constant and the gradient is computed as a small rotation ('rotation' terms). This can be seen by looking directly at the terms in Eq 5.10. Using the product rule for differentiation, it follows that a small perturbation of either term in this equation has the form

$$
\mathrm{d}\left(\mathcal{R}_i{}^{\mathrm{T}}\mathbf{N}_i^{\pm\,\mathrm{T}}\mathbf{F}_i\right) = \mathrm{d}\left(\mathcal{R}_i{}^{\mathrm{T}}\mathbf{N}_i^{\pm\,\mathrm{T}}\mathbf{F}_i\right)_{\text{fixed } \theta_i, \varphi_i, \gamma_i} + \left(\mathrm{d}\mathcal{R}_i{}^{\mathrm{T}}\right)\mathbf{N}_i^{\pm\,\mathrm{T}}\mathbf{F}_i \qquad (5.12)
$$

in which the member force matrix $\mathbf{F}_i$ is held fixed.

The stretch terms are most easy to obtain. These are derived from the first term of the right hand side of Eq. 5.12 as the gradient, $\nabla\left(\mathcal{R}_i{}^{\mathrm{T}}\mathbf{N}_i^{\pm\,\mathrm{T}}\mathbf{F}_i\right)_{\text{fixed } \theta_i, \varphi_i, \gamma_i}$. Note that the matrices $\mathbf{N}_i^{\pm}$ only contain constants and the term L_i^{-1} which arises when member shears are obtained from moments by dividing by the member length. Clearly, L_i depends only on the coordinates of the end of the member. The gradient of L_i^{-1} then follows directly as

$$
\begin{aligned}
\frac{\partial L_i^{-1}}{\partial x_{\mathrm{A}}} &= -(\mathbf{n}_i)_x L_i^{-2}; \quad \frac{\partial L_i^{-1}}{\partial x_{\mathrm{C}}} = -\frac{\partial L_i^{-1}}{\partial x_{\mathrm{A}}} \\
\frac{\partial L_i^{-1}}{\partial y_{\mathrm{A}}} &= -(\mathbf{n}_i)_y L_i^{-2}; \quad \frac{\partial L_i^{-1}}{\partial y_{\mathrm{C}}} = -\frac{\partial L_i^{-1}}{\partial y_{\mathrm{A}}} \\
\frac{\partial L_i^{-1}}{\partial z_{\mathrm{A}}} &= -(\mathbf{n}_i)_z L_i^{-2}; \quad \frac{\partial L_i^{-1}}{\partial z_{\mathrm{C}}} = -\frac{\partial L_i^{-1}}{\partial z_{\mathrm{A}}}
\end{aligned}
\tag{5.13}
$$

Let each $(\mathbf{K}_{\mathrm{G}})_{iIJ}$ term in Eq. 5.11 be written as the sum of a stretch term and a rotation term according to Eq. 5.12

$$
(\mathbf{K}_{\mathrm{G}})_{iIJ} = (\mathbf{K}_{\mathrm{G}})_{iIJ}^{\mathrm{S}} + (\mathbf{K}_{\mathrm{G}})_{iIJ}^{\mathrm{R}}
$$

The stretch term can be written directly as

$$
(\mathbf{K}_{\mathrm{G}})_{i11}^{\mathrm{S}} = (\mathbf{K}_{\mathrm{G}})_{i22}^{\mathrm{S}} = \begin{bmatrix}
k_{\mathrm{G}11} & k_{\mathrm{G}12} & k_{\mathrm{G}13} & -k_{\mathrm{G}11} & -k_{\mathrm{G}12} & -k_{\mathrm{G}13} \\
k_{\mathrm{G}21} & k_{\mathrm{G}22} & k_{\mathrm{G}23} & -k_{\mathrm{G}21} & -k_{\mathrm{G}22} & -k_{\mathrm{G}23} \\
k_{\mathrm{G}31} & k_{\mathrm{G}23} & k_{\mathrm{G}33} & -k_{\mathrm{G}31} & -k_{\mathrm{G}32} & -k_{\mathrm{G}33} \\
0 & 0 & 0 & 0 & 0 & 0 \\
0 & 0 & 0 & 0 & 0 & 0 \\
0 & 0 & 0 & 0 & 0 & 0
\end{bmatrix}
\tag{5.14}
$$

where

$$
\begin{aligned}
k_{\mathrm{G}11} &= -f_1^+(\mathbf{n}_i)_x L_i^{-1} \quad k_{\mathrm{G}21} = -f_2^+(\mathbf{n}_i)_x L_i^{-1} \quad k_{\mathrm{G}31} = -f_3^+(\mathbf{n}_i)_x L_i^{-1} \\
k_{\mathrm{G}12} &= -f_1^+(\mathbf{n}_i)_y L_i^{-1} \quad k_{\mathrm{G}22} = -f_2^+(\mathbf{n}_i)_y L_i^{-1} \quad k_{\mathrm{G}32} = -f_3^+(\mathbf{n}_i)_y L_i^{-1} \\
k_{\mathrm{G}13} &= -f_1^+(\mathbf{n}_i)_z L_i^{-1} \quad k_{\mathrm{G}23} = -f_2^+(\mathbf{n}_i)_z L_i^{-1} \quad k_{\mathrm{G}33} = -f_3^+(\mathbf{n}_i)_z L_i^{-1}
\end{aligned}
\tag{5.15}
$$

In this equation $\mathbf{f}_i^+$ is the 6×1 column matrix that originates from the nodal equilibrium vector $\mathcal{R}_i^{\mathrm{T}} \mathbf{N}_i^{+\mathrm{T}} \mathbf{F}_i$ with the thrust term set equal to zero. The term $f_j^+, j = 1 \rightarrow 6$ is then the j th element of that column matrix. The result is a nonsymmetric contribution. Clearly

$$(\mathbf{K}_{\mathrm{G}})_{i12}^{\mathrm{S}} = (\mathbf{K}_{\mathrm{G}})_{i21}^{\mathrm{S}} = -(\mathbf{K}_{\mathrm{G}})_{i11}^{\mathrm{S}} \tag{5.16}$$

Before proceeding with the rotation term it is instructive to demonstrate the derivation of one of the terms of Eq. 5.15, say, the term $k_{\mathrm{G}13}$. This term is evaluated from the partial derivative with respect to x_{A} of the 3rd term of Eq. 5.10. In expanded form it becomes

$$k_{\mathrm{G}13} = \left(0 + R_{i23}\frac{-(\mathbf{M}_i^+)_{z^i} - (\mathbf{M}_i^-)_{z^i}}{L_i} + R_{i33}\frac{(\mathbf{M}_i^+)_{y^i} + (\mathbf{M}_i^-)_{y^i}}{L_i} \right) \frac{-(\mathbf{n}_i)_z}{L_i} \tag{5.17}$$

and appears in Eq. 5.15.

The rotation term in the geometric stiffness which comes from $\left(\mathrm{d}\mathcal{R}_i^{\mathrm{T}}\right)\mathbf{N}_i^{+\mathrm{T}}\mathbf{F}_i$, the second term on the right hand side of Eq. 5.12, can be computed using the relationship from rigid body mechanics (Goldstein, 1957) which states that the change of a vector $\mathbf{G}$ under a small rigid body rotation that is described by the small rotation vector $\boldsymbol{\omega}$ can be written as

$$\mathrm{d}\mathbf{G} = \boldsymbol{\omega} \times \mathbf{G} = -\mathbf{G} \times \boldsymbol{\omega} \tag{5.18}$$

This is a much more simple procedure to use than computing the required derivatives by the direct use of the gradient.

In matrix form the cross product of Eq 5.18 becomes:

$$\mathrm{d}\mathbf{G} = \overline{\mathbf{G}}\boldsymbol{\omega}, \tag{5.19}$$

or

$$\begin{Bmatrix} (\mathrm{d}\mathbf{G})_x \\ (\mathrm{d}\mathbf{G})_y \\ (\mathrm{d}\mathbf{G})_z \end{Bmatrix} = -\begin{bmatrix} 0 & -(\mathbf{G})_z & (\mathbf{G})_y \\ (\mathbf{G})_z & 0 & -(\mathbf{G})_x \\ -(\mathbf{G})_y & (\mathbf{G})_x & 0 \end{bmatrix} \begin{Bmatrix} (\boldsymbol{\omega})_x \\ (\boldsymbol{\omega})_y \\ (\boldsymbol{\omega})_z \end{Bmatrix}$$

In this discussion $\mathbf{G}$ plays the role of a force vector and a moment vector in the 'plus' and 'minus' ends of the member respectively, i.e. $\mathbf{G}_{force}^+$, $\mathbf{G}_{force}^-$, $\mathbf{G}_{moment}^+$ and $\mathbf{G}_{moment}^-$.

Several steps are required to find the 'rotation' terms of the geometric stiffness matrix. First of all it is necessary to write the rotation vector $\boldsymbol{\omega}$ in terms of the joint displacements. This is typically done by working first in the local coordinate system of member i and then transforming these results into the global coordinate system. Finally, the cross product of Eq 5.18 is written as a matrix product.

Let $\boldsymbol{\omega}_i^{'}$ be the 3×1 column matrix representing the small rotations in the local coordinate system. The components of $\boldsymbol{\omega}_i^{'}$ are simply

$$(\boldsymbol{\omega}_i)_1^{'} = (\theta_{\mathrm{C}})_{x^i}$$

$$(\boldsymbol{\omega}_i)_2^{'} = -\left[(\boldsymbol{\delta}_{\mathrm{A}})_{z^i} - (\boldsymbol{\delta}_{\mathrm{C}})_{z^i}\right]/\,L_i \tag{5.20}$$

$$(\boldsymbol{\omega}_i)_3^{'} = \left[(\boldsymbol{\delta}_{\mathrm{A}})_{y^i} - (\boldsymbol{\delta}_{\mathrm{C}})_{y^i}\right]/\,L_i$$

Note that the first of Eq 5.20 comes from the arbitrary assumption of fixing the rigid body rotation of the negative end C to the rotation of joint C. Put into matrix form Eq 5.20 becomes

$$\boldsymbol{\omega}_i^{'} = \mathbf{A}_{\mathrm{A}}^{'}\boldsymbol{\delta}_{\mathrm{A}}^{'} + \mathbf{A}_{\mathrm{C}}^{'}\boldsymbol{\delta}_{\mathrm{C}}^{'}$$

where

$$\mathbf{A}_{\mathrm{A}}^{'} = \begin{bmatrix} 0 & 0 & 0 & 0 & 0 & 0 \\ 0 & 0 & -L_i^{-1} & 0 & 0 & 0 \\ 0 & +L_i^{-1} & 0 & 0 & 0 & 0 \end{bmatrix}; \quad \mathbf{A}_{\mathrm{C}}^{'} = \begin{bmatrix} 0 & 0 & 0 & 1 & 0 & 0 \\ 0 & 0 & +L_i^{-1} & 0 & 0 & 0 \\ 0 & L_i^{-1} & 0 & 0 & 0 & 0 \end{bmatrix} \tag{5.21}$$

These matrices are transformed into the global coordinate system as

$$\mathbf{A}_{\mathrm{A}} = \mathcal{R}_i^{\mathrm{T}} \begin{bmatrix} \mathbf{A}_{\mathrm{A}}^{'} \\ \mathbf{A}_{\mathrm{A}}^{'} \end{bmatrix} \mathcal{R}_i; \quad \mathbf{A}_{\mathrm{C}} = \mathcal{R}_i^{\mathrm{T}} \begin{bmatrix} \mathbf{A}_{\mathrm{C}}^{'} \\ \mathbf{A}_{\mathrm{C}}^{'} \end{bmatrix} \mathcal{R}_i \tag{5.22}$$

The small rotations vector $\boldsymbol{\omega}_i$, can now be transformed into global coordinates as

$$\boldsymbol{\Omega} = \begin{Bmatrix} \boldsymbol{\omega}_i \\ \boldsymbol{\omega}_i \end{Bmatrix} = \mathbf{A}_{\mathrm{A}}\boldsymbol{\delta}_{\mathrm{A}} + \mathbf{A}_{\mathrm{C}}\boldsymbol{\delta}_{\mathrm{C}} \tag{5.23}$$

In this equation the matrices of Eq 5.21 are "stacked" so that they can operate simultaneously on two vectors.

The "rotation" terms of the geometric stiffness matrix can now be completed. These rotation terms that stem from $\left(d\mathcal{R}_i^{\mathrm{T}}\right)\mathbf{N}_i^{\pm\mathrm{T}}\mathbf{F}_i$ are in essence the change in nodal force and moment equilibrium equations of member i (Eq. 5.10) due to the small rotations $\boldsymbol{\omega}_i$. Using Eq. 5.19 to describe the changes, the rotation terms emerge from

$$
\begin{aligned}
\left(d\mathcal{R}_i^{\mathrm{T}}\right)\mathbf{N}_i^{\pm\mathrm{T}}\mathbf{F}_i &= d\begin{Bmatrix} \mathcal{R}_i^{\mathrm{T}}\mathbf{N}_i^{+\mathrm{T}}\mathbf{F}_i \\ \mathcal{R}_i^{\mathrm{T}}\mathbf{N}_i^{-\mathrm{T}}\mathbf{F}_i \end{Bmatrix} \\
&= \begin{bmatrix} \overline{\mathbf{G}}^{+} & \mathbf{0} \\ \mathbf{0} & \overline{\mathbf{G}}^{-} \end{bmatrix}\begin{Bmatrix} \boldsymbol{\Omega} \\ \boldsymbol{\Omega} \end{Bmatrix} = \begin{bmatrix} \overline{\mathbf{G}}^{+}\mathbf{A}_{\mathrm{A}} & \overline{\mathbf{G}}^{+}\mathbf{A}_{\mathrm{C}} \\ \overline{\mathbf{G}}^{-}\mathbf{A}_{\mathrm{A}} & \overline{\mathbf{G}}^{-}\mathbf{A}_{\mathrm{C}} \end{bmatrix}\begin{Bmatrix} \boldsymbol{\delta}_{\mathrm{A}} \\ \boldsymbol{\delta}_{\mathrm{C}} \end{Bmatrix}
\end{aligned}
\tag{5.24}
$$

as

$$
\begin{aligned}
(\mathbf{K}_{\mathrm{G}})_{i11}^{\mathrm{R}} &= \overline{\mathbf{G}}^{+}\mathbf{A}_{\mathrm{A}} \\
(\mathbf{K}_{\mathrm{G}})_{i12}^{\mathrm{R}} &= \overline{\mathbf{G}}^{+}\mathbf{A}_{\mathrm{C}} \\
(\mathbf{K}_{\mathrm{G}})_{i21}^{\mathrm{R}} &= \overline{\mathbf{G}}^{-}\mathbf{A}_{\mathrm{A}} \\
(\mathbf{K}_{\mathrm{G}})_{i22}^{\mathrm{R}} &= \overline{\mathbf{G}}^{-}\mathbf{A}_{\mathrm{C}}
\end{aligned}
\tag{5.25}
$$

where

$$
\overline{\mathbf{G}}^{+} = \begin{bmatrix} \mathbf{G}_{force}^{+} & \\ & \mathbf{G}_{moment}^{+} \end{bmatrix}; \quad \overline{\mathbf{G}}^{-} = \begin{bmatrix} \mathbf{G}_{force}^{-} & \\ & \mathbf{G}_{moment}^{-} \end{bmatrix}
$$

and

$$
\mathbf{G}_{force}^{+} = -\begin{bmatrix} 0 & -g_3^{+} & g_2^{+} \\ g_3^{+} & 0 & -g_1^{+} \\ -g_2^{+} & g_1^{+} & 0 \end{bmatrix}; \quad \mathbf{G}_{moment}^{+} = -\begin{bmatrix} 0 & -g_6^{+} & g_5^{+} \\ g_6^{+} & 0 & -g_4^{+} \\ -g_5^{+} & g_4^{+} & 0 \end{bmatrix}
$$

$$\mathbf{G}_{force}^{-} = -\begin{bmatrix} 0 & -g_3^- & g_2^- \\ g_3^- & 0 & -g_1^- \\ -g_2^- & g_1^- & 0 \end{bmatrix}; \quad \mathbf{G}_{moment}^{+} = -\begin{bmatrix} 0 & -g_6^- & g_5^- \\ g_6^- & 0 & -g_4^- \\ -g_5^- & g_4^- & 0 \end{bmatrix}$$

Here $\mathbf{g}_i^+$ is the 6 × 1 column matrix obtained by computing the product $\mathcal{R}_i^{\mathrm{T}} \mathbf{N}_i^{+\mathrm{T}} \mathbf{F}_i$. The term g_i^+ is then the i-th element of that column matrix. Moreover $\mathbf{g}_i^-$ is the 6 × 1 column matrix obtained by computing the product $\mathcal{R}_i^{\mathrm{T}} \mathbf{N}_i^{-\mathrm{T}} \mathbf{F}_i$. The term g_i^- is then the i-th element of that column matrix. The result is a nonsymmetric contribution.

Displayed below is a 12×12 layout of the geometric stiffness matrix with a mapping of the stretch terms, S, and the rotation terms, R. Columns 10-12 are due to $(\boldsymbol{\omega}_i)_1^{'} = (\theta_C)_{x^i}$ that takes the value of "1" inside the matrix $\mathbf{A}_C^{'}$ (Eq. 5.21). When $\mathbf{A}_C^{'}$ is transformed to global coordinates it fills out these columns.

	1	2	3	4	5	6	7	8	9	10	11	12
1	RS	RS	RS	S	S	S	RS	RS	RS	RS	RS	RS
2	RS	RS	RS	S	S	S	RS	RS	RS	RS	RS	RS
3	RS	RS	RS	S	S	S	RS	RS	RS	RS	RS	RS
4	R	R	R	0	0	0	R	R	R	R	R	R
5	R	R	R	0	0	0	R	R	R	R	R	R
6	R	R	R	0	0	0	R	R	R	R	R	R
7	RS	RS	RS	S	S	S	RS	RS	RS	RS	RS	RS
8	RS	RS	RS	S	S	S	RS	RS	RS	RS	RS	RS
9	RS	RS	RS	S	S	S	RS	RS	RS	RS	RS	RS
10	R	R	R	0	0	0	R	R	R	R	R	R
11	R	R	R	0	0	0	R	R	R	R	R	R
12	R	R	R	0	0	0	R	R	R	R	R	R

5.6 Computer Programs P10-FR3DNL2.FOR, P11-FR3DNL3.FOR and P12-FR3DNLSR.FOR

FR3DNL2.FOR is a three dimensional version of the nonlinear plane frame program incorporating the nonlinear effects and the geometric stiffness matrix as described in the previous sections. It also uses the two dimensional beam-column equations in its member stiffness matrix. This is a matter of expediency. It is proper to use the three-dimensional beam-column of Chapter 8 below when computing the member stiffness matrix for three-dimensional frames and that is done here in program P11-FR3DNL3.FOR. Unfortunately, the program which uses the finite difference solution of Chapter 8 is computationally intensive and runs too slowly for practical applications on a PC. The program P12-FR3DNLSR uses a power series solution and is much faster. The compromise of program P10-FR3DNL2.FOR is to use two-dimensional beam column effects in the member stiffness together with short enough member lengths to mitigate the absence of three-dimensional beam column effects such as lateral buckling.

5.6.1 Example 5.3 Lateral Torsional Buckling

This example is a good test for computer programs for space frames because it fails in torsional buckling with the out-of-plane bending around the weak axis. The particular parameters used in this case come from an experiment (Spillers *et al*,, 1993) for which the buckling load was measured to be about 25 lbs. (See Figure 5.7.) With regard to input data, this program requires coordinates to be specified and therefore only one angle of orientation is necessary. That angle is γ. If the other angles are also specified they will be automatically replaced with calculated angles for reasons of internal consistency. The input file for this example is FR3DNL53.DAT.

The output file FR3DNL53.OUT is partially listed below. First an echo of the data is returned. Then with each iteration after the displacements are calculated both nodal and member quantities are updated. Updated nodal coordinates are listed under `NEW COORDINATES`. For members there is the `INCREMENTAL ELASTIC RESPONSE` that provides member deformations in the first line and member forces in the second line for the current iteration and load step. `UPDATED NONLINEAR RESPONSE` provides the accumulated deformations and 'forces'. The accumulated member forces are assembled and listed again under `CURRENT MEMBER FORCES`.

The output file is scanned for the word NEG which appears when the system becomes singular. In this case the system is stable at load step 7 and unstable at load step 8 which implies that the buckling load lies between 8/10×28=22.4 lbs and 9/10×28=25.2 lbs which agrees with the observed buckling load.

Interestingly, this problem exhibits two different bifurcation loads, depending on the direction of loading. For the upward loading, buckling of the frame was reached at 62% of that for the downward loading.

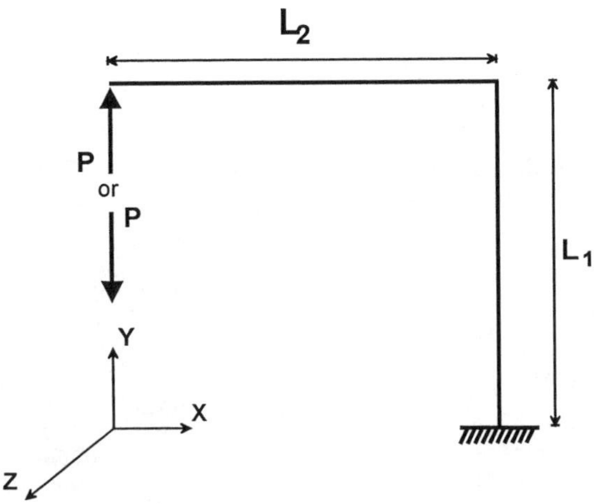

Figure 5.7 L-frame in lateral torsional buckling.

Input file FR3DNL53.DAT

```
10     11     1    10     5
                   28.
 0
 0
 0
 0
 0
 0
 0
 0
 0
 0
 0
 0
 0
 0
 0
 0
 0
 0
 0
.12       2.1       .000576 .000144 .01         2    1 0.
.12       2.1       .000576 .000144 .01         3    2
.12       2.1       .000576 .000144 .01         4    3
.12       2.1       .000576 .000144 .01         5    4
.12       2.1       .000576 .000144 .01         6    5
.12       2.1       .000576 .000144 .01         7    6           -90.
.12       2.1       .000576 .000144 .01         8    7           -90.
.12       2.1       .000576 .000144 .01         9    8           -90.
.12       2.1       .000576 .000144 .01        10    9           -90.
.12       2.1       .000576 .000144 .01        11   10           -90.
0.          10.64
2.1         10.64
4.2         10.64
6.2         10.64
8.4         10.64
```

```
10.5      10.64
10.5       8.512
10.5       6.384
10.5       4.256
10.5       2.128
10.5
```

Output file FR3DNL53.OUT

```
  10  NO. MEMBERS
  11  NO. NODES
   1  NO. SUPPORTS
  10  NO. LOAD STEPS
   5  NO. ITERATIONS PER STEP

APPLIED LOADS-MOMENTS
            PX          PY          PZ           MX           MY           MZ
 1 .00000D+00 .28000D+02 .00000D+00  .00000D+00  .00000D+00  .00000D+00
 2 .00000D+00 .00000D+00 .00000D+00  .00000D+00  .00000D+00  .00000D+00
 3 .00000D+00 .00000D+00 .00000D+00  .00000D+00  .00000D+00  .00000D+00
 4 .00000D+00 .00000D+00 .00000D+00  .00000D+00  .00000D+00  .00000D+00
 5 .00000D+00 .00000D+00 .00000D+00  .00000D+00  .00000D+00  .00000D+00
 6 .00000D+00 .00000D+00 .00000D+00  .00000D+00  .00000D+00  .00000D+00
 7 .00000D+00 .00000D+00 .00000D+00  .00000D+00  .00000D+00  .00000D+00
 8 .00000D+00 .00000D+00 .00000D+00  .00000D+00  .00000D+00  .00000D+00
 9 .00000D+00 .00000D+00 .00000D+00  .00000D+00  .00000D+00  .00000D+00
10 .00000D+00 .00000D+00 .00000D+00  .00000D+00  .00000D+00  .00000D+00

NODE COORDINATES
             X           Y           Z         THX         THY         THZ
 1  .00000E+00  .10640E+02  .00000E+00 .00000E+00 .00000E+00 .00000E+00
 2  .21000E+01  .10640E+02  .00000E+00 .00000E+00 .00000E+00 .00000E+00
 3  .42000E+01  .10640E+02  .00000E+00 .00000E+00 .00000E+00 .00000E+00
 4  .62000E+01  .10640E+02  .00000E+00 .00000E+00 .00000E+00 .00000E+00
 5  .84000E+01  .10640E+02  .00000E+00 .00000E+00 .00000E+00 .00000E+00
 6  .10500E+02  .10640E+02  .00000E+00 .00000E+00 .00000E+00 .00000E+00
 7  .10500E+02  .85120E+01  .00000E+00 .00000E+00 .00000E+00 .00000E+00
 8  .10500E+02  .63840E+01  .00000E+00 .00000E+00 .00000E+00 .00000E+00
 9  .10500E+02  .42560E+01  .00000E+00 .00000E+00 .00000E+00 .00000E+00
10  .10500E+02  .21280E+01  .00000E+00 .00000E+00 .00000E+00 .00000E+00
11  .10500E+02  .00000E+00  .00000E+00 .00000E+00 .00000E+00 .00000E+00

 MEMBER PROPERTIES
    AREA LENGTH       IX       IY        IZ +END -END  THETA   PHI GAMMA
 1  0.12   2.10 0.000576 0.000144 0.010000    2    1   0.00  0.00  0.00
 2  0.12   2.10 0.000576 0.000144 0.010000    3    2   0.00  0.00  0.00
 3  0.12   2.00 0.000576 0.000144 0.010000    4    3   0.00  0.00  0.00
 4  0.12   2.20 0.000576 0.000144 0.010000    5    4   0.00  0.00  0.00
 5  0.12   2.10 0.000576 0.000144 0.010000    6    5   0.00  0.00  0.00
 6  0.12   2.13 0.000576 0.000144 0.010000    7    6 -90.00  0.00  0.00
 7  0.12   2.13 0.000576 0.000144 0.010000    8    7 -90.00  0.00  0.00
 8  0.12   2.13 0.000576 0.000144 0.010000    9    8 -90.00  0.00  0.00
 9  0.12   2.13 0.000576 0.000144 0.010000   10    9 -90.00  0.00  0.00
10  0.12   2.13 0.000576 0.000144 0.010000   11   10 -90.00  0.00  0.00

******** NEW LOAD STEP ***********

CURRENT MEMBER FORCES
       THRUST     TORQUE       MY(+)        MZ(+)        MY(-)        MZ(-)
 1 .00000E+00 00000E+00 .00000E+00  .00000E+00  .00000E+00  .00000E+00
 2 .00000E+00 00000E+00 .00000E+00  .00000E+00  .00000E+00  .00000E+00
 3 .00000E+00 00000E+00 .00000E+00  .00000E+00  .00000E+00  .00000E+00
 4 .00000E+00 00000E+00 .00000E+00  .00000E+00  .00000E+00  .00000E+00
```

```
 5 .00000E+00 00000E+00 .00000E+00  .00000E+00  .00000E+00  .00000E+00
 6 .00000E+00 00000E+00 .00000E+00  .00000E+00  .00000E+00  .00000E+00
 7 .00000E+00 00000E+00 .00000E+00  .00000E+00  .00000E+00  .00000E+00
 8 .00000E+00 00000E+00 .00000E+00  .00000E+00  .00000E+00  .00000E+00
 9 .00000E+00 00000E+00 .00000E+00  .00000E+00  .00000E+00  .00000E+00
10 .00000E+00 00000E+00 .00000E+00  .00000E+00  .00000E+00  .00000E+00

*** ERROR =          .56000000E+01
    LOAD STEP             1
    ITERATION NUMBER      1
APPLIED LOADS-MOMENTS
           PX          PY          PZ          MX          MY          MZ
 1 .00000D+00 .56000D+01 .00000D+00  .00000D+00  .00000D+00  .00000D+00
 2 .00000D+00 .00000D+00 .00000D+00  .00000D+00  .00000D+00  .00000D+00
 3 .00000D+00 .00000D+00 .00000D+00  .00000D+00  .00000D+00  .00000D+00
 4 .00000D+00 .00000D+00 .00000D+00  .00000D+00  .00000D+00  .00000D+00
 5 .00000D+00 .00000D+00 .00000D+00  .00000D+00  .00000D+00  .00000D+00
 6 .00000D+00 .00000D+00 .00000D+00  .00000D+00  .00000D+00  .00000D+00
 7 .00000D+00 .00000D+00 .00000D+00  .00000D+00  .00000D+00  .00000D+00
 8 .00000D+00 .00000D+00 .00000D+00  .00000D+00  .00000D+00  .00000D+00
 9 .00000D+00 .00000D+00 .00000D+00  .00000D+00  .00000D+00  .00000D+00
10 .00000D+00 .00000D+00 .00000D+00  .00000D+00  .00000D+00  .00000D+00

JOINT DISPLACEMENTS
           X           Y           Z        TH X        TH Y        TH Z
 1 .33284E-01 .87350E-01 .00000E+00  .00000E+00  .00000E+00 -.93433E-02
 2 .33284E-01 .67815E-01 .00000E+00  .00000E+00  .00000E+00 -.92198E-02
 3 .33284E-01 .48800E-01 .00000E+00  .00000E+00  .00000E+00 -.88494E-02
 4 .33284E-01 .31646E-01 .00000E+00  .00000E+00  .00000E+00 -.82670E-02
 5 .33284E-01 .14398E-01 .00000E+00  .00000E+00  .00000E+00 -.73676E-02
 6 .33284E-01 .49652E-04 .00000E+00  .00000E+00  .00000E+00 -.62563E-02
 7 .21302E-01 .39722E-04 .00000E+00  .00000E+00  .00000E+00 -.50051E-02
 8 .11982E-01 .29791E-04 .00000E+00  .00000E+00  .00000E+00 -.37538E-02
 9 .53254E-02 .19861E-04 .00000E+00  .00000E+00  .00000E+00 -.25025E-02
10 .13313E-02 .99306E-05 .00000E+00  .00000E+00  .00000E+00 -.12513E-02

NEW JOINT COORDINATES
           X           Y           Z        TH X        TH Y        TH Z
 1 .33284E-01 .10727E+02 .00000E+00  .00000E+00  .00000E+00 -.93433E-02
 2 .21333E+01 .10708E+02 .00000E+00  .00000E+00  .00000E+00 -.92198E-02
 3 .42333E+01 .10689E+02 .00000E+00  .00000E+00  .00000E+00 -.88494E-02
 4 .62333E+01 .10672E+02 .00000E+00  .00000E+00  .00000E+00 -.82670E-02
 5 .84333E+01 .10654E+02 .00000E+00  .00000E+00  .00000E+00 -.73676E-02
 6 .10533E+02 .10640E+02 .00000E+00  .00000E+00  .00000E+00 -.62563E-02
 7 .10521E+02 .85120E+01 .00000E+00  .00000E+00  .00000E+00 -.50051E-02
 8 .10512E+02 .63840E+01 .00000E+00  .00000E+00  .00000E+00 -.37538E-02
 9 .10505E+02 .42560E+01 .00000E+00  .00000E+00  .00000E+00 -.25025E-02
10 .10501E+02 .21280E+01 .00000E+00  .00000E+00  .00000E+00 -.12513E-02
11 .10500E+02 .00000E+00 .00000E+00  .00000E+00  .00000E+00  .00000E+00

INCREMENTAL ELASTIC RESPONSE
 1 .00000E+00 .00000E+00 .00000E+00  .82320E-04  .00000E+00 -.41160E-04
 1 .00000E+00 .00000E+00 .00000E+00  .11760E+02  .00000E+00 -.15102E-11
UPDATED NONLINEAR RESPONSE
 1 .90855E-04 .00000E+00 .00000E+00  .82052E-04  .00000E+00 -.41428E-04
 1 .51917E+02 .00000E+00 .00000E+00  .11683E+02  .00000E+00 -.76655E-01

INCREMENTAL ELASTIC RESPONSE
 2 .00000E+00 .00000E+00 .00000E+00  .20580E-03  .00000E+00 -.16464E-03
 2 .00000E+00 .00000E+00 .00000E+00  .23520E+02  .00000E+00 -.11760E+02
UPDATED NONLINEAR RESPONSE
 2 .86095E-04 .00000E+00 .00000E+00  .20555E-03  .00000E+00 -.16489E-03
 2 .49197E+02 .00000E+00 .00000E+00  .23449E+02  .00000E+00 -.11831E+02

 ⋮
```

```
INCREMENTAL ELASTIC RESPONSE
10 .99307E-05 .00000E+00 .00000E+00  .62563E-03  .00000E+00 -.62563E-03
10 .56000E+01 .00000E+00 .00000E+00  .58800E+02  .00000E+00 -.58800E+02
UPDATED NONLINEAR RESPONSE
10 .10347E-04 .00000E+00 .00000E+00  .62559E-03  .00000E+00 -.62568E-03
10 .58348E+01 .00000E+00 .00000E+00  .58787E+02  .00000E+00 -.58813E+02

CURRENT MEMBER FORCES
       THRUST      TORQUE       MY(+)        MZ(+)        MY(-)        MZ(-)
 1 .51917E+02 .00000E+00 .00000E+00  .11683E+02  .00000E+00 -.76655E-01
 2 .49197E+02 .00000E+00 .00000E+00  .23449E+02  .00000E+00 -.11831E+02
 3 .44137E+02 .00000E+00 .00000E+00  .34657E+02  .00000E+00 -.23583E+02
 4 .36878E+02 .00000E+00 .00000E+00  .46996E+02  .00000E+00 -.34764E+02
 5 .28010E+02 .00000E+00 .00000E+00  .58770E+02  .00000E+00 -.47070E+02
 6 .24622E+02 .00000E+00 .00000E+00  .58763E+02  .00000E+00 -.58837E+02
 7 .17107E+02 .00000E+00 .00000E+00  .58774E+02  .00000E+00 -.58826E+02
 8 .11471E+02 .00000E+00 .00000E+00  .58781E+02  .00000E+00 -.58819E+02
 9 .77135E+01 .00000E+00 .00000E+00  .58785E+02  .00000E+00 -.58815E+02
10 .58348E+01 .00000E+00 .00000E+00  .58787E+02  .00000E+00 -.58813E+02

*** ERROR =         .64062560E+02


MEMBER PROPERTIES
    AREA LENGTH        IX        IY        IZ +END -END  THETA   PHI GAMMA
 1  0.12    2.10 0.000576 0.000144 0.010000    2    1    -.53  0.00  0.00
 2  0.12    2.10 0.000576 0.000144 0.010000    3    2    -.52  0.00  0.00
 3  0.12    2.00 0.000576 0.000144 0.010000    4    3    -.49  0.00  0.00
 4  0.12    2.20 0.000576 0.000144 0.010000    5    4    -.45  0.00  0.00
 5  0.12    2.10 0.000576 0.000144 0.010000    6    5    -.39  0.00  0.00
 6  0.12    2.13 0.000576 0.000144 0.010000    7    6  -90.32  0.00  0.00
 7  0.12    2.13 0.000576 0.000144 0.010000    8    7  -90.25  0.00  0.00
 8  0.12    2.13 0.000576 0.000144 0.010000    9    8  -90.18  0.00  0.00
 9  0.12    2.13 0.000576 0.000144 0.010000   10    9  -90.11  0.00  0.00
10  0.12    2.13 0.000576 0.000144 0.010000   11   10  -90.04  0.00  0.00

:

    LOAD STEP             7
    ITERATION NUMBER      5

JOINT DISPLACEMENTS
             X            Y            Z         TH X         TH Y         TH Z
 1 -.13128E-08 -.75124E-08  .00000E+00 .00000E+00 .00000E+00 .92972E-09
 2 -.12410E-08 -.55601E-08  .00000E+00 .00000E+00 .00000E+00 .93289E-09
 3 -.11744E-08 -.36984E-08  .00000E+00 .00000E+00 .00000E+00 .83103E-09
 4 -.11238E-08 -.22078E-08  .00000E+00 .00000E+00 .00000E+00 .63826E-09
 5 -.10820E-08 -.86045E-09  .00000E+00 .00000E+00 .00000E+00 .53634E-09
 6 -.10583E-08  .16148E-10  .00000E+00 .00000E+00 .00000E+00 .30153E-09
 7 -.58958E-09  .61334E-11  .00000E+00 .00000E+00 .00000E+00 .17268E-09
 8 -.25959E-09  .30372E-12  .00000E+00 .00000E+00 .00000E+00 .13391E-09
 9 -.44756E-10 -.22361E-11  .00000E+00 .00000E+00 .00000E+00 .59862E-10
10  .57997E-11 -.19451E-11  .00000E+00 .00000E+00 .00000E+00 .80584E-12

NEW JOINT COORDINATES
           X           Y           Z         TH X          TH Y         TH Z
 1 .13745E+00 .10986E+02 .00000E+00  .00000E+00  .00000E+00 -.37143E-01
 2 .22360E+01 .10908E+02 .00000E+00  .00000E+00  .00000E+00 -.36649E-01
 3 .43347E+01 .10833E+02 .00000E+00  .00000E+00  .00000E+00 -.35168E-01
 4 .63335E+01 .10765E+02 .00000E+00  .00000E+00  .00000E+00 -.32840E-01
 5 .85324E+01 .10696E+02 .00000E+00  .00000E+00  .00000E+00 -.29245E-01
 6 .10632E+02 .10639E+02 .00000E+00  .00000E+00  .00000E+00 -.24802E-01
 7 .10584E+02 .85116E+01 .00000E+00  .00000E+00  .00000E+00 -.19812E-01
 8 .10547E+02 .63839E+01 .00000E+00  .00000E+00  .00000E+00 -.14841E-01
```

```
 9 .10521E+02 .42560E+01 .00000E+00  .00000E+00  .00000E+00 -.98857E-02
10 .10505E+02 .21280E+01 .00000E+00  .00000E+00  .00000E+00 -.49404E-02
11 .10500E+02 .00000E+00 .00000E+00  .00000E+00  .00000E+00  .00000E+00

CURRENT MEMBER FORCES
        THRUST      TORQUE        MY(+)         MZ(+)         MY(-)         MZ(-)
 1 .82812E+00 .00000E+00 .00000E+00  .47008E+02  .00000E+00  .19273E-07
 2 .80602E+00 .00000E+00 .00000E+00  .94017E+02  .00000E+00 -.47008E+02
 3 .76322E+00 .00000E+00 .00000E+00  .13879E+03  .00000E+00 -.94017E+02
 4 .69726E+00 .00000E+00 .00000E+00  .18805E+03  .00000E+00 -.13879E+03
 5 .60710E+00 .00000E+00 .00000E+00  .23507E+03  .00000E+00 -.18805E+03
 6 .22394E+02 .00000E+00 .00000E+00  .23401E+03  .00000E+00 -.23507E+03
 7 .22397E+02 .00000E+00 .00000E+00  .23318E+03  .00000E+00 -.23401E+03
 8 .22398E+02 .00000E+00 .00000E+00  .23259E+03  .00000E+00 -.23318E+03
 9 .22399E+02 .00000E+00 .00000E+00  .23224E+03  .00000E+00 -.23259E+03
10 .22400E+02 .00000E+00 .00000E+00  .23212E+03  .00000E+00 -.23224E+03

*** ERROR =          .15142980E-04

MEMBER PROPERTIES
     AREA LENGTH        IX        IY         IZ +END -END  THETA    PHI GAMMA
 1  0.12    2.10 0.000576 0.000144 0.010000    2    1  -2.12   0.00  0.00
 2  0.12    2.10 0.000576 0.000144 0.010000    3    2  -2.06   0.00  0.00
 3  0.12    2.00 0.000576 0.000144 0.010000    4    3  -1.95   0.00  0.00
 4  0.12    2.20 0.000576 0.000144 0.010000    5    4  -1.78   0.00  0.00
 5  0.12    2.10 0.000576 0.000144 0.010000    6    5  -1.55   0.00  0.00
 6  0.12    2.13 0.000576 0.000144 0.010000    7    6 -91.28   0.00  0.00
 7  0.12    2.13 0.000576 0.000144 0.010000    8    7 -90.99   0.00  0.00
 8  0.12    2.13 0.000576 0.000144 0.010000    9    8 -90.71   0.00  0.00
 9  0.12    2.13 0.000576 0.000144 0.010000   10    9 -90.42   0.00  0.00
10  0.12    2.13 0.000576 0.000144 0.010000   11   10 -90.14   0.00  0.00

:

     LOAD STEP           8
     ITERATION NUMBER    3
NEG TERM ON THE DIAGONAL AT ROW    58 $$$$

:

Stop - Program terminated.
```

5.7 Problems

1. Prove that the relative rotation of joint A with respect to joint C is a compound rotation given by the matrix product $\mathbf{R}_{\mathrm{A}}^{\mathrm{T}}\mathbf{R}_{\mathrm{C}}$.

2. Redo Example 4.4 as a three dimensional beam.

3. Introduce a small lateral load in the y-direction in Example 5.2 and plot the coordinates of the tip as P is incremented.

CHAPTER 6

NONLINEAR ANALYSIS OF MEMBRANES

6.1 Introduction

This chapter is an addition to the sequence of discussions of discrete systems. It is included here for several reasons. In terms of applications, if you are going to discuss fabric structures you need access to a membrane finite element. The plane stress/strain element which follows is probably the most simple finite element which can be discussed. Since the functional dependence of the element upon the node coordinates is so simple, it is an easy matter to apply perturbation methods to it.

Subsequently the book discusses three-dimensional beam-columns as continuous systems rather than discrete systems. This membrane finite element is in these terms something of a hybrid. Membranes are, of course, continuous in the sense that they are properly described by their field equations. The most direct way to come up with a nonlinear membrane finite element would be to first of all derive the appropriate field equations and then formulate some discrete solution of them such as the finite element solution. In the material below an alternative approach is taken. That is, the discretization of the finite element method for the linear membrane problem is accepted as a starting point and then perturbation methods applied to it. Certainly the procedure is simple.

The work below proceeds as follows. First, Zienkiewicz's plane stress constant strain finite element is introduced and its corresponding geometrically nonlinear finite element is derived. This is a two-dimensional result. It is then a simple matter to construct the three-dimensional finite element using the rotation matrix. An application is included with the computer program which is discussed subsequently.

6.2 The Geometric Stiffness Matrix of the Plane Stress Triangular Finite Element

For the purpose of this book, the triangular plane stress (constant stress) finite element (Fig. 6.1) described by Zienkiewicz (1977) is used. This section derives its element's nodal force vector in the 'global' two-dimensional coordinate system of the element's plane. This global coordinate system makes sense in problems of two-dimensional elasticity only. It reduces to a member's local coordinate system for the three-dimensional membrane shells considered herein.

Let the nodal displacement vector be defined as

$$\mathbf{a}^e = \begin{Bmatrix} u_i \\ v_i \\ u_j \\ v_j \\ u_m \\ v_m \end{Bmatrix} \tag{6.1}$$

and the displacement field be approximated using *linear* shape functions $G_r(x, y),\ r = i, j, m$ in both the x and y directions such that

$$\mathbf{u}(x, y) = \begin{Bmatrix} u(x, y) \\ v(x, y) \end{Bmatrix} = \mathbf{G}\mathbf{a}^e = \left[\mathbf{I}G_i(x, y) \quad \mathbf{I}G_j(x, y) \quad \mathbf{I}G_m(x, y)\right]\mathbf{a}^e \tag{6.2}$$

The shape function coefficients are determined using the condition that for a given degree of freedom its shape function will attain a value of unity and the shape functions that are linked to the other degrees of freedom will attain a value of zero. The shape functions may, thus, be written as

$$G_r(x, y) = A_r + B_r x + C_r y \qquad (r = i, j, m) \tag{6.3}$$

and each is required to satisfy the following three conditions

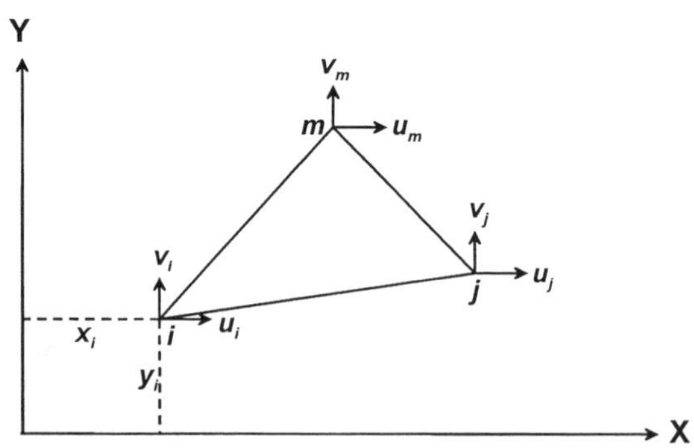

Figure 6.1 Plane stress triangular finite element.

$$G_r(x_s, y_s) = \begin{Bmatrix} 1 & \text{if} \quad s = r \\ 0 & \text{otherwise} \end{Bmatrix} \quad s = i, j, m \tag{6.4}$$

for a given r. When Eq. 6.3 is expanded for the case of r=i and s=i, j, m it becomes

$$\begin{aligned} A_i + B_i x_i + C_i y_i &= 1 \\ A_i + B_i x_j + C_i y_j &= 0 \Rightarrow \\ A_i + B_i x_m + C_i y_m &= 0 \end{aligned} \begin{bmatrix} 1 & x_i & y_i \\ 1 & x_j & y_j \\ 1 & x_m & y_m \end{bmatrix} \begin{Bmatrix} A_i \\ B_i \\ C_i \end{Bmatrix} = \begin{Bmatrix} 1 \\ 0 \\ 0 \end{Bmatrix} \tag{6.5}$$

and yields the following shape function coefficients:

$$A_i = \frac{x_j y_m - x_m y_j}{2A}; \; B_i = \frac{y_j - y_m}{2A}; \; C_i = \frac{x_m - x_j}{2A} \tag{6.6}$$

where A is the area of the triangle ijm so that

$$2A = \det \begin{bmatrix} 1 & x_i & y_i \\ 1 & x_j & y_j \\ 1 & x_m & y_m \end{bmatrix} \tag{6.7}$$

Proceeding in the same manner and calculating the rest of the shape function coefficients by introducing the remaining two conditions into Eq. 6.3, i.e.

$$G_j(x_s, y_s) = \begin{Bmatrix} 1 & \text{if} \quad s = j \\ 0 & \text{otherwise} \end{Bmatrix}; \; G_m(x_s, y_s) = \begin{Bmatrix} 1 & \text{if} \quad s = m \\ 0 & \text{otherwise} \end{Bmatrix}$$

and changing variables so that $a_r = 2AA_r$; $b_r = 2AB_r$; $c_r = 2AC_r$; $(r = i, j, m)$ the full set of shape function coefficients becomes:

$$\begin{aligned} a_i &= x_j y_m - x_m y_j; \; b_i = y_j - y_m = y_{jm}; \; c_i = x_m - x_j = x_{mi}; \\ a_j &= x_m y_i - x_i y_m; \; b_j = y_m - y_i = y_{mi}; \; c_j = x_i - x_m = x_{im}; \\ a_m &= x_i y_j - x_j y_i; \; b_m = y_i - y_j = y_{ij}; \; c_m = x_j - x_i = x_{ji} \end{aligned} \tag{6.8}$$

The strains are described using the following linear relationship

$$\boldsymbol{\varepsilon}(x,y)=\begin{Bmatrix}\varepsilon_x(x,y)\\ \varepsilon_y(x,y)\\ \gamma_{xy}(x,y)\end{Bmatrix}=\begin{bmatrix}\dfrac{\partial}{\partial x} & 0\\ 0 & \dfrac{\partial}{\partial y}\\ \dfrac{\partial}{\partial y} & \dfrac{\partial}{\partial x}\end{bmatrix}\begin{Bmatrix}u(x,y)\\ v(x,y)\end{Bmatrix}=\mathbf{Lu}(x,y) \tag{6.9}$$

Substitution for $\mathbf{u}(x,y)$ from Eq. 6.2 yields *constant strains* since

$$\boldsymbol{\varepsilon}(x,y)=\mathbf{LG}(x,y)\mathbf{a}^e=\mathbf{L}\left[\mathbf{I}G_i(x,y)\quad \mathbf{I}G_j(x,y)\quad \mathbf{I}G_m(x,y)\right]\mathbf{a}^e \tag{6.10}$$

or

$$\boldsymbol{\varepsilon}(x,y)=\mathbf{B}(x,y)\mathbf{a}^e=\left[\mathbf{B}_i\quad \mathbf{B}_j\quad \mathbf{B}_m\right]\mathbf{a}^e \tag{6.11}$$

where

$$\mathbf{B}_r=\mathbf{LI}G_r=\begin{bmatrix}\dfrac{\partial G_r}{\partial x} & 0\\ 0 & \dfrac{\partial G_r}{\partial y}\\ \dfrac{\partial G_r}{\partial y} & \dfrac{\partial G_r}{\partial x}\end{bmatrix}=\frac{1}{2A}\begin{bmatrix}b_r & 0\\ 0 & c_r\\ c_r & b_r\end{bmatrix}\qquad r=i,j,m \tag{6.12}$$

Now since $\mathbf{B}_r$ is constant it makes the strains $\boldsymbol{\varepsilon}(x,y)=\mathbf{B}(x,y)=\textit{constant}$.

Stresses are obtained, using the usual linear constitutive rule, as

$$\boldsymbol{\sigma}(x,y)=\begin{Bmatrix}\sigma_x(x,y)\\ \sigma_y(x,y)\\ \tau_{xy}(x,y)\end{Bmatrix}=\frac{E}{1-\nu^2}\begin{bmatrix}1 & \nu & 0\\ \nu & 1 & 0\\ 0 & 0 & \dfrac{1-\nu}{2}\end{bmatrix}\boldsymbol{\varepsilon}(x,y)=\mathbf{D}\boldsymbol{\varepsilon}(x,y) \tag{6.13}$$

where E is Young's modulus and ν is Poisson's ratio and, $\sigma_x(x,y)$, $\sigma_y(x,y)$ and $\tau_{xy}(x,y)$ are the usual stresses of plane elasticity. Substitution of Eq. 6.11 into Eq. 6.13 yields *constant stresses* of

$$\boldsymbol{\sigma} = \mathbf{DBa}^e \tag{6.14}$$

The linear elastic stiffness matrix which takes the general form of

$$\mathbf{K}_{\mathrm{E}}^{e} = \int_V \mathbf{B}^{\mathrm{T}}\mathbf{DB}dV \tag{6.15}$$

reduces to a constant matrix, for the case of the constant stress triangular element,

$$\mathbf{K}_{\mathrm{E}}^{e} = \int_V \mathbf{B}^{\mathrm{T}}\mathbf{DB}t\mathrm{d}x\mathrm{d}y = \mathbf{B}^{\mathrm{T}}\mathbf{DB}tA \tag{6.16}$$

where t is the thickness of the element. Equation 6.16 may be expanded to yield a 6×6 two-dimensional stiffness matrix

$$\left(\mathbf{K}_{\mathrm{E}}^{e}\right) = \begin{bmatrix} \left(\mathbf{K}_{\mathrm{E}}^{e}\right)_{ii} & \left(\mathbf{K}_{\mathrm{E}}^{e}\right)_{ij} & \left(\mathbf{K}_{\mathrm{E}}^{e}\right)_{im} \\ \left(\mathbf{K}_{\mathrm{E}}^{e}\right)_{ji} & \left(\mathbf{K}_{\mathrm{E}}^{e}\right)_{jj} & \left(\mathbf{K}_{\mathrm{E}}^{e}\right)_{jm} \\ \left(\mathbf{K}_{\mathrm{E}}^{e}\right)_{mi} & \left(\mathbf{K}_{\mathrm{E}}^{e}\right)_{mj} & \left(\mathbf{K}_{\mathrm{E}}^{e}\right)_{mm} \end{bmatrix} \tag{6.17}$$

with

$$\left(\mathbf{K}_{\mathrm{E}}^{e}\right)_{rs} = \frac{t}{4A}\cdot\frac{E}{1-\nu^2}\begin{bmatrix} b_r b_s + \frac{1-\nu}{2}c_r c_s & \nu b_r b_s + \frac{1-\nu}{2}c_r b_s \\ \nu b_r b_s + \frac{1-\nu}{2}b_r c_s & c_r c_s + \frac{1-\nu}{2}b_r b_s \end{bmatrix} \tag{6.18}$$

for $r,s = i, j, m$.

The element's nodal force vector in the local coordinate system is defined as

$$\mathbf{F}^e = \mathbf{K}_{\mathrm{E}}^{e}\mathbf{a}^e \tag{6.19}$$

and when Eq. (6.16) is substituted it becomes $\mathbf{F}^e = tA\mathbf{B}^T\mathbf{DBa}^e$ or

$$\mathbf{F}^e = tA\mathbf{B}^T\boldsymbol{\sigma} \tag{6.20}$$

More explicitly the two-dimensional nodal force vector (Eq. 6.20) in the element's plane coordinate system may be written as

$$\mathbf{F}^e = \frac{t}{2}\begin{Bmatrix} b_i\sigma_x + c_i\tau_{xy} \\ c_i\sigma_y + b_i\tau_{xy} \\ \vdots \\ b_j\sigma_x + c_j\tau_{xy} \\ c_j\sigma_y + b_j\tau_{xy} \\ \vdots \\ b_m\sigma_x + c_m\tau_{xy} \\ c_m\sigma_y + b_m\tau_{xy} \end{Bmatrix} = \frac{t}{2}\begin{Bmatrix} (\mathbf{F}_i)_x \\ (\mathbf{F}_i)_y \\ \vdots \\ (\mathbf{F}_j)_x \\ (\mathbf{F}_j)_y \\ \vdots \\ (\mathbf{F}_m)_x \\ (\mathbf{F}_m)_y \end{Bmatrix} \begin{matrix} \text{row } i \\ \\ \text{row } j \\ \\ \text{row } m \end{matrix} \tag{6.21}$$

As was done in the previous chapters the geometric stiffness matrix is derived as the gradient of the nodal force vector, $\mathbf{F}^e$ (Eq. 6.21). The gradient of $\mathbf{F}^e$ has a particularly simple form. It comprises the geometric stiffness matrix having 9 sub matrices (instead of 4 in two noded elements like trusses and beams) arranged in an antisymmetric manner as

$$\mathbf{K}_G^e = \begin{bmatrix} \mathbf{0} & \mathbf{A} & -\mathbf{A} \\ -\mathbf{A} & \mathbf{0} & \mathbf{A} \\ \mathbf{A} & -\mathbf{A} & \mathbf{0} \end{bmatrix} \tag{6.22}$$

where

$$\mathbf{A} = \frac{t}{2}\begin{bmatrix} -\tau_{xy} & \sigma_x \\ -\sigma_y & \tau_{xy} \end{bmatrix} \tag{6.23}$$

The gradient has been taken with respect to the node coordinates x_i, y_i; x_j, y_j; x_m, y_m; while holding down the stresses $\sigma_x, \sigma_y, \tau_{xy}$ fixed. Equations 6.18 and 6.22 give $\mathbf{K}_E^e$ and $\mathbf{K}_G^e$ explicitly for plane elasticity problems. An additional step is required to deal with three dimensional membranes.

6.3. The Geometric Stiffness Matrix of Three-dimensional Membranes

A plane stress problem defines a two-dimensional (x, y) displacement field while a three-dimensional membrane in general defines a three-dimensional (x,y,z) displacement field. For linear elastic problems, a membrane has no out-of-plane stiffness so that the 3-*D* element stiffness can be constructed by simply "rearranging" terms from Eq. 6.18 and Eq. 6.23 in a 3-*D* array i.e. expanding the 2×2 matrices to 3×3 by inserting a zero third row and a zero third column. That is not true in the nonlinear case. Just as the string (truss bar) has a "geometric" stiffness component normal to its axes, the nonlinear membrane has an out-of-plane "geometric" stiffness. That effect will be described below using the relationship from mechanics (that was introduced in Chapter 5) which gives the change **dF** of a vector force **F** which is subjected to a small rigid body rotation vector $\boldsymbol{\omega}$ as

$$\mathbf{dF} = \boldsymbol{\omega} \times \mathbf{F} = -\mathbf{F} \times \boldsymbol{\omega} \tag{6.24}$$

This approach is possible since within small strain theory an out-of-plane displacement produces no strain in a membrane. Certainly this is a simple procedure to use and will in general produce rather accurate results in incremental nonlinear analysis where equilibrium is satisfied at each increment of load.

Equation 6.24 expands for local quantities, into

$$\begin{Bmatrix} (\mathbf{dF})_{x'} \\ (\mathbf{dF})_{y'} \\ (\mathbf{dF})_{z'} \end{Bmatrix} = -\begin{bmatrix} 0 & -(\mathbf{F})_{z'} & (\mathbf{F})_{y'} \\ (\mathbf{F})_{z'} & 0 & -(\mathbf{F})_{x'} \\ -(\mathbf{F})_{y'} & (\mathbf{F})_{x'} & 0 \end{bmatrix} \begin{Bmatrix} \omega_{x'} \\ \omega_{y'} \\ \omega_{z'} \end{Bmatrix} \tag{6.25}$$

First of all, it should be clear that small, in-plane (z component) rotations are included within the geometric stiffness matrix shown in Eq. 6.22. It is the effect of out-of-plane rotations (x, y, components only) which must be added to the existing formulation to complete the 3-dimensional model. Given the x, y, z displacement components of the nodes of a typical finite element (Figure 6.2) the out-of-plane rotation components $\omega_{x'}$, $\omega_{y'}$ in the local coordinate system, (x', y', z') are

$$\omega_{x'} = \frac{1}{a}\left[(\boldsymbol{\delta}_m)_{z'} - (\boldsymbol{\delta}_i)_{z'}\left(1 - \frac{c}{e}\right) - (\boldsymbol{\delta}_j)_{z'} \frac{c}{e} \right] \tag{6.26}$$

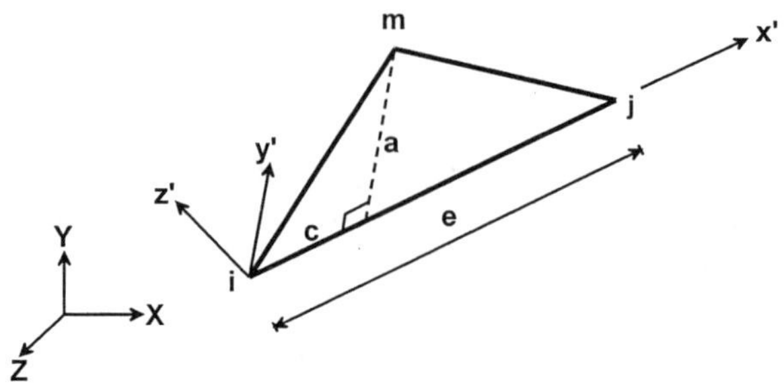

Figure 6.2 A triangular finite element in its local coordinate system.

$$\omega_{y'} = -\frac{1}{e}\left[(\boldsymbol{\delta}_j)_{z'} - (\boldsymbol{\delta}_i)_{z'}\right] \tag{6.27}$$

These rotations are obtained from Figure 6.3 as

$$\theta = \omega_{x'} = \frac{\delta_{mi} - \frac{c}{e}\delta_{ji}}{a} \text{ and } -\phi = \omega_{y'} = -\frac{\delta_{ji}}{e} \tag{6.28}$$

where

$$\delta_{mi} = (\boldsymbol{\delta}_m)_{z'} - (\boldsymbol{\delta}_i)_{z'} \text{ and } \delta_{ji} = (\boldsymbol{\delta}_j)_{z'} - (\boldsymbol{\delta}_i)_{z'} \tag{6.29}$$

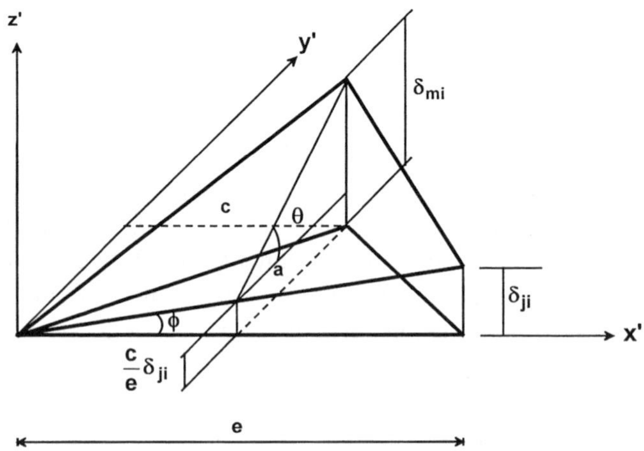

Figure 6.3 Out-of-plane rotations.

Now since the rotation components are linear in displacements (Eqs. 6.26, 6.27), the rotation vector $\boldsymbol{\omega}$ can be written as

$$\boldsymbol{\omega} = \mathbf{A}^{*}\boldsymbol{\delta}^{e} \tag{6.30}$$

or

$$\begin{Bmatrix} \omega_{x'} \\ \omega_{y'} \\ \omega_{z'} \end{Bmatrix} = \begin{bmatrix} 0 & 0 & -\dfrac{e-c}{ae} & 0 & 0 & -\dfrac{c}{ae} & 0 & 0 & 1/a \\ 0 & 0 & 1/e & 0 & 0 & -1/e & 0 & 0 & 0 \\ 0 & 0 & 0 & 0 & 0 & 0 & 0 & 0 & 0 \end{bmatrix} \begin{Bmatrix} (\boldsymbol{\delta}_i)_{x'} \\ (\boldsymbol{\delta}_i)_{y'} \\ (\boldsymbol{\delta}_i)_{z'} \\ (\boldsymbol{\delta}_j)_{x'} \\ (\boldsymbol{\delta}_j)_{y'} \\ (\boldsymbol{\delta}_j)_{z'} \\ (\boldsymbol{\delta}_m)_{x'} \\ (\boldsymbol{\delta}_m)_{y'} \\ (\boldsymbol{\delta}_m)_{z'} \end{Bmatrix} \tag{6.31}$$

It only remains to construct a matrix representation of the incremental forces Eq. 6.24, produced by these rotations acting upon the element nodal forces of Eq. 6.21. Noting that out-of plane components of nodal forces are zero i.e. $(\mathbf{F}_r)_{z'} = 0$ for $r = i, j, m$, the incremental forces (Eq. 6.24) for each node may be written as

$$\begin{Bmatrix} (\mathrm{d}\mathbf{F}_i)_{x'} \\ (\mathrm{d}\mathbf{F}_i)_{y'} \\ (\mathrm{d}\mathbf{F}_i)_{z'} \end{Bmatrix} = -\begin{bmatrix} 0 & 0 & (\mathbf{F}_i)_{y'} \\ 0 & 0 & -(\mathbf{F}_i)_{x'} \\ -(\mathbf{F}_i)_{y'} & (\mathbf{F}_i)_{x'} & 0 \end{bmatrix} \begin{Bmatrix} \omega_{x'} \\ \omega_{y'} \\ \omega_{z'} \end{Bmatrix} \tag{6.32}$$

$$\begin{Bmatrix} (\mathrm{d}\mathbf{F}_j)_{x'} \\ (\mathrm{d}\mathbf{F}_j)_{y'} \\ (\mathrm{d}\mathbf{F}_j)_{z'} \end{Bmatrix} = -\begin{bmatrix} 0 & 0 & (\mathbf{F}_j)_{y'} \\ 0 & 0 & -(\mathbf{F}_j)_{x'} \\ -(\mathbf{F}_j)_{y'} & (\mathbf{F}_j)_{x'} & 0 \end{bmatrix} \begin{Bmatrix} \omega_{x'} \\ \omega_{y'} \\ \omega_{z'} \end{Bmatrix} \tag{6.33}$$

$$\begin{Bmatrix} (\mathbf{dF}_m)_{x'} \\ (\mathbf{dF}_m)_{y'} \\ (\mathbf{dF}_m)_{z'} \end{Bmatrix} = -\begin{bmatrix} 0 & 0 & (\mathbf{F}_m)_{y'} \\ 0 & 0 & -(\mathbf{F}_m)_{x'} \\ -(\mathbf{F}_m)_{y'} & (\mathbf{F}_m)_{x'} & 0 \end{bmatrix} \begin{Bmatrix} \omega_{x'} \\ \omega_{y'} \\ \omega_{z'} \end{Bmatrix} \tag{6.34}$$

and the incremental force vector for member e becomes

$$\mathbf{dF}^e = -\mathbf{F}^* \boldsymbol{\omega} \tag{6.35}$$

where $\mathbf{F}^*$ is a 9×3 matrix obtained by stacking the 3×3 matrices of Eqs. 6.32-6.34. Substitution of Eq. 6.31 into Eq. 6.35 yields

$$\mathbf{dF}^e = -\mathbf{F}^* \mathbf{A}^* \boldsymbol{\delta}^{\mathrm{e}} = \mathbf{K}_{\mathrm{G}}^{e*} \boldsymbol{\delta}^{\mathrm{e}} \tag{6.36}$$

The required out-of-plane contribution, $\mathbf{K}_{\mathrm{G}}^{e*}$, to the geometric stiffness matrix, in the local coordinate system is then simply

$$\mathbf{K}_{\mathrm{G}}^{e*} = -\mathbf{F}^* \mathbf{A}^* \tag{6.37}$$

or more explicitly

$$\mathbf{K}_{\mathrm{G}}^{e*} = \frac{t}{2} \begin{bmatrix} \mathbf{K}_{\mathrm{G}_{ii}}^{e*} & \mathbf{K}_{\mathrm{G}_{ji}}^{e*} & \mathbf{K}_{\mathrm{G}_{mi}}^{e*} \\ \mathbf{K}_{\mathrm{G}_{ij}}^{e*} & \mathbf{K}_{\mathrm{G}_{jj}}^{e*} & \mathbf{K}_{\mathrm{G}_{mj}}^{e*} \\ \mathbf{K}_{\mathrm{G}_{im}}^{e*} & \mathbf{K}_{\mathrm{G}_{jm}}^{e*} & \mathbf{K}_{\mathrm{G}_{mm}}^{e*} \end{bmatrix} \tag{6.38}$$

where

$$\mathbf{K}_{\mathrm{G}_{ir}}^{e*} = \begin{bmatrix} 0 & 0 & 0 \\ 0 & 0 & 0 \\ 0 & 0 & \alpha_r \end{bmatrix}; \; \mathbf{K}_{\mathrm{G}_{jr}}^{e*} = \begin{bmatrix} 0 & 0 & 0 \\ 0 & 0 & 0 \\ 0 & 0 & \beta_r \end{bmatrix}; \; \mathbf{K}_{\mathrm{G}_{mr}}^{e*} = \begin{bmatrix} 0 & 0 & 0 \\ 0 & 0 & 0 \\ 0 & 0 & \lambda_r \end{bmatrix} \tag{6.39}$$

where

$$\alpha_r = -\frac{e-c}{ae}(\mathbf{F}_r)_{y'} - \frac{1}{e}(\mathbf{F}_r)_{x'}; \; r = i, j, m \tag{6.40}$$

$$\beta_r = -\frac{c}{ae}(\mathbf{F}_r)_{y'} + \frac{1}{e}(\mathbf{F}_r)_{x'};\ r = i, j, m \tag{6.41}$$

$$\lambda_r = \frac{1}{a}(\mathbf{F}_r)_{y'};\ r = i, j, m \tag{6.42}$$

6.4 A Direct Alternative Derivation of the Geometric Stiffness Matrix of Three-Dimensional Membranes

As was previously noted the geometric stiffness matrix is the gradient of the element nodal force vector in global coordinates, i.e.

$$\mathbf{K}^e_{\mathrm{G\,3D}} = \nabla\left(\mathbf{F}^e_{\mathrm{global}}\right) = tA\nabla\left(\mathcal{R}^e\mathbf{B}^{\mathrm{T}}\boldsymbol{\sigma}\right) \tag{6.43}$$

The matrix $\mathcal{R}^e$ in Eq. 6.43 is formed from the three-dimensional rotation matrix $\mathbf{R}^e$ which describes the transformation of member e from global to local coordinates as follows:

$$\mathcal{R}^e = \begin{bmatrix} \mathbf{R}^e & & \\ & \mathbf{R}^e & \\ & & \mathbf{R}^e \end{bmatrix} \tag{6.44}$$

Here, the rotation matrix $\mathbf{R}^e$ is assembled from

$$\mathbf{R}^e = \begin{bmatrix} (\mathbf{i}')_x & (\mathbf{i}')_y & (\mathbf{i}')_z \\ (\mathbf{j}')_x & (\mathbf{j}')_y & (\mathbf{j}')_z \\ (\mathbf{k}')_x & (\mathbf{k}')_y & (\mathbf{k}')_z \end{bmatrix} = \begin{Bmatrix} \mathbf{i}' \\ \mathbf{j}' \\ \mathbf{k}' \end{Bmatrix} \tag{6.45}$$

where $\mathbf{i}'$, $\mathbf{j}'$, $\mathbf{k}'$ are the unit vectors in the **x'**, **y'**, **z'** directions of the local coordinate system which is described in Figure 6.2.

No attempt is made here to perform the gradient of Eq. 6.43 but for the sake of completeness of the discussion of the previous section of in-plane and out-of-plane effects, the gradient is split in two to yield

$$\nabla\left(\mathcal{R}^{e}\mathbf{B}^{\mathrm{T}}\boldsymbol{\sigma}\right)=\nabla\left(\mathcal{R}^{e^{\mathrm{T}}}\mathbf{B}^{\mathrm{T}}\boldsymbol{\sigma}\right)_{\text{fixed }\mathcal{R}_{ij}^{e}}+\left(\nabla\mathcal{R}^{e^{\mathrm{T}}}\mathbf{B}^{\mathrm{T}}\boldsymbol{\sigma}\right)_{\text{fixed }\mathbf{B}_{ij}} \tag{6.45}$$

The first term on the right-hand side of Eq. 6.45 denotes the 'in-plane' effect of the geometric stiffness matrix, which requires only the gradients of member length, i.e., b_r, c_r. The second term on the right-hand side of Eq. 6.45 relates to the 'out-of-plane' effect of the matrix, which stems from the gradient of the rotation matrix. It is the calculation of this term that poses some hardship and the previous section was devoted to circumventing this difficulty by deriving a geometric stiffness matrix that assumes small rigid body rotations.

6.5 Computer Programs

6.5.1 Program P13-FEMPS.FOR

This program is one of the most simple finite element analysis programs. It is the constant stress finite element program carefully described in Zienkiewicz's book and is in fact quite similar to the other programs described in this book. In terms of loads and coordinates this program is identical to the earlier programs; it also includes the same solver used in these earlier programs. The coefficients *b* (here B) and *c* (here CZ) are those of Eq. 6.8.

6.5.2 Program P14-MEMBRN.FOR

This program is a step along the way toward a computer program for fabric structures or membranes. It is a linear analysis program which does not take into account changes in geometry. The elements themselves only have in-plane stiffness so that you end up with a structure which can only transmit axial forces and in-plane shear which is of course the membrane shell.

It is well known that to construct a membrane shell finite element program it is simply necessary to start with a plane stress finite element and rotate the element until it has the proper geometric orientation in space. That is what is done here and also the reason that much of the code which follows is taken from program P13.

This program has the structure of the other analysis programs in the book. It uses the following subroutines:

•`ABC`. This subroutine computes the coefficients *b* and *c* and the surface area A.

•`COMPKT`. This subroutine collects terms from a global array and places them in a local array.

•`PCOORD` sets up so-called "plane coordinates" and determines the rotation matrix from the element coordinates.

•`PLSTR` sets the member stiffness for a plane stress/strain element.

•`FORCES` computes the member stresses/strains given the member displacements.

•`TRANS` rotates a vector from global to local coordinates.

•`RASM` rotates the member stiffness to global coordinates.

6.5.3 Program P15-MEMNL.FOR

This program is simply program P14 with geometrical nonlinearities added. Truss bars are also included as an added capability since they would be needed for a typical fabric structure application. New subroutines in this program include:

•`MGSTIF` generates the geometric stiffness matrix.

•`UPDATE` calculates the new member forces once rigid body rotations have been removed.

This program uses the same input as program P14.

6.6 Examples

6.6.1 Example 6.1. A Deep Beam

This example is a "deep beam", that is a beam in which the depth is not small in comparison to its span. In this case, the problem data is generated by a simple computer program FEMPSDAT.FOR. This program generates a rectangular finite element grid which is subsequently divided into triangles. The variables in this program include: `NX` = number of grid lines in the *x* direction; NY = number of grid lines in the y direction; T = beam thickness; H = beam height and AL = beam length.

It this case the program places a unit load at the end of the beam. The easiest way to run these programs is to compile both the data generation program, FEMPSDAT.FOR, and the finite element analysis program, FEMPS.FOR, and then use the DOS pipeline command to concatenate their execution FEMPSDAT|FEMPS>FEMPS.OUT. The beam is 30 inches long and 28 inches deep and 1 inch thick. Young's modulus is 29,000,000 psi and Poison's ratio is *.32*. Figure 6.4 shows the structural analysis model.

The program reads joint coordinates R, loads P, member incidence nodes NP, and thickness S. It computes joint displacements P, strains AZ and stresses, AI. The output file is listed below. Assessment of the results as representative of the phenomenon is left as an exercise to the reader.

Output file FEMPS.OUT

```
             COORDINATES                          LOADS
                X                Y                 PX                PY
  1    .30000000D+02    .00000000D+00    .00000000D+00   -.10000000D+01
  2    .30000000D+02    .93333330D+01    .00000000D+00    .00000000D+00
  3    .30000000D+02    .18666670D+02    .00000000D+00    .00000000D+00
  4    .30000000D+02    .28000000D+02    .00000000D+00    .00000000D+00
  5    .22500000D+02    .00000000D+00    .00000000D+00    .00000000D+00
  6    .22500000D+02    .93333330D+01    .00000000D+00    .00000000D+00
  7    .22500000D+02    .18666670D+02    .00000000D+00    .00000000D+00
  8    .22500000D+02    .28000000D+02    .00000000D+00    .00000000D+00
  9    .15000000D+02    .00000000D+00    .00000000D+00    .00000000D+00
 10    .15000000D+02    .93333330D+01    .00000000D+00    .00000000D+00
 11    .15000000D+02    .18666670D+02    .00000000D+00    .00000000D+00
 12    .15000000D+02    .28000000D+02    .00000000D+00    .00000000D+00
 13    .75000000D+01    .00000000D+00    .00000000D+00    .00000000D+00
 14    .75000000D+01    .93333330D+01    .00000000D+00    .00000000D+00
 15    .75000000D+01    .18666670D+02    .00000000D+00    .00000000D+00
 16    .75000000D+01    .28000000D+02    .00000000D+00    .00000000D+00
 17    .00000000D+00    .00000000D+00    .00000000D+00    .00000000D+00
 18    .00000000D+00    .93333330D+01    .00000000D+00    .00000000D+00
 19    .00000000D+00    .18666670D+02    .00000000D+00    .00000000D+00
 20    .00000000D+00    .28000000D+02    .00000000D+00    .00000000D+00

   ELEMENT            ELEMENT NODES                      THICKNESS
         1            5         1         2        .10000000E+01
         2            5         2         6        .10000000E+01
         3            6         2         3        .10000000E+01
         4            6         3         7        .10000000E+01
         5            7         3         4        .10000000E+01
         6            7         4         8        .10000000E+01
         7            9         5         6        .10000000E+01
         8            9         6        10        .10000000E+01
         9           10         6         7        .10000000E+01
        10           10         7        11        .10000000E+01
        11           11         7         8        .10000000E+01
        12           11         8        12        .10000000E+01
        13           13         9        10        .10000000E+01
        14           13        10        14        .10000000E+01
        15           14        10        11        .10000000E+01
        16           14        11        15        .10000000E+01
        17           15        11        12        .10000000E+01
        18           15        12        16        .10000000E+01
```

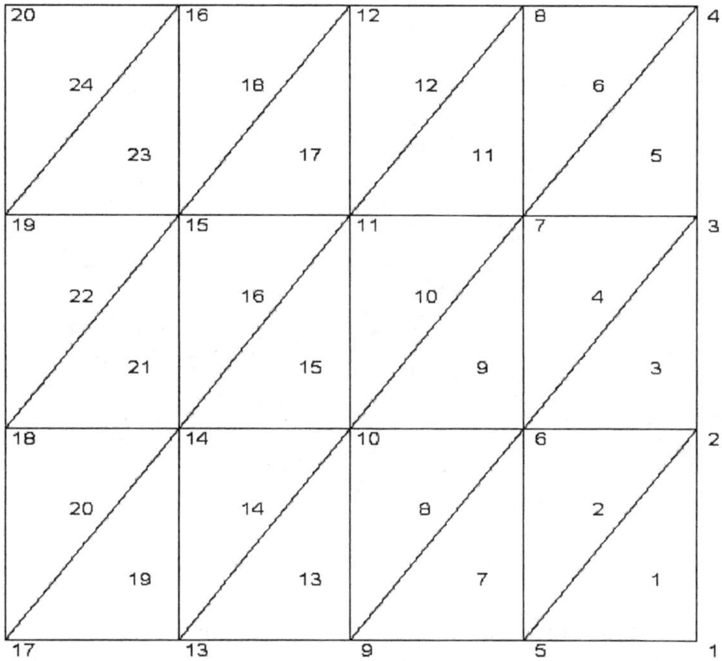

Figure 6.4 The plane stress cantilever beam.

```
  19        17        13        14       .10000000E+01
  20        17        14        18       .10000000E+01
  21        18        14        15       .10000000E+01
  22        18        15        19       .10000000E+01
  23        19        15        16       .10000000E+01
  24        19        16        20       .10000000E+01

DISPLACEMENTS
                          X                    Y
   1      -.12561651D-06       -.29653664D-06
   2      -.24175047D-07       -.23195584D-06
   3       .25915595D-07       -.20460964D-06
   4       .79456403D-07       -.19862350D-06
   5      -.96539144D-07       -.16937614D-06
   6      -.20086171D-07       -.15520392D-06
   7       .24816674D-07       -.14673469D-06
   8       .75709248D-07       -.14552769D-06
   9      -.72706518D-07       -.95010275D-07
  10      -.15657549D-07       -.86800959D-07
  11       .19421555D-07       -.84794679D-07
  12       .61803372D-07       -.87771887D-07
  13      -.41962270D-07       -.40817802D-07
  14      -.80423984D-08       -.32318339D-07
  15       .10651986D-07       -.32806949D-07
  16       .35986177D-07       -.37874193D-07

        STRAINS                              STRESSES
          EX          EY           GA          SX          SY         TAU
1-.38770D-08 .69194D-08-.60860D-08 -.53722D-01  .18347D+00 -.66854D-01
2-.54518D-09 .15185D-08-.20422D-08 -.19152D-02  .43422D-01 -.22433D-01
```

```
 3-.54518D-09 .29299D-08-.48667D-08  .12678D-01  .89025D-01 -.53460D-01
 4 .14652D-09 .90742D-09-.29056D-08  .14115D-01  .30832D-01 -.31918D-01
 5 .14652D-09 .64137D-09-.19801D-08  .11365D-01  .22237D-01 -.21752D-01
 6 .49962D-09 .12932D-09-.16267D-08  .17479D-01  .93436D-02 -.17869D-01
 7 .31777D-08 .15185D-08-.17241D-08 -.86967D-01  .16206D-01 -.18939D-01
 8 .59048D-09 .87957D-09-.30080D-08 -.99840D-02  .22313D-01 -.33042D-01
 9 .59048D-09 .90742D-09-.43094D-08 -.96960D-02  .23212D-01 -.47338D-01
10 .71935D-09 .21496D-09-.45002D-08  .25463D-01  .14382D-01 -.49434D-01
11 .71935D-09 .12932D-09-.28059D-08  .24578D-01  .11615D-01 -.30822D-01
12 .18541D-08-.31899D-09-.31599D-08  .56606D-01  .88632D-02 -.34711D-01
13-.40992D-08 .87957D-09-.11133D-08 -.12335D+00 -.13963D-01 -.12229D-01
14 .10154D-08 .91066D-09-.36301D-08 -.23389D-01  .18924D-01 -.39876D-01
15-.10154D-08 .21496D-09-.35059D-08 -.30582D-01 -.35525D-02 -.38512D-01
16 .11693D-08-.52351D-10-.49287D-08  .37236D-01  .10397D-01 -.54141D-01
17 .11693D-08-.31899D-09-.23908D-08  .34479D-01  .17828D-02 -.26262D-01
18 .34423D-08-.54292D-09-.39386D-08  .10560D+00  .18048D-01 -.43265D-01
19 .55950D-08 .91066D-09-.18081D-08 -.17135D+00 -.28423D-01 -.19862D-01
20 .10723D-08 .00000D+00-.43091D-08 -.34645D-01 -.11086D-01 -.47335D-01
21-.10723D-08-.52351D-10-.23061D-08 -.35186D-01 -.12778D-01 -.25333D-01
22 .14203D-08 .00000D+00-.43743D-08  .45886D-01  .14684D-01 -.48051D-01
23 .14203D-08-.54292D-09-.16599D-08  .40273D-01 -.28572D-02 -.18234D-01
24 .47982D-08 .00000D+00-.50499D-08  .15502D+00  .49607D-01 -.55472D-01
Stop - Program terminated.
```

6.6.2 Example 6.2. A Spherical Cap

This example analyses a spherical cap for membrane stresses. Figure 6.5 shows a plan view of the analysis model. Again, the data is generated using a computer program (MEMBRNDAT.FOR) with the following variables: `RAD` = shell radius; `THICK` = shell thickness; `PHI` = shell central angle; `NSEG` = number of sectors taken around the shell and `NRING` = number of levels into which the shell is subdivided. Again, both the data program and the analysis program must first be compiled. At run time they can be concatenated using the DOS pipeline command MEMBRNDAT|MEMBRN>MEMBRN.OUT.

The cap has a radius of 4.76 inches, a central angle of 10.9 degrees and a thickness of 0.01576 inches. Young's modulus and Poisson's ratio are 10,000,000 psi and 0.3 respectively. The output for this example is listed below.

Output file MEMBRN.OUT

```
NUMBER OF BARS                     0
NUMBER OF NODES                   25
NUMBER OF SUPPORTS                 8
NUMBER OF CONSTRAINTS              0
NUMBER OF ITERATIONS               1
NUMBER OF FINITE ELEMENTS         40

            COORDINATES                         LOADS
            X           Y           Z           PX          PY          PZ
 1 .00000D+00 .00000D+00 .47600D+01  .00000D+00  .00000D+00 -.10000D+01
 2 .30165D+00 .00000D+00 .47504D+01  .00000D+00  .00000D+00  .00000D+00
 3 .21330D+00 .21330D+00 .47504D+01  .00000D+00  .00000D+00  .00000D+00
 4 .38236D-06 .30165D+00 .47504D+01  .00000D+00  .00000D+00  .00000D+00
 5-.21330D+00 .21330D+00 .47504D+01  .00000D+00  .00000D+00  .00000D+00
```

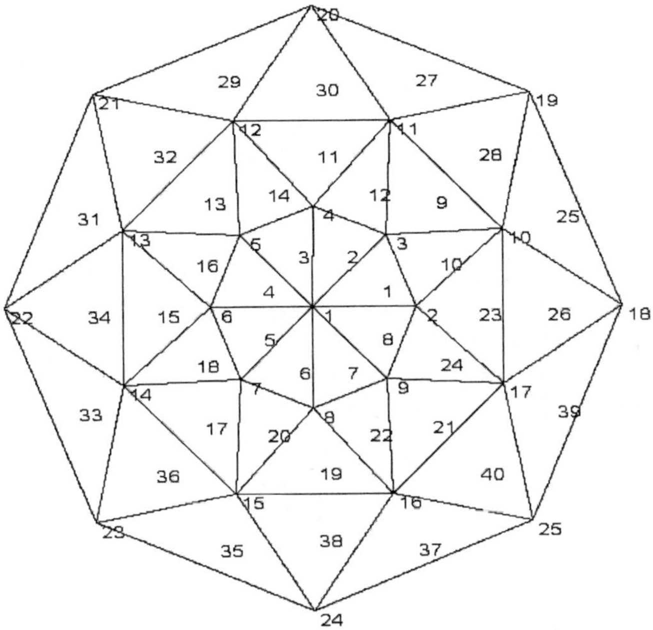

Figure 6.5 The spherical membrane shell.

```
 6-.30165D+00 .76473D-06 .47504D+01  .00000D+00  .00000D+00  .00000D+00
 7-.21330D+00-.21330D+00 .47504D+01  .00000D+00  .00000D+00  .00000D+00
 8-.11471D-05-.30165D+00 .47504D+01  .00000D+00  .00000D+00  .00000D+00
 9 .21330D+00-.21330D+00 .47504D+01  .00000D+00  .00000D+00  .00000D+00
10 .55625D+00 .23041D+00 .47218D+01  .00000D+00  .00000D+00  .00000D+00
11 .23041D+00 .55625D+00 .47218D+01  .00000D+00  .00000D+00  .00000D+00
12-.23041D+00 .55625D+00 .47218D+01  .00000D+00  .00000D+00  .00000D+00
13-.55625D+00 .23041D+00 .47218D+01  .00000D+00  .00000D+00  .00000D+00
14-.55625D+00-.23040D+00 .47218D+01  .00000D+00  .00000D+00  .00000D+00
15-.23041D+00-.55625D+00 .47218D+01  .00000D+00  .00000D+00  .00000D+00
16 .23040D+00-.55625D+00 .47218D+01  .00000D+00  .00000D+00  .00000D+00
17 .55625D+00-.23041D+00 .47218D+01  .00000D+00  .00000D+00  .00000D+00
18 .90009D+00 .00000D+00 .46741D+01  .00000D+00  .00000D+00  .00000D+00
19 .63646D+00 .63646D+00 .46741D+01  .00000D+00  .00000D+00  .00000D+00
20 .11410D-05 .90009D+00 .46741D+01  .00000D+00  .00000D+00  .00000D+00
21-.63646D+00 .63646D+00 .46741D+01  .00000D+00  .00000D+00  .00000D+00
22-.90009D+00 .22819D-05 .46741D+01  .00000D+00  .00000D+00  .00000D+00
23-.63646D+00-.63646D+00 .46741D+01  .00000D+00  .00000D+00  .00000D+00
24-.34229D-05-.90009D+00 .46741D+01  .00000D+00  .00000D+00  .00000D+00
25 .63646D+00-.63647D+00 .46741D+01  .00000D+00  .00000D+00  .00000D+00
```

```
FINITE ELEMENT DATA
 NO.         NODES             THICKNESS
             I    J    M
  1          1    2    3       .1576E-01
  2          1    3    4       .1576E-01
  3          1    4    5       .1576E-01
  4          1    5    6       .1576E-01
  5          1    6    7       .1576E-01
  6          1    7    8       .1576E-01
  7          1    8    9       .1576E-01
```

8	1	9	2	.1576E-01
9	10	11	3	.1576E-01
10	2	10	3	.1576E-01
11	11	12	4	.1576E-01
12	3	11	4	.1576E-01
13	12	13	5	.1576E-01
14	4	12	5	.1576E-01
15	13	14	6	.1576E-01
16	5	13	6	.1576E-01
17	14	15	7	.1576E-01
18	6	14	7	.1576E-01
19	15	16	8	.1576E-01
20	7	15	8	.1576E-01
21	16	17	9	.1576E-01
22	8	16	9	.1576E-01
23	17	10	2	.1576E-01
24	9	17	2	.1576E-01
25	18	19	10	.1576E-01
26	17	18	10	.1576E-01
27	19	20	11	.1576E-01
28	10	19	11	.1576E-01
29	20	21	12	.1576E-01
30	11	20	12	.1576E-01
31	21	22	13	.1576E-01
32	12	21	13	.1576E-01
33	22	23	14	.1576E-01
34	13	22	14	.1576E-01
35	23	24	15	.1576E-01
36	14	23	15	.1576E-01
37	24	25	16	.1576E-01
38	15	24	16	.1576E-01
39	25	18	17	.1576E-01
40	16	25	17	.1576E-01

DISPLACEMENTS

	X	Y	Z
1	-.28711713D-10	.17348089D-10	-.23013107D-02
2	.26981282D-04	-.58449789D-11	.12403418D-03
3	.19078607D-04	.19078568D-04	.12403315D-03
4	.28488593D-10	.26981224D-04	.12403335D-03
5	-.19078584D-04	.19078653D-04	.12403344D-03
6	-.26981239D-04	.74573549D-10	.12403344D-03
7	-.19078664D-04	-.19078536D-04	.12403328D-03
8	-.85858794D-10	-.26981189D-04	.12403299D-03
9	.19078543D-04	-.19078664D-04	.12403294D-03
10	.82616315D-05	.34220653D-05	.34118350D-04
11	.34220896D-05	.82616316D-05	.34118449D-04
12	-.34220756D-05	.82616545D-05	.34118522D-04
13	-.82616462D-05	.34221082D-05	.34118544D-04
14	-.82616561D-05	-.34220602D-05	.34118502D-04
15	-.34221060D-05	-.82616211D-05	.34118430D-04
16	.34220644D-05	-.82616785D-05	.34118600D-04
17	.82616260D-05	-.34221213D-05	.34118451D-04

MEMBER ANALYSIS

	STRAINS			STRESSES		
	EX	EY	GA	SX	SY	TAU
1	-.16540E-03	.45773E-04	-.21100E-03	-.16667E+04	-.42271E+02	-.81153E+03
2	-.16540E-03	.45773E-04	-.21100E-03	-.16667E+04	-.42271E+02	-.81153E+03
3	-.16540E-03	.45773E-04	-.21100E-03	-.16667E+04	-.42271E+02	-.81153E+03
4	-.16540E-03	.45773E-04	-.21100E-03	-.16667E+04	-.42271E+02	-.81153E+03
5	-.16540E-03	.45773E-04	-.21100E-03	-.16667E+04	-.42271E+02	-.81153E+03
6	-.16540E-03	.45773E-04	-.21100E-03	-.16667E+04	-.42270E+02	-.81153E+03
7	-.16540E-03	.45773E-04	-.21100E-03	-.16667E+04	-.42270E+02	-.81153E+03

```
 8-.16540E-03 .45773E-04-.21100E-03 -.16667E+04 -.42277E+02 -.81153E+03
 9 .14852E-04-.33342E-04-.30056E-10  .53294E+02 -.31744E+03 -.11560E-03
10-.11794E-04 .76648E-04 .71993E-04  .12309E+03  .80341E+03  .27689E+03
11 .14852E-04-.33342E-04-.30224E-11  .53294E+02 -.31744E+03 -.11625E-04
12-.11794E-04 .76648E-04 .71993E-04  .12309E+03  .80341E+03  .27689E+03
13 .14852E-04-.33342E-04 .18448E-11  .53294E+02 -.31744E+03  .70953E-05
14-.11794E-04 .76648E-04 .71993E-04  .12309E+03  .80341E+03  .27689E+03
15 .14852E-04-.33342E-04-.56942E-11  .53294E+02 -.31744E+03 -.21901E-04
16-.11794E-04 .76648E-04 .71993E-04  .12309E+03  .80341E+03  .27689E+03
17 .14852E-04-.33342E-04 .13423E-11  .53294E+02 -.31744E+03  .51628E-05
18-.11794E-04 .76648E-04 .71993E-04  .12309E+03  .80341E+03  .27689E+03
19 .14852E-04-.33342E-04-.28137E-11  .53294E+02 -.31744E+03 -.10822E-04
20-.11794E-04 .76648E-04 .71993E-04  .12309E+03  .80341E+03  .27689E+03
21 .14852E-04-.33342E-04 .16695E-09  .53293E+02 -.31744E+03  .64210E-03
22-.11794E-04 .76648E-04 .71992E-04  .12309E+03  .80341E+03  .27689E+03
23 .14852E-04-.33342E-04 .22670E-09  .53294E+02 -.31743E+03  .87192E-03
24-.11794E-04 .76648E-04 .71992E-04  .12308E+03  .80341E+03  .27689E+03
25 .00000E+00-.77662E-05-.44590E-10 -.25603E+02 -.85343E+02 -.17150E-03
26-.24579E-05 .72260E-05 .22979E-04 -.31876E+01  .71304E+02  .88383E+02
27 .00000E+00-.77662E-05-.24496E-11 -.25603E+02 -.85343E+02 -.94214E-05
28-.24580E-05 .72261E-05 .22979E-04 -.31882E+01  .71305E+02  .88382E+02
29 .00000E+00-.77662E-05 .61971E-11 -.25603E+02 -.85343E+02  .23835E-04
30-.24580E-05 .72262E-05 .22979E-04 -.31883E+01  .71305E+02  .88382E+02
31 .00000E+00-.77662E-05-.33468E-11 -.25603E+02 -.85343E+02 -.12872E-04
32-.24580E-05 .72262E-05 .22979E-04 -.31883E+01  .71305E+02  .88382E+02
33 .00000E+00-.77662E-05-.84355E-11 -.25603E+02 -.85343E+02 -.32444E-04
34-.24580E-05 .72261E-05 .22979E-04 -.31883E+01  .71305E+02  .88382E+02
35 .00000E+00-.77662E-05 .12291E-10 -.25603E+02 -.85343E+02  .47275E-04
36-.24580E-05 .72262E-05 .22979E-04 -.31882E+01  .71305E+02  .88382E+02
37 .00000E+00-.77662E-05 .19779E-10 -.25603E+02 -.85343E+02  .76073E-04
38-.24580E-05 .72262E-05 .22979E-04 -.31883E+01  .71305E+02  .88382E+02
39 .00000E+00-.77661E-05-.88205E-10 -.25603E+02 -.85342E+02 -.33925E-03
40-.24580E-05 .72262E-05 .22979E-04 -.31884E+01  .71305E+02  .88382E+02
Stop - Program terminated.
```

6.6.3 Example 6.3. A Flat Stretched Membrane

The transversely loaded flat cable-net of Example 3.2 is transformed to a flat stretched membrane ending a sequence of examples that started with Biot's prestressed 2-bar truss. The symmetric analysis model of Figure 3.3 is turned to a symmetric finite element analysis model. (See Figure 6.6.) The thickness of the membrane is taken as the cross-sectional area of a single cable divided by one span i.e. $\frac{0.25}{5\times 12} = 0.004167 \text{ in}$. Similarly a uniform stress of $\frac{20{,}000}{5\times 12} \div \frac{0.25}{5\times 12} = 80{,}000 \text{ psi}$ is applied all around the membrane. Young's modulus and Poisson's ratio are taken as 30,000,000 psi and 0.3 respectively. Input file MEMNL.DAT is invoked by MEMNL.FOR and the results are partially listed in MEMNL.OUT. Actually only the echo of the input file and the computations of the last iteration of the last load step are listed. It is important to mention here that `LINEAR INCREMENTAL ANALYSIS` lists the strains and stresses that are computed from the usual plane nodal displacements/plane strain relations of the plane stress triangular element i.e.

nodal displacements are transformed to local coordinates and the z-coordinate discarded. INCREMENTAL NONLINEAR ANALYSIS, however, lists strains and stresses which are computed after rigid body rotations have been removed. (See problem 6.2.) UPDATED PRESTRESS simply adds the stresses from the current increment to those of the previous increment.

It might be of interest to observe that the central deflection here 6.62 inches whereas the central deflection of the cable net is 7.11 inches. But then again this is to be expected because of the shear in the membrane that tends to stiffen it.

Output file MEMNL.OUT

```
NUMBER OF BARS                      0
NUMBER OF NODES                    25
NUMBER OF SUPPORTS                 16
NUMBER OF CONSTRAINTS               0
NUMBER OF ITERATIONS               10
NUMBER OF FINITE ELEMENTS          32
NUMBER OF LOAD STEPS                5

            COORDINATES                          LOADS
          X            Y            Z            PX           PY           PZ
  1-.60000E+02 .60000E+02 .00000E+00  .00000E+00  .00000E+00  .00000E+00
```

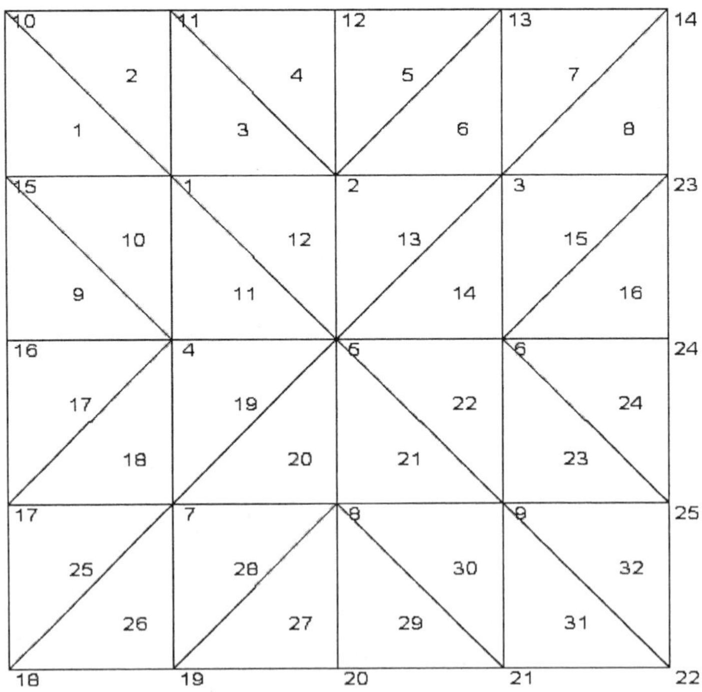

Figure 6.6 The transversely loaded stretched membrane.

```
 2 .00000E+00 .60000E+02 .00000E+00  .00000E+00  .00000E+00  .00000E+00
 3 .60000E+02 .60000E+02 .00000E+00  .00000E+00  .00000E+00  .00000E+00
 4-.60000E+02 .00000E+00 .00000E+00  .00000E+00  .00000E+00  .00000E+00
 5 .00000E+00 .00000E+00 .00000E+00  .00000E+00  .00000E+00 -.10000E+05
 6 .60000E+02 .00000E+00 .00000E+00  .00000E+00  .00000E+00  .00000E+00
 7-.60000E+02-.60000E+02 .00000E+00  .00000E+00  .00000E+00  .00000E+00
 8 .00000E+00-.60000E+02 .00000E+00  .00000E+00  .00000E+00  .00000E+00
 9 .60000E+02-.60000E+02 .00000E+00  .00000E+00  .00000E+00  .00000E+00
10-.12000E+03 .12000E+03 .00000E+00  .00000E+00  .00000E+00  .00000E+00
11-.60000E+02 .12000E+03 .00000E+00  .00000E+00  .00000E+00  .00000E+00
12 .00000E+00 .12000E+03 .00000E+00  .00000E+00  .00000E+00  .00000E+00
13 .60000E+02 .12000E+03 .00000E+00  .00000E+00  .00000E+00  .00000E+00
14 .12000E+03 .12000E+03 .00000E+00  .00000E+00  .00000E+00  .00000E+00
15-.12000E+03 .60000E+02 .00000E+00  .00000E+00  .00000E+00  .00000E+00
16-.12000E+03 .00000E+00 .00000E+00  .00000E+00  .00000E+00  .00000E+00
17-.12000E+03-.60000E+02 .00000E+00  .00000E+00  .00000E+00  .00000E+00
18-.12000E+03-.12000E+03 .00000E+00  .00000E+00  .00000E+00  .00000E+00
19-.60000E+02-.12000E+03 .00000E+00  .00000E+00  .00000E+00  .00000E+00
20 .00000E+00-.12000E+03 .00000E+00  .00000E+00  .00000E+00  .00000E+00
21 .60000E+02-.12000E+03 .00000E+00  .00000E+00  .00000E+00  .00000E+00
22 .12000E+03-.12000E+03 .00000E+00  .00000E+00  .00000E+00  .00000E+00
23 .12000E+03 .60000E+02 .00000E+00  .00000E+00  .00000E+00  .00000E+00
24 .12000E+03 .00000E+00 .00000E+00  .00000E+00  .00000E+00  .00000E+00
25 .12000E+03-.60000E+02 .00000E+00  .00000E+00  .00000E+00  .00000E+00

FINITE ELEMENT DATA
  NO.    NODES           THICKNESS          PRESTRESS
         I     J     M                      SIGX        SIGY        TAU
    1   15     1    10  .41600E-02  .80000E+05  .80000E+05  .00000E+00
    2   11    10     1  .41600E-02  .80000E+05  .80000E+05  .00000E+00
    3    1     2    11  .41600E-02  .80000E+05  .80000E+05  .00000E+00
    4   12    11     2  .41600E-02  .80000E+05  .80000E+05  .00000E+00
    5   13    12     2  .41600E-02  .80000E+05  .80000E+05  .00000E+00
    6    2     3    13  .41600E-02  .80000E+05  .80000E+05  .00000E+00
    7   14    13     3  .41600E-02  .80000E+05  .80000E+05  .00000E+00
    8    3    23    14  .41600E-02  .80000E+05  .80000E+05  .00000E+00
    9   16     4    15  .41600E-02  .80000E+05  .80000E+05  .00000E+00
   10    1    15     4  .41600E-02  .80000E+05  .80000E+05  .00000E+00
   11    4     5     1  .41600E-02  .80000E+05  .80000E+05  .00000E+00
   12    2     1     5  .41600E-02  .80000E+05  .80000E+05  .00000E+00
   13    3     2     5  .41600E-02  .80000E+05  .80000E+05  .00000E+00
   14    5     6     3  .41600E-02  .80000E+05  .80000E+05  .00000E+00
   15   23     3     6  .41600E-02  .80000E+05  .80000E+05  .00000E+00
   16    6    24    23  .41600E-02  .80000E+05  .80000E+05  .00000E+00
   17    4    16    17  .41600E-02  .80000E+05  .80000E+05  .00000E+00
   18   17     7     4  .41600E-02  .80000E+05  .80000E+05  .00000E+00
   19    5     4     7  .41600E-02  .80000E+05  .80000E+05  .00000E+00
   20    7     8     5  .41600E-02  .80000E+05  .80000E+05  .00000E+00
   21    8     9     5  .41600E-02  .80000E+05  .80000E+05  .00000E+00
   22    6     5     9  .41600E-02  .80000E+05  .80000E+05  .00000E+00
   23    9    25     6  .41600E-02  .80000E+05  .80000E+05  .00000E+00
   24   24     6    25  .41600E-02  .80000E+05  .80000E+05  .00000E+00
   25    7    17    18  .41600E-02  .80000E+05  .80000E+05  .00000E+00
   26   18    19     7  .41600E-02  .80000E+05  .80000E+05  .00000E+00
   27   19    20     8  .41600E-02  .80000E+05  .80000E+05  .00000E+00
   28    8     7    19  .41600E-02  .80000E+05  .80000E+05  .00000E+00
   29   20    21     8  .41600E-02  .80000E+05  .80000E+05  .00000E+00
   30    9     8    21  .41600E-02  .80000E+05  .80000E+05  .00000E+00
   31   21    22     9  .41600E-02  .80000E+05  .80000E+05  .00000E+00
   32   25     9    22  .41600E-02  .80000E+05  .80000E+05  .00000E+00

⋮

****ITERATION NUMBER  10
    LOAD STEP          5
```

```
           COORDINATES                              LOADS
              X             Y            Z           PX           PY           PZ
  1-.59985E+02 .59985E+02-.14309E+01  .00000E+00  .00000E+00  .00000E+00
  2-.25607E-15 .59983E+02-.26046E+01  .00000E+00  .00000E+00  .00000E+00
  3 .59985E+02 .59985E+02-.14309E+01  .00000E+00  .00000E+00  .00000E+00
  4-.59983E+02-.82747E-15-.26045E+01  .00000E+00  .00000E+00  .00000E+00
  5 .39495E-16 .36615E-16-.66418E+01  .00000E+00  .00000E+00 -.10000E+05
  6 .59983E+02 .82668E-15-.26045E+01  .00000E+00  .00000E+00  .00000E+00
  7-.59985E+02-.59985E+02-.14309E+01  .00000E+00  .00000E+00  .00000E+00
  8 .27357E-15-.59983E+02-.26046E+01  .00000E+00  .00000E+00  .00000E+00
  9 .59985E+02-.59985E+02-.14309E+01  .00000E+00  .00000E+00  .00000E+00
 10-.12000E+03 .12000E+03 .00000E+00  .00000E+00  .00000E+00  .00000E+00
 11-.60000E+02 .12000E+03 .00000E+00  .00000E+00  .00000E+00  .00000E+00
 12 .00000E+00 .12000E+03 .00000E+00  .00000E+00  .00000E+00  .00000E+00
 13 .60000E+02 .12000E+03 .00000E+00  .00000E+00  .00000E+00  .00000E+00
 14 .12000E+03 .12000E+03 .00000E+00  .00000E+00  .00000E+00  .00000E+00
 15-.12000E+03 .60000E+02 .00000E+00  .00000E+00  .00000E+00  .00000E+00
 16-.12000E+03 .00000E+00 .00000E+00  .00000E+00  .00000E+00  .00000E+00
 17-.12000E+03-.60000E+02 .00000E+00  .00000E+00  .00000E+00  .00000E+00
 18-.12000E+03-.12000E+03 .00000E+00  .00000E+00  .00000E+00  .00000E+00
 19-.60000E+02-.12000E+03 .00000E+00  .00000E+00  .00000E+00  .00000E+00
 20 .00000E+00-.12000E+03 .00000E+00  .00000E+00  .00000E+00  .00000E+00
 21 .60000E+02-.12000E+03 .00000E+00  .00000E+00  .00000E+00  .00000E+00
 22 .12000E+03-.12000E+03 .00000E+00  .00000E+00  .00000E+00  .00000E+00
 23 .12000E+03 .60000E+02 .00000E+00  .00000E+00  .00000E+00  .00000E+00
 24 .12000E+03 .00000E+00 .00000E+00  .00000E+00  .00000E+00  .00000E+00
 25 .12000E+03-.60000E+02 .00000E+00  .00000E+00  .00000E+00  .00000E+00

  ERROR =          .12389175D-02

 DISPLACEMENTS
                          X                   Y                   Z
        1       .52016588D-09      -.10951429D-08      -.46234717D-09
        2       .11365291D-15      -.21571404D-08      -.16859462D-09
        3      -.52016577D-09      -.10951434D-08      -.46234508D-09
        4      -.29205406D-09       .72583523D-15       .28383451D-09
        5      -.12545721D-16      -.14985675D-16       .45354019D-09
        6       .29205406D-09      -.72999134D-15       .28383420D-09
        7       .52016577D-09       .10951434D-08      -.46234496D-09
        8      -.12116067D-15       .21571404D-08      -.16859486D-09
        9      -.52016589D-09       .10951429D-08      -.46234720D-09

 MEMBER ANALYSIS
 MEMBRANE FINITE ELEMENTS

  ELEMENT      1
  NODES       15     1    10
 LINEAR INCREMENTAL ANALYSIS
             EX           EY          GA           SX           SY          TAU
  1 .88504E-11-.44298E-14-.18238E-10  .29173E-03  .87385E-04 -.21044E-03
 INCREMENTAL NONLINEAR ANALYSIS
             EX           EY          GA           SX           SY          TAU
  1 .88505E-11-.43817E-14-.18239E-10  .29173E-03  .87388E-04 -.21044E-03
 UPDATED PRESTRESS
                        SX                SY                TAU
          .9737767E+05       .8521242E+05      -.2801543E+04

 ELEMENT      2
 NODES       11    10     1
 LINEAR INCREMENTAL ANALYSIS
             EX           EY          GA           SX           SY          TAU
  2 .00000E+00 .18421E-10-.86649E-11  .18219E-03  .60729E-03 -.99979E-04
 INCREMENTAL NONLINEAR ANALYSIS
             EX           EY          GA           SX           SY          TAU
```

```
 2 .00000E+00 .18421E-10-.86649E-11  .18219E-03  .60729E-03 -.99979E-04
UPDATED PRESTRESS
                  SX                SY                TAU
         .8521326E+05      .9737754E+05     -.2802885E+04

ELEMENT      3
NODES        1    2   11
LINEAR INCREMENTAL ANALYSIS
           EX          EY          GA          SX          SY          TAU
 3-.87633E-11 .18403E-10-.26405E-10 -.10689E-03  .52004E-03 -.30467E-03
INCREMENTAL NONLINEAR ANALYSIS
           EX          EY          GA          SX          SY          TAU
 3-.87633E-11 .18403E-10-.26405E-10 -.10689E-03  .52004E-03 -.30467E-03
UPDATED PRESTRESS
                  SX                SY                TAU
         .8351018E+05      .9685909E+05     -.8657109E+04

ELEMENT      4
NODES       12   11    2
LINEAR INCREMENTAL ANALYSIS
           EX          EY          GA          SX          SY          TAU
 4 .00000E+00 .35996E-10-.18919E-17  .35601E-03  .11867E-02 -.21830E-10
INCREMENTAL NONLINEAR ANALYSIS
           EX          EY          GA          SX          SY          TAU
 4 .00000E+00 .35996E-10-.18919E-17  .35601E-03  .11867E-02 -.21830E-10
UPDATED PRESTRESS
                  SX                SY                TAU
         .9211395E+05      .1203798E+06      .2734976E-10

ELEMENT      5
NODES       13   12    2
LINEAR INCREMENTAL ANALYSIS
           EX          EY          GA          SX          SY          TAU
 5 .00000E+00 .35996E-10-.18919E-17 .35601E-03  .11867E-02 -.21830E-10
INCREMENTAL NONLINEAR ANALYSIS
           EX          EY          GA          SX          SY          TAU
 5 .00000E+00 .35996E-10 .00000E+00  .35601E-03  .11867E-02  .00000E+00
UPDATED PRESTRESS
                  SX                SY                TAU
         .9211395E+05      .1203798E+06      .0000000E+00

ELEMENT      6
NODES        2    3   13
LINEAR INCREMENTAL ANALYSIS
           EX          EY          GA          SX          SY          TAU
 6-.87633E-11 .18403E-10 .26405E-10 -.10689E-03  .52004E-03  .30467E-03
INCREMENTAL NONLINEAR ANALYSIS
           EX          EY          GA          SX          SY          TAU
 6-.87633E-11 .18403E-10 .26405E-10 -.10689E-03  .52004E-03  .30467E-03
UPDATED PRESTRESS
                  SX                SY                TAU
         .8351018E+05      .9685909E+05      .8657109E+04

ELEMENT      7
NODES       14   13    3
LINEAR INCREMENTAL ANALYSIS
           EX          EY          GA          SX          SY          TAU
 7 .00000E+00 .18421E-10 .86649E-11  .18219E-03  .60729E-03  .99979E-04
INCREMENTAL NONLINEAR ANALYSIS
           EX          EY          GA          SX          SY          TAU
 7 .00000E+00 .18421E-10 .86649E-11  .18219E-03  .60729E-03  .99979E-04
UPDATED PRESTRESS
                  SX                SY                TAU
         .8521326E+05      .9737754E+05      .2802885E+04
```

```
ELEMENT      8
NODES         3   23   14
LINEAR INCREMENTAL ANALYSIS
          EX            EY            GA             SX             SY            TAU
 8 .88504E-11-.44298E-14 .18238E-10  .29173E-03  .87385E-04  .21044E-03
INCREMENTAL NONLINEAR ANALYSIS
          EX            EY            GA             SX             SY            TAU
 8 .88505E-11-.43817E-14 .18238E-10  .29173E-03  .87388E-04  .21044E-03
UPDATED PRESTRESS
                    SX                    SY                    TAU
        .9737767E+05       .8521242E+05       .2801543E+04

ELEMENT      9
NODES        16    4   15
LINEAR INCREMENTAL ANALYSIS
          EX            EY            GA             SX             SY            TAU
 9-.50619E-11 .16643E-33 .12082E-16 -.16688E-03 -.50063E-04  .13941E-09
INCREMENTAL NONLINEAR ANALYSIS
          EX            EY            GA             SX             SY            TAU
 9-.50619E-11 .00000E+00 .12082E-16 -.16687E-03 -.50062E-04  .13941E-09
UPDATED PRESTRESS
                    SX                    SY                    TAU
        .1203582E+06       .9210745E+05      -.1955792E-10

ELEMENT     10
NODES         1   15    4
LINEAR INCREMENTAL ANALYSIS
          EX            EY            GA             SX             SY            TAU
10 .88504E-11-.18497E-10-.45423E-11  .10883E-03 -.52226E-03 -.52411E-04
INCREMENTAL NONLINEAR ANALYSIS
          EX            EY            GA             SX             SY            TAU
10 .88505E-11-.18497E-10-.45423E-11  .10884E-03 -.52226E-03 -.52411E-04
UPDATED PRESTRESS
                    SX                    SY                    TAU
        .9686449E+05       .8350182E+05      -.8644788E+04

ELEMENT     11
NODES         4    5    1
LINEAR INCREMENTAL ANALYSIS
          EX            EY            GA             SX             SY            TAU
11 .46574E-11-.18474E-10 .14410E-10 -.29171E-04 -.56298E-03  .16627E-03
INCREMENTAL NONLINEAR ANALYSIS
          EX            EY            GA             SX             SY            TAU
11 .46575E-11-.18474E-10 .14410E-10 -.29168E-04 -.56298E-03  .16627E-03
UPDATED PRESTRESS
                    SX                    SY                    TAU
        .1446910E+06       .9783068E+05      -.1561558E+05

ELEMENT     12
NODES         2    1    5
LINEAR INCREMENTAL ANALYSIS
          EX            EY            GA             SX             SY            TAU
12-.87633E-11-.36519E-10-.17152E-10 -.65007E-03 -.12906E-02 -.19791E-03
INCREMENTAL NONLINEAR ANALYSIS
          EX            EY            GA             SX             SY            TAU
12-.87633E-11-.36519E-10-.17152E-10 -.65008E-03 -.12906E-02 -.19791E-03
UPDATED PRESTRESS
                    SX                    SY                    TAU
        .9784664E+05       .1446473E+06      -.1565376E+05

ELEMENT     13
NODES         3    2    5
LINEAR INCREMENTAL ANALYSIS
          EX            EY            GA             SX             SY            TAU
13-.87633E-11-.36519E-10 .17152E-10 -.65007E-03 -.12906E-02  .19791E-03
```

```
INCREMENTAL NONLINEAR ANALYSIS
          EX          EY          GA          SX          SY          TAU
13-.87633E-11-.36519E-10 .17152E-10 -.65008E-03 -.12906E-02  .19791E-03
UPDATED PRESTRESS
                 SX               SY               TAU
        .9784664E+05     .1446473E+06     .1565376E+05

ELEMENT     14
NODES        5    6    3
LINEAR INCREMENTAL ANALYSIS
          EX          EY          GA          SX          SY          TAU
14 .46574E-11-.18474E-10-.14410E-10 -.29171E-04 -.56298E-03 -.16627E-03
INCREMENTAL NONLINEAR ANALYSIS
          EX          EY          GA          SX          SY          TAU
14 .46575E-11-.18474E-10-.14410E-10 -.29168E-04 -.56298E-03 -.16627E-03
UPDATED PRESTRESS
                 SX               SY               TAU
        .1446910E+06     .9783068E+05     .1561558E+05

ELEMENT     15
NODES       23    3    6
LINEAR INCREMENTAL ANALYSIS
          EX          EY          GA          SX          SY          TAU
15 .88504E-11-.18497E-10 .45423E-11  .10883E-03 -.52226E-03  .52411E-04
INCREMENTAL NONLINEAR ANALYSIS
          EX          EY          GA          SX          SY          TAU
15 .88505E-11-.18497E-10 .45423E-11  .10884E-03 -.52226E-03  .52411E-04
UPDATED PRESTRESS
                 SX               SY               TAU
        .9686449E+05     .8350182E+05     .8644788E+04

ELEMENT     16
NODES        6   24   23
LINEAR INCREMENTAL ANALYSIS
          EX          EY          GA          SX          SY          TAU
16-.50619E-11 .14390E-32 .12152E-16 -.16688E-03 -.50063E-04  .14021E-09
INCREMENTAL NONLINEAR ANALYSIS
          EX          EY          GA          SX          SY          TAU
16-.50619E-11 .00000E+00-.59973E-27 -.16687E-03 -.50062E-04 -.69199E-20
UPDATED PRESTRESS
                 SX               SY               TAU
        .1203582E+06     .9210745E+05    -.1366110E-08

ELEMENT     17
NODES        4   16   17
LINEAR INCREMENTAL ANALYSIS
          EX          EY          GA          SX          SY          TAU
17-.50619E-11 .14309E-32 .12082E-16 -.16688E-03 -.50063E-04  .13941E-09
INCREMENTAL NONLINEAR ANALYSIS
          EX          EY          GA          SX          SY          TAU
17-.50619E-11 .00000E+00-.59973E-27 -.16687E-03 -.50062E-04 -.69199E-20
UPDATED PRESTRESS
                 SX               SY               TAU
        .1203582E+06     .9210745E+05    -.1366110E-08

ELEMENT     18
NODES       17    7    4
LINEAR INCREMENTAL ANALYSIS
          EX          EY          GA          SX          SY          TAU
18 .88504E-11-.18497E-10 .45423E-11  .10883E-03 -.52226E-03  .52411E-04
INCREMENTAL NONLINEAR ANALYSIS
          EX          EY          GA          SX          SY          TAU
18 .88505E-11-.18497E-10 .45423E-11  .10884E-03 -.52226E-03  .52411E-04
UPDATED PRESTRESS
                 SX               SY               TAU
```

```
         .9686449E+05      .8350182E+05      .8644788E+04

ELEMENT     19
NODES        5    4    7
LINEAR INCREMENTAL ANALYSIS
           EX         EY          GA           SX           SY           TAU
19 .46574E-11-.18474E-10-.14410E-10 -.29171E-04 -.56298E-03 -.16627E-03
INCREMENTAL NONLINEAR ANALYSIS
           EX         EY          GA           SX           SY           TAU
19 .46575E-11-.18474E-10-.14410E-10 -.29168E-04 -.56298E-03 -.16627E-03
UPDATED PRESTRESS
                  SX                SY                TAU
         .1446910E+06      .9783068E+05      .1561558E+05

ELEMENT     20
NODES        7    8    5
LINEAR INCREMENTAL ANALYSIS
           EX         EY          GA           SX           SY           TAU
20-.87633E-11-.36519E-10 .17152E-10 -.65007E-03 -.12906E-02  .19791E-03
INCREMENTAL NONLINEAR ANALYSIS
           EX         EY          GA           SX           SY           TAU
20-.87633E-11-.36519E-10 .17152E-10 -.65008E-03 -.12906E-02  .19791E-03
UPDATED PRESTRESS
                  SX                SY                TAU
         .9784664E+05      .1446473E+06      .1565376E+05

ELEMENT     21
NODES        8    9    5
LINEAR INCREMENTAL ANALYSIS
           EX         EY          GA           SX           SY           TAU
21-.87633E-11-.36519E-10-.17152E-10 -.65007E-03 -.12906E-02 -.19791E-03
INCREMENTAL NONLINEAR ANALYSIS
           EX         EY          GA           SX           SY           TAU
21-.87633E-11-.36519E-10-.17152E-10 -.65008E-03 -.12906E-02 -.19791E-03
UPDATED PRESTRESS
                  SX                SY                TAU
         .9784664E+05      .1446473E+06     -.1565376E+05

ELEMENT     22
NODES        6    5    9
LINEAR INCREMENTAL ANALYSIS
           EX         EY          GA           SX           SY           TAU
22 .46574E-11-.18474E-10 .14410E-10 -.29171E-04 -.56298E-03  .16627E-03
INCREMENTAL NONLINEAR ANALYSIS
           EX         EY          GA           SX           SY           TAU
22 .46575E-11-.18474E-10 .14410E-10 -.29168E-04 -.56298E-03  .16627E-03
UPDATED PRESTRESS
                  SX                SY                TAU
         .1446910E+06      .9783068E+05     -.1561558E+05

ELEMENT     23
NODES        9   25    6
LINEAR INCREMENTAL ANALYSIS
           EX         EY          GA           SX           SY           TAU
23 .88504E-11-.18497E-10-.45423E-11  .10883E-03 -.52226E-03 -.52411E-04
INCREMENTAL NONLINEAR ANALYSIS
           EX         EY          GA           SX           SY           TAU
23 .88505E-11-.18497E-10-.45423E-11  .10884E-03 -.52226E-03 -.52411E-04
UPDATED PRESTRESS
                  SX                SY                TAU
         .9686449E+05      .8350182E+05     -.8644788E+04

ELEMENT     24
NODES       24    6   25
LINEAR INCREMENTAL ANALYSIS
```

```
          EX          EY          GA          SX          SY          TAU
24-.50619E-11 .16722E-33 .12152E-16 -.16688E-03 -.50063E-04  .14021E-09
INCREMENTAL NONLINEAR ANALYSIS
          EX          EY          GA          SX          SY          TAU
24-.50619E-11 .00000E+00 .12152E-16 -.16687E-03 -.50062E-04  .14021E-09
UPDATED PRESTRESS
                   SX                SY                TAU
        .1203582E+06      .9210745E+05      -.1860152E-10

ELEMENT    25
NODES       7   17   18
LINEAR INCREMENTAL ANALYSIS
          EX          EY          GA          SX          SY          TAU
25 .88504E-11-.44298E-14 .18238E-10  .29173E-03  .87385E-04  .21044E-03
INCREMENTAL NONLINEAR ANALYSIS
          EX          EY          GA          SX          SY          TAU
25 .88505E-11-.43817E-14 .18238E-10  .29173E-03  .87388E-04  .21044E-03
UPDATED PRESTRESS
                   SX                SY                TAU
        .9737767E+05      .8521242E+05      .2801543E+04

ELEMENT    26
NODES      18   19    7
LINEAR INCREMENTAL ANALYSIS
          EX          EY          GA          SX          SY          TAU
26 .00000E+00 .18421E-10 .86649E-11  .18219E-03  .60729E-03  .99979E-04
INCREMENTAL NONLINEAR ANALYSIS
          EX          EY          GA          SX          SY          TAU
26 .00000E+00 .18421E-10 .86649E-11  .18219E-03  .60729E-03  .99979E-04
UPDATED PRESTRESS
                   SX                SY                TAU
        .8521326E+05      .9737754E+05      .2802885E+04

ELEMENT    27
NODES      19   20    8
LINEAR INCREMENTAL ANALYSIS
          EX          EY          GA          SX          SY          TAU
27 .00000E+00 .35996E-10-.20169E-17  .35601E-03  .11867E-02 -.23272E-10
INCREMENTAL NONLINEAR ANALYSIS
          EX          EY          GA          SX          SY          TAU
27 .00000E+00 .35996E-10 .00000E+00  .35601E-03  .11867E-02  .00000E+00
UPDATED PRESTRESS
                   SX                SY                TAU
        .9211395E+05      .1203798E+06      .0000000E+00

ELEMENT    28
NODES       8    7   19
LINEAR INCREMENTAL ANALYSIS
          EX          EY          GA          SX          SY          TAU
28-.87633E-11 .18403E-10 .26405E-10 -.10689E-03  .52004E-03  .30467E-03
INCREMENTAL NONLINEAR ANALYSIS
          EX          EY          GA          SX          SY          TAU
28-.87633E-11 .18403E-10 .26405E-10 -.10689E-03  .52004E-03  .30467E-03
UPDATED PRESTRESS
                   SX                SY                TAU
        .8351018E+05      .9685909E+05      .8657109E+04

ELEMENT    29
NODES      20   21    8
LINEAR INCREMENTAL ANALYSIS
          EX          EY          GA          SX          SY          TAU
29 .00000E+00 .35996E-10-.20169E-17  .35601E-03  .11867E-02 -.23272E-10
INCREMENTAL NONLINEAR ANALYSIS
          EX          EY          GA          SX          SY          TAU
29 .00000E+00 .35996E-10-.20169E-17  .35601E-03  .11867E-02 -.23272E-10
```

```
UPDATED PRESTRESS
                     SX                SY                TAU
          .9211395E+05       .1203798E+06      .2927084E-10

ELEMENT     30
NODES        9     8    21
LINEAR INCREMENTAL ANALYSIS
            EX          EY          GA          SX          SY         TAU
30-.87633E-11 .18403E-10-.26405E-10 -.10689E-03  .52004E-03 -.30467E-03
INCREMENTAL NONLINEAR ANALYSIS
            EX          EY          GA          SX          SY         TAU
30-.87633E-11 .18403E-10-.26405E-10 -.10689E-03  .52004E-03 -.30467E-03
UPDATED PRESTRESS
                     SX                SY                TAU
          .8351018E+05       .9685909E+05     -.8657109E+04

ELEMENT     31
NODES       21    22     9
LINEAR INCREMENTAL ANALYSIS
            EX          EY          GA          SX          SY         TAU
31 .00000E+00 .18421E-10-.86649E-11  .18219E-03  .60729E-03 -.99979E-04
INCREMENTAL NONLINEAR ANALYSIS
            EX          EY          GA          SX          SY         TAU
31 .00000E+00 .18421E-10-.86649E-11  .18219E-03  .60729E-03 -.99979E-04
UPDATED PRESTRESS
                     SX                SY                TAU
          .8521326E+05       .9737754E+05     -.2802885E+04

ELEMENT     32
NODES       25     9    22
LINEAR INCREMENTAL ANALYSIS
            EX          EY          GA          SX          SY         TAU
32 .88504E-11-.44298E-14-.18238E-10  .29173E-03  .87385E-04 -.21044E-03
INCREMENTAL NONLINEAR ANALYSIS
            EX          EY          GA          SX          SY         TAU
32 .88505E-11-.43817E-14-.18239E-10  .29173E-03  .87388E-04 -.21044E-03
UPDATED PRESTRESS
                     SX                SY                TAU
          .9737767E+05       .8521242E+05     -.2801543E+04
Stop - Program terminated.
```

6.7 Problems

1. For the plane stress cantilever Example 6.1 study the effect of changing the grid size (element density) in an attempt to obtain reasonable deep beam stresses. (Timoshenko and Goodier, 1951, pp. 361 ff)

2. Discuss the removal of rigid body rotations for the cases of a truss bar and a plane stress triangular element. (Hint: for the truss bar see figures 1.13 and 1.16)

3. An aluminum ($E = 10 \times 10^6$ psi, $\nu = 0.3$) prestressed sheet is shown in Figure 6.7. The sheet is $L=4$ inches square, *0.1* inches thick and there is a uniform initial prestress of *1000* psi. Using the method of load incrementing find the buckling load.(Ans: between *430* lbs and *440* lbs.)

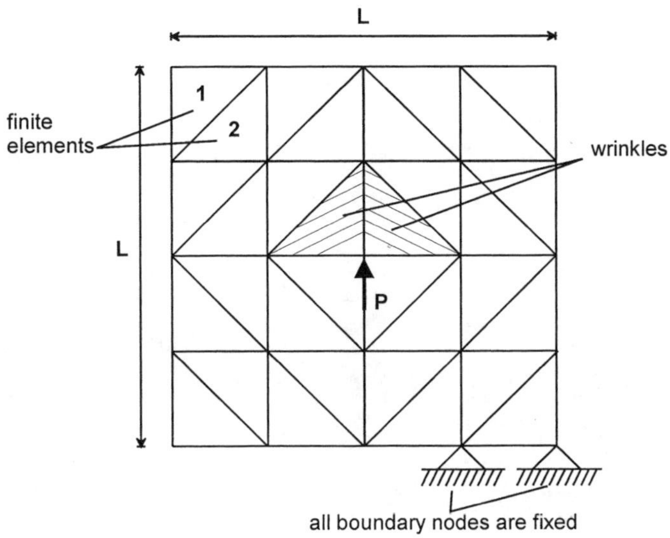

Figure 6.7 Finite element grid for the prestressed sheet.

4. For fabric structures applications, rather than buckling, fabric "wrinkling" would be the appropriate phenomenon. Fabric wrinkling can be modeled by monitoring the principal stresses as the structure is loaded. When the initial prestress which is of course tension is taken to zero in terms of either principal stress during the load application, the isotropic finite element stiffness can be changed to an anisotropic one with only stiffness in the direction of the principal tensile stress. (This is the "tension field" concept used in plate girders.) Modify MEMNL.FOR to include principal stresses and monitoring of the loss of prestress and find for the prestressed sheet of Figure 6.7, at what load do wrinkles occur and their angle of inclination. (Ans: P=240 lbs at an angle of 27° to the horizontal. Up to a load of 600 lbs wrinkles do not form in the other triangular facets.)

5. Model the 24-bar truss of Example 3.6 using nonlinear membrane plane stress elements and find the buckling load. Comment on the expected and attained closeness of the membrane results vs. truss results.

CHAPTER 7

CABLE NETS AND FABRIC STRUCTURES

7.1 Introduction

This chapter is something of a digression from the other chapters of this text which are concerned strictly with the analysis of specific types of structures. Here the concern is with how cable nets and fabric structures are designed. This in turn raises the question of how shape is to be determined which in turn returns to the question of the application of some methods of nonlinear analysis. But no new analysis topics are introduced in this chapter.

In the material which follows there will not always be a clear distinction made between cable nets and fabric structures. Obviously, they are physically quite similar but the matter goes deeper than this. The most simple model of a piece of fabric is one with fibers running in two directions (a cable net). Arguments against this model usually revolve around the fact that a cable net can not transmit shear. The counter argument is of course that fabric is not good at transmitting shear anyway and that the cable net model of fabric is physically intuitive and simple. The simple cable net also does not model to warp/woof load interchange found in a piece of fabric when originally straight fibers become deformed under biaxial load (Stubbs and Fluss, 1979).

The design of cable nets and fabric structures can be described in terms of three events: shape finding, analysis, and patterning. In the process of shape finding, the designer specifies a set of parameters and then computes other parameters finally resolving the details of the shape of the structure. Under analysis, loads are applied to a structure whose shape is known and the response to these loads computed. Patterning is concerned with how a curved surface is to be formed from rolls of fabric.

This chapter is first concerned with the process of finding the details of the shape of a structure and later with how a structure is to be patterned. It is argued that methods of analysis have already been discussed in earlier chapters. Finally, only occasional references will be provided below as part of the text. The more ambitious reader may wish to consult the excellent book of Leonard (1988) on tension structures.

7.2 Basic Methods of Shape Finding

The process of shape finding can be thought of roughly in terms of stretching fabric over a frame of arbitrary shape. (For example, in the skylight problem of Figure 7.1 , the geometry is fixed along both crossed arches and the base.) Clearly the fabric must follow the frame at the boundaries and certain tensions can be specified on these boundaries. But just as clearly, the locations of the fabric points within the frame must be determined from the equations of equilibrium and in some cases the material parameters. Finding the locations of these internal points is the process of shape finding.

In the early days (Frei Otto, 1973) and in the absence of the computer, physical models were commonly used in the design of fabric structures and cable nets. It is now conventional wisdom that small scale models are not sufficiently accurate either for the prediction of forces or the patterning of the fabric. In the following some computer based approaches which have been used to find shape are discussed.

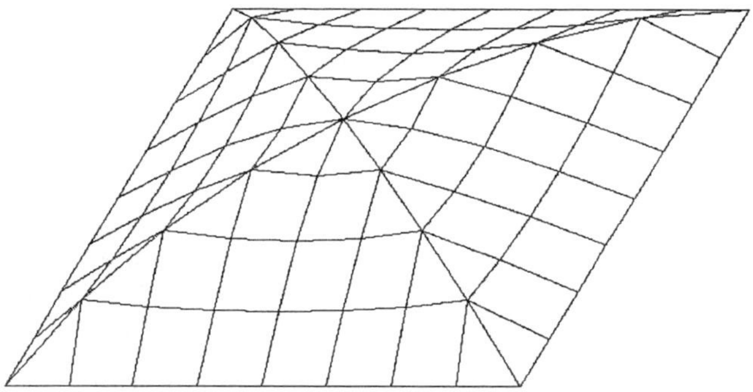

Figure 7.1 Skylight example: crossed circular arches on structural frame.

7.2.1 Deformed Shape

If you need to find a shape which is in equilibrium, this can be done by applying loads to, for example, a stretched elastic sheet and then using the deformed sheet or a scaled version of it as the shape. There is nothing wrong with doing so but care is required since loads applied to a sheet may introduce stress concentrations which may not be desired in the structure under design. The basic reference to this method is Argyris, *et al.* (1964).

Any computer program for nonlinear structural analysis can be used to achieve shape in this manner but it does not appear common to do so. Pertinent to this is the fact that the fabrics now commonly used in permanent structures can not tolerate large strains without tearing.

7.2.2 Force Density Method

It has been noted by Schek (1974) that if the ratio of the bar force to its length is held constant in a cable net, the associated geometry can be found by solving a system of linear equations. (A similar statement can be made for a finite element fabric model.) This approach is frequently used in the design of fabric structures.

The force density method is based on the fact that the force on the end of a truss bar can be represented by the product of the bar force and a unit vector in the direction of the bar as shown in Figure 7.2. Here $\mathbf{n}_i$ is the unit vector of member *i*, F_i is the bar force of member *i*, and L_i is the length of member *i*. The components of the force vector can be written as

$$\begin{aligned}(\mathbf{F}_i)_x &= \frac{F_i}{L_i}(X_A - X_C)\\ (\mathbf{F}_i)_y &= \frac{F_i}{L_i}(Y_A - Y_C)\\ (\mathbf{F}_i)_z &= \frac{F_i}{L_i}(Z_A - Z_C)\end{aligned} \tag{7.1}$$

Clearly, if the "force density", F_i/L_i the ratio of the bar force to its bar length is some known constant, the force at the end of the bar becomes a linear function of the coordinates X_A, Y_A, Z_A and X_C, Y_C, Z_C of the ends of the bar. What this means is that when the equilibrium equations are written, they are linear in the node coordinates. Thus problems can be formulated in which the node coordinates (shape) are easily found by solving a system of linear equations.

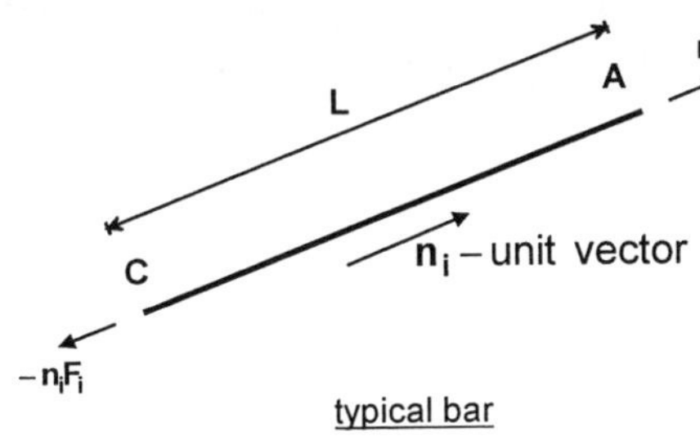

Figure 7.2 The force density method for a bar element.

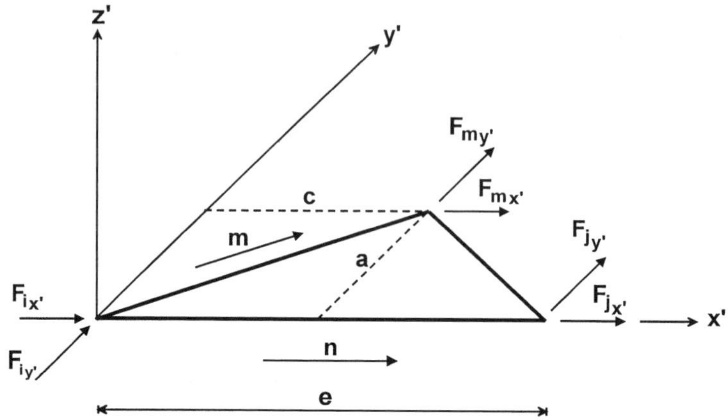

Figure 7.3 Force density method for a membrane element.

One of the shortcomings, of course, with this method is that it is usually preferable to specify the member force F_i rather than the force density F_i/L_i. In fact, L_i is not really known until the node coordinates have been computed by some shape finding algorithm.

As Figure 7.3 indicates, similar arguments can be made for a simple membrane finite element. Roughly, given a simple finite element such as Zienkiewicz's plane stress element (Zienkiewicz, 1977) and an existing state of stress, nodal forces can be computed. (They depend upon both the state of stress and the geometry of the finite element.) These nodal forces can be expressed as linear combinations of the unit vectors **n** and **m** (Figure 7.3) which describe the slopes of two sides of the finite element. As was done for the bar, these unit vectors can be factored into a term linear in the node coordinates and whatever remains. Obviously, when the coefficients of these linear terms are specified as in the case of the bar, the equilibrium equations become linear in the coordinates of the elements.

7.2.3. Grid Method

Siev and Eidelman (1964) showed that if equilibrium is satisfied over a grid in the horizontal plane (see Figure 7.4), vertical equilibrium can be used to compute the elevation (shape) at the grid points. This is probably the most simple way to find shape and will be discussed at some length below.

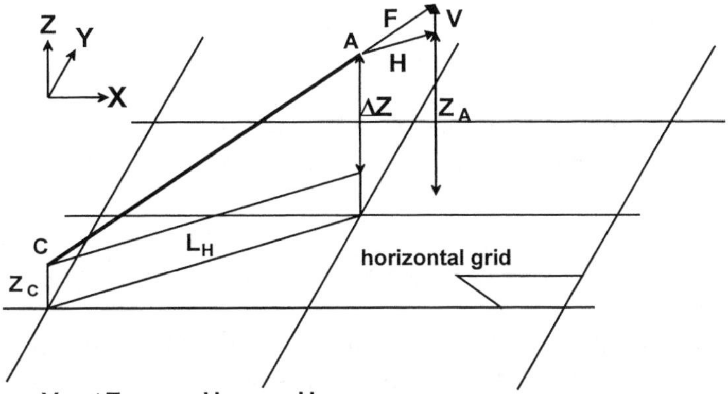

$$\frac{V}{H} = \frac{\Delta Z}{L_H} \Rightarrow V = \frac{H}{L_H}\Delta Z = \frac{H}{L_H}(Z_A - Z_C)$$

Figure 7.4 The grid method.

7.2.4 Smoothing

Given a shape, a computer program for nonlinear analysis can be used (Haber and Abel, 1982) to smooth out the forces or stresses sometimes resulting in a state of uniform stress. How this can be done is described below along with variations which allow shape to be found directly using computer programs for nonlinear analysis.

In the material which follows, a path will be described from the grid method, through smoothing, and on to a procedure which allows shape to be determined using computer programs for nonlinear analysis which are of course central to this text.

7.3 The Grid Method

The grid method is probably the most simple way to find shape. It requires, first of all, a grid in the horizontal plane but any numerical method for finding shape requires some kind of a grid of points. Member forces must be assumed or determined so that the cable net or finite element system is in equilibrium in the horizontal plane but this too can be a trivial step. That is clearly the case in the simple example of Figure 7.6 below where it is assumed that the horizontal cable force components are all 1,000 lbs. (Constant force components over a rectangular grid are clearly in equilibrium in the horizontal plane.) If only the rise of the circular arches has been specified, it is also required to compute the elevations of the points of attachment along these arches as is done below. The final step is to write the equations of equilibrium in the vertical direction and then solve a system of linear equations for the unknown elevations at these points. Following are two examples of shape finding using the grid method. The first, which has the form of part of a skylight, is solved by hand whereas the second skylight example is solved using the computer program LAYOUT.FOR.

7.3.1 Example 7.1 A Piece of a Cross Arched Skylight

This example is a piece (type B in Figure 7.5) of a cross arched skylight of dimensions 20ft × 40ft. Since the base of this skylight is not square, it would be necessary to find the shape of another piece (type A in Figure 7.5) to complete its design.

The piece of skylight is taken as a 9-member cable net laid over a right angled triangular structural base having two edges of 10 ft and 20 ft as shown in Figure 7.6. It is supported at nodes 4,5,6 by the third edge which is a circular arch having a rise of 5 ft. With horizontal force components of 1,000 lbs in each of the cables it is required to find the vertical coordinates of nodes 1,2,3 so that equilibrium is maintained.

The elevations of nodes 5 and 6 are first established from the arch geometry. The radius of the arch is obtained from:

$$\theta = 2 \times \arctan\left(\frac{5}{22.3607}\right) = 25.2087^\circ$$

$$R = \frac{22.3607}{\sin 25.2087^\circ} = 52.500 \text{ ft.}$$

Now using the equation for a circle, $x^2 + y^2 = R^2$, the z-coordinates of nodes 5 and 6 are obtained as

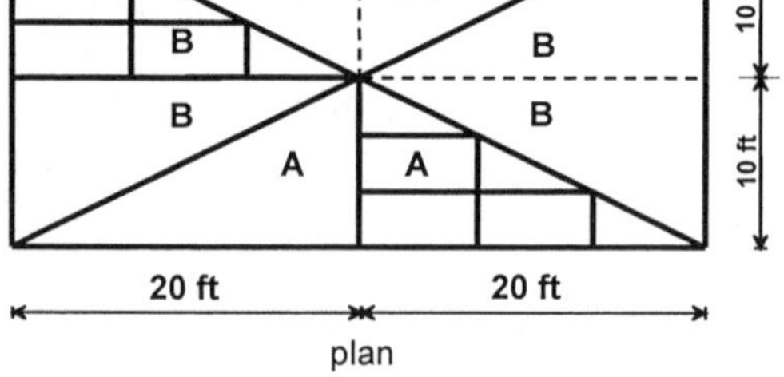

Figure 7.5 Rectangular skylight.

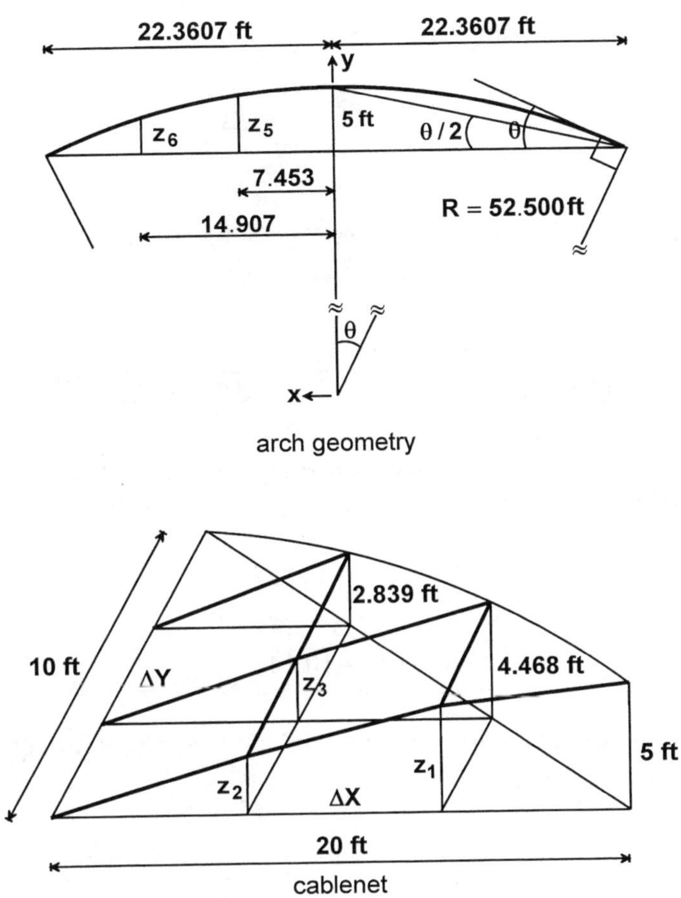

Figure 7.6 A simple cable net.

$$z_5 = \sqrt{R^2 - 7.453^2} - R\cos\theta = 4.4682 \text{ ft}$$

$$z_6 = \sqrt{R^2 - 14.907^2} - R\cos\theta = 2.8391 \text{ ft}$$

The next step is to write vertical joint equilibrium for each free joint by adding the contribution of each member. (Remember that the grid method starts with the satisfaction of equilibrium in the horizontal plane.) The simple relation of Figure 7.4 is used for that purpose, i.e.

$$V = \frac{H}{L_H}\left(Z_A - Z_C\right) \tag{7.2}$$

Here H=1,000lbs and L_H has only two values which are $\Delta Y = 3.333$ ft and $\Delta X = 6.667$ ft. Considering vertical equilibrium:

node 1: $$H\left(\frac{5-z_1}{\Delta X}+2\times\frac{4.468-z_1}{\Delta Y}-\frac{z_1-z_2}{\Delta X}\right)=0$$

node 2: $$H\left(\frac{z_1-z_2}{\Delta X}+2\times\frac{z_3-z_2}{\Delta Y}-\frac{z_2}{\Delta X}\right)=0$$

node 3: $$H\left(\frac{2.839-z_3}{\Delta Y}+\frac{4.468-z_3}{\Delta X}-\frac{z_3-z_2}{\Delta Y}-\frac{z_3}{\Delta X}\right)=0$$

Observing that $\Delta X/\Delta Y = 2$ and canceling H the above equations are rewritten and solved as

$$\begin{aligned} -6z_1+z_2+22.872&=0\\ z_1-6z_2+4z_3&=0\\ 2z_2-6z_3+10.146&=0 \end{aligned}$$

Substituting z_2 from the first equation and z_3 from the third equation into the second equation yields

$$z_1-6(-22.872+6z_1)+4\left(\frac{10.146+2(-22.872+6z_1)}{6}\right)=0$$

$\Rightarrow -27z_1+113.5=0 \Rightarrow z_1=4.203$ ft and $z_2=2.350$ ft; $z_3=2.474$ ft. This completes the example since an initial shape in equilibrium has been found.

7.3.2 Example 7.2 A Squared Base Skylight

This example, which is described by Figure 7.7, is in fact the skylight of Figure 7.1. This time the computer program LAYOUT.FOR will be used to determine shape. Figure 7.8 shows a node map of 1/8 of this structure which is all that it is required for design due to symmetry.

Following are the calculations which are necessary to determine the elevations of the fixed points on the arch (nodes 5, 9, 12, 14 and 15) having a specified rise of 30 inches. Here the base of the skylight is 172 inches square and the cables are equally spaced in the horizontal plane. The elevations z_5 and z_{15} are given as 30 inches and 0 inches respectively. The angle θ and the radius of the arch are obtained as

$$\frac{\theta}{2} = \tan^{-1}\frac{30}{121.622} = 13.856° \Rightarrow \theta = 27.7125°$$

$$R\sin\theta = 121.622 \Rightarrow R = 261.533 \text{ in}$$

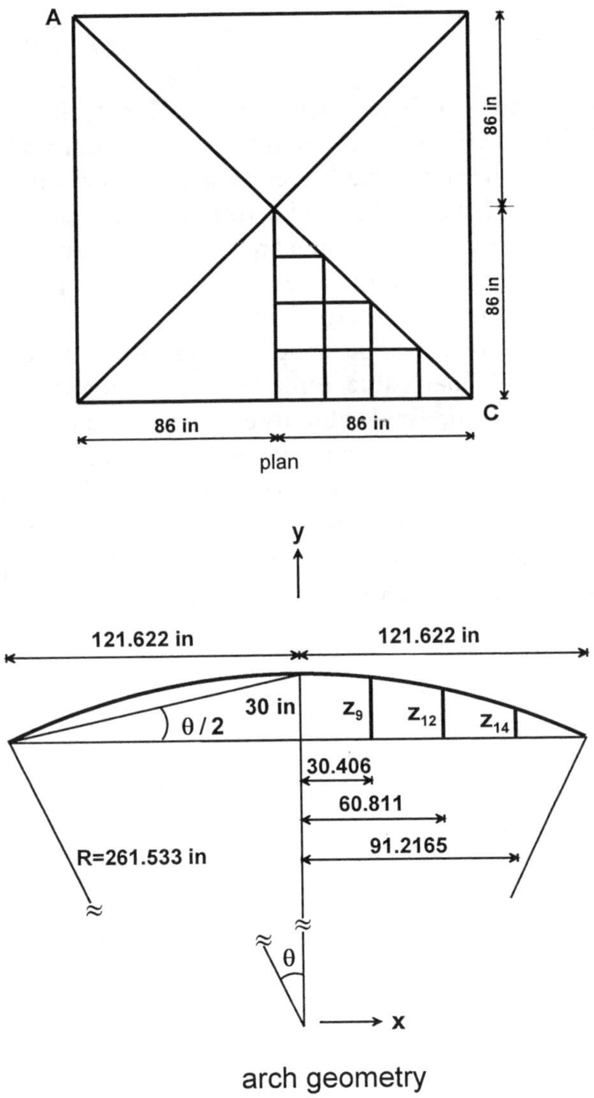

Figure 7.7 Skylight dimensions for example 7.2.

Using the equation of a circle $x^2 + y^2 = R^2$ the elevations for nodes 9, 12 and 14 become

$$z_9 = \sqrt{R^2 - 30.406^2} - R\cos\theta = 28.2265 \text{ in}$$

$$z_{12} = \sqrt{R^2 - 60.811^2} - R\cos\theta = 22.832 \text{ in}$$

$$z_{14} = \sqrt{R^2 - 90.2165^2} - R\cos\theta = 13.578 \text{ in}$$

DOME which is listed below is the input file required by the program LAYOUT.FOR. First the number of bars and nodes is specified. Then a set of lines describes the bar ends and the horizontal component of the force in each cable here taken to be 1000 lbs; the second part of the file describes the node coordinates (which are taken to be zero when they are not known), some symmetry conditions to be discussed later, and the member connectivity. This connectivity is simply a list of the nodes which are connected by cables to the node under consideration. Once the shape has been determined the final cable forces can be computed as indicated in Figure 7.9. Obviously, the final cable forces can not be equal since the cable slopes are not equal. The output file FORT8 lists for each node the coordinates and the `FORCE BALANCE` (from which reactions can be computed) and for each member its prestress. Fictitious unbalanced components indicated at points of symmetry are simply a product of the manner by which LAYOUT.FOR deals with symmetry. See Section 7.8 for more details regarding the running of the computer program LAYOUT.FOR.

Input file DOME

```
   24   15
1000.          1    2
1000.          2    3
1000.          3    4
1000.          4    5
1000.          6    7
1000.          7    8
1000.          8    9
1000.         10   11
1000.         11   12
1000.         13   14
1000.         15   14
1000.         14   12
1000.          2    7
1000.          7   11
1000.         11   14
1000.          3    8
1000.          8   12
1000.          4    9
```

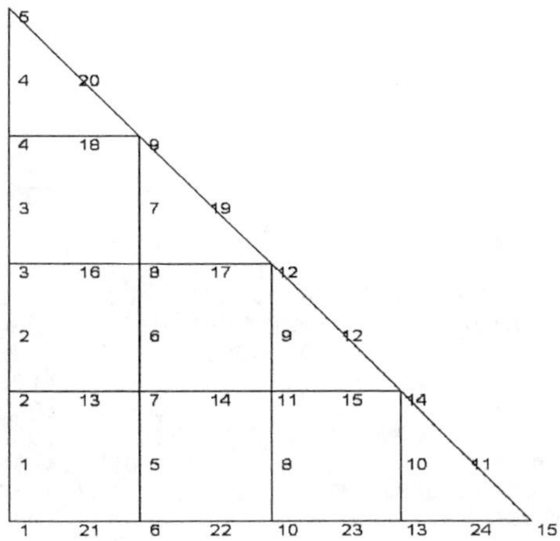

Figure 7.8 Node map of 1/8 of the skylight structure.

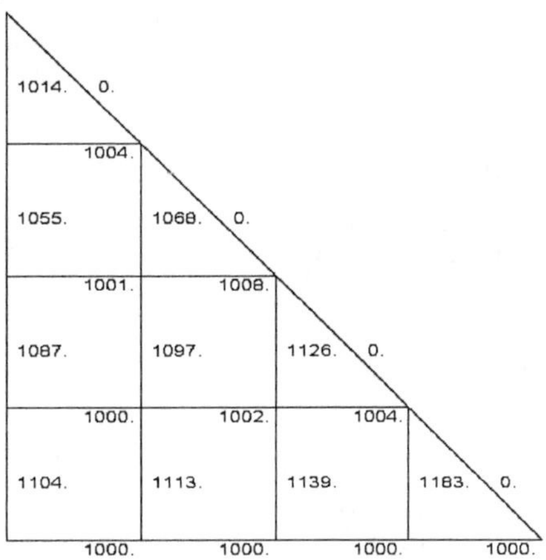

Figure 7.9 Computed prestress on 1/8 skylight structure.

```
1000.        12     9
1000.         9     5
1000.         1     6
1000.         6    10
1000.        10    13
1000.        13    15
0.        0.        0.          1      2     6     6
0.        21.5      0.                 1     7     3     7
0.        43.       0.                 2     8     4     8
0.        64.5                         3     9     5     9
0.        86.       30.         1 4    4
21.5                0.          1      1     7    10
21.5      21.5      0.                 2     6     8    11
21.5      43.       0.                 3     7     9    12
21.5      64.5      28.227      1 4    8     4
43.       0.        0.          1      6    11    13
```

```
43.        21.5      0.                7   10   14   12
43.        43.       22.832     1 4    8   11
64.5       0.        0.         1     10   14   15
64.5       21.5      13.577     1 4   13
86.         0.       0.         1 4
```

Output file FORT8

```
                COORDINATES                             FORCE BALANCE
                X            Y            Z            X           Y            Z
   1 .00000E+00 .00000E+00 .00000E+00 .20000E+04  .10000E+04  .46736E+03
   2 .00000E+00 .21500E+02 .10048E+02 .20000E+04 -.23820E-04 -.11462E-03
   3 .00000E+00 .43000E+02 .19211E+02 .20000E+04 -.11285E-03 -.80640E-04
   4 .00000E+00 .64500E+02 .26416E+02 .20000E+04  .46334E-04 -.16394E-03
   5 .00000E+00 .86000E+02 .30000E+02 .00000E+00 -.10000E+04 -.16668E+03
   6 .21500E+02 .00000E+00 .00000E+00 .00000E+00  .10000E+04  .48795E+03
   7 .21500E+02 .21500E+02 .10491E+02-.14171E-05  .44862E-04 -.65020E-04
   8 .21500E+02 .43000E+02 .20190E+02-.12679E-04 -.42546E-04 -.24605E-03
   9 .21500E+02 .64500E+02 .28227E+02-.10000E+04 -.10000E+04 -.45802E+03
  10 .43000E+02 .00000E+00 .00000E+00 .00000E+00  .10000E+04  .54535E+03
  11 .43000E+02 .21500E+02 .11725E+02-.25455E-04  .15764E-04  .32629E-04
  12 .43000E+02 .43000E+02 .22832E+02-.10000E+04 -.10000E+04 -.63948E+03
  13 .64500E+02 .00000E+00 .00000E+00 .00000E+00  .10000E+04  .63149E+03
  14 .64500E+02 .21500E+02 .13577E+02 .00000E+00 -.10000E+04 -.63149E+03
  15 .86000E+02 .00000E+00 .00000E+00 .00000E+00  .00000E+00  .00000E+00
```

```
MEMBER OUTPUT
        +END  -END              FORCE
    1     1     2          .11038236E+04
    2     2     3          .10870308E+04
    3     3     4          .10546581E+04
    4     4     5          .10137963E+04
    5     6     7          .11126964E+04
    6     7     8          .10970520E+04
    7     8     9          .10675797E+04
    8    10    11          .11390364E+04
    9    11    12          .11255585E+04
   10    13    14          .11826992E+04
   11    15    14          .00000000E+00
   12    14    12          .00000000E+00
   13     2     7          .10002119E+04
   14     7    11          .10016460E+04
   15    11    14          .10037032E+04
   16     3     8          .10010362E+04
   17     8    12          .10075203E+04
   18     4     9          .10035400E+04
   19    12     9          .00000000E+00
   20     9     5          .00000000E+00
   21     1     6          .10000000E+04
   22     6    10          .10000000E+04
   23    10    13          .10000000E+04
   24    13    15          .10000000E+04
```

The shape of the 1/8 model is shown in Figure 7.10. For practical reasons it is desirable to look at the entire structure rather than simply 1/8 of it. That is done here by running two graphics routines: LPLOT1.FOR takes the geometry of 1/8 of the structure and reflects it about the diagonal to get a view of 1/4 of the structure as shown in Figure 7.11; the routine LAYOUTPLOT.FOR can then be used to produce a picture of the entire structure of Figure 7.1. (See also Section 7.8 below.)

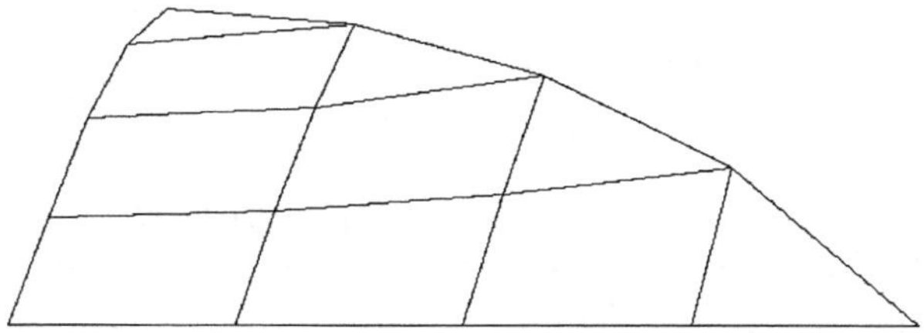

Figure 7.10 Initial equilibrium shape of 1/8 skylight structure.

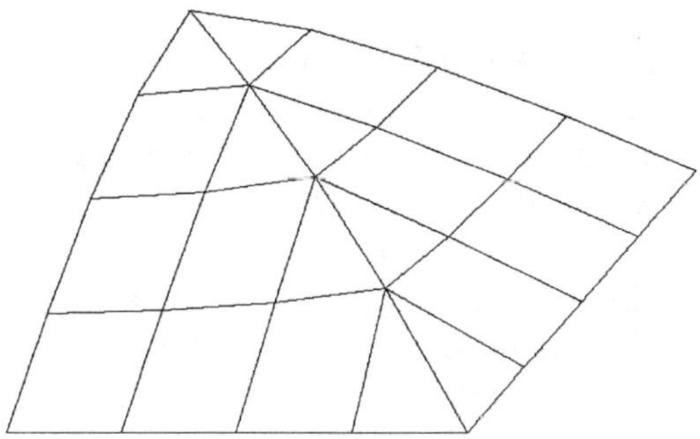

Figure 7.11 Initial equilibrium shape of 1/4 skylight example.

7.4 Smoothing

Having obtained a shape using in this case the grid method, it is natural to hope that a better shape perhaps close to the given shape might be obtained. For example, there are many reasons such as efficiency in the use of material to look for a shape in which the bar forces or the membrane stresses are constant.

When such a shape exists, it may be possible to generate it using a nonlinear structural analysis computer program. This is the idea of "smoothing" which has been described by Haber and Abel (1982). There are two steps to the smoothing procedure:

1) The output of the shape finding program is used as input to the nonlinear analysis program except that the desired cable forces, here taken to be 1,000 lbs rather than the computed cable forces are used. Clearly this initial configuration is not in equilibrium.

2) A fictitious, small value of E, Young's modulus, is used during the subsequent analysis. This essentially disables the elastic stiffness matrix

and prevents the cable forces from changing. When the nonlinear analysis converges, a shape has been obtained for which the cable forces are constant. (See Figure 7.12) In this case the cables have moved to form geodesics in the surface of the structure.

A final note with regard to this example. It is, in this case, possible to start from the beginning using the analysis program avoiding the grid method entirely. Doing so allows a shape to be achieved directly using an analysis program while specifying the cable forces. The three dimensional nonlinear analysis program TR3DNL.FOR was slightly modified to find an equilibrium shape given a state of prestress. This modified version, TR3DSM.FOR basically paralyzes the elastic stiffness matrix by inserting zero for Young's modulus leaving the task of shape finding to the geometric stiffness matrix.

The input file TRDOMESM.DAT is used with TR3DSM.FOR to resolve Example 7.2 above for the case of constant cable forces (1000 lbs). Two output files, FORT60 and FORT600, are saved but only FORT600, listed below, provides information on the number of bars, the number of nodes, the number of supports, the number of load steps, and the number of iterations. Member forces, connectivity, and area follow in subsequent lines. The second part of the output gives the final coordinates of the nodes. The input file is the same as DOME except for the information on symmetry which is not supplied here. Instead, 'constraints' are specified for all degrees of freedom: the number 1 indicates a constrained degree of freedom whereas the number 0 indicates a free degree of freedom.

Input file TRDOMESM.DAT

```
  24   15
1000.             1    2
1000.             2    3
1000.             3    4
1000.             4    5
1000.             6    7
1000.             7    8
1000.             8    9
1000.            10   11
1000.            11   12
1000.            13   14
1000.            15   14
1000.            14   12
1000.             2    7
1000.             7   11
1000.            11   14
1000.             3    8
1000.             8   12
1000.             4    9
1000.            12    9
1000.             9    5
1000.             1    6
1000.             6   10
1000.            10   13
```

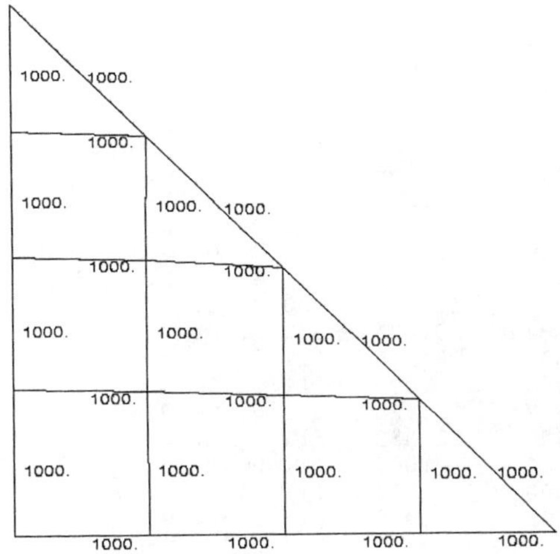

Figure 7.12 Skylight structure with constant cable forces.

```
1000.          13    15
0.          0.          0.          111
0.          21.5        0.          1
0.          43.         0.          1
0.          64.5                    1
0.          86.         30.         111
21.5                    0.          111
21.5        21.5        0.
21.5        43.         0.
21.5        64.5        28.227      111
43.         0.          0.          111
43.         21.5        0.
43.         43.         22.832      111
64.5        0.          0.          111
64.5        21.5        13.578      111
86.          0.         0.          111
```

Output file FORT600

```
  24    15      0     1    10
 1000.000      1     2          .006
 1000.000      2     3          .006
 1000.000      3     4          .006
 1000.000      4     5          .006
 1000.000      6     7          .006
 1000.000      7     8          .006
 1000.000      8     9          .006
 1000.000     10    11          .006
 1000.000     11    12          .006
 1000.000     13    14          .006
 1000.000     15    14          .006
 1000.000     14    12          .006
 1000.000      2     7          .006
 1000.000      7    11          .006
 1000.000     11    14          .006
 1000.000      3     8          .006
 1000.000      8    12          .006
```

```
 1000.000     4     9       .006
 1000.000    12     9       .006
 1000.000     9     5       .006
 1000.000     1     6       .006
 1000.000     6    10       .006
 1000.000    10    13       .006
 1000.000    13    15       .006
     .000        .000       .000   111 0. 0. 0.
     .000      23.335     10.558   100 0. 0. 0.
     .000      44.720     19.276   100 0. 0. 0.
     .000      65.121     25.835   100 0. 0. 0.
     .000      86.000     30.000   111 0. 0. 0.
   21.500        .000       .000   111 0. 0. 0.
   21.405      23.015     11.302   000 0. 0. 0.
   21.363      44.164     20.805   000 0. 0. 0.
   21.500      64.500     28.227   111 0. 0. 0.
   43.000        .000       .000   111 0. 0. 0.
   42.977      22.377     12.723   000 0. 0. 0.
   43.000      43.000     22.832   111 0. 0. 0.
   64.500        .000       .000   111 0. 0. 0.
   64.500      21.500     13.578   111 0. 0. 0.
   86.000        .000       .000   111 0. 0. 0.
```

7.5 A More Complex Example

Figures 7.13 (a)-(b) show a cable net which looks something like a hyperbolic paraboloid; it is supported at four points and bounded by four edge cables on four sides. This structure is more complex than the example above because of its four edge cables. A node map has been created and displayed in Fig. 7.14. (Note that cable node 15, which defines the "sag" of the edge cable has been forced to lie on a grid point.)

In order to get started with the grid method in this case it is first necessary to go off and solve the (horizontal) cable problem by hand. Figure 7.14 is a free body diagram of half of one edge cable. Symmetry requires that the force components at the exposed end be equal. Moment equilibrium then provides *R* (see below) after which individual cable forces and reactions may be calculated using force equilibrium starting with cable 32 and moving upwards up to cable 29. Following are the required calculations:

Σ moments about top $\Rightarrow 1000\times2+1000\times4+1000\times6+1000\times2+R\times4-R\times6=0 \Rightarrow$ $14000=2R \Rightarrow R=7000$.

force in cable 32 is $\sqrt{6000^2+7000^2}=9219$ lbs

force in cable 31 is $\sqrt{5000^2+7000^2}=8602$ lbs

force in cable 30 is $\sqrt{5000^2+8000^2}=9433$ lbs

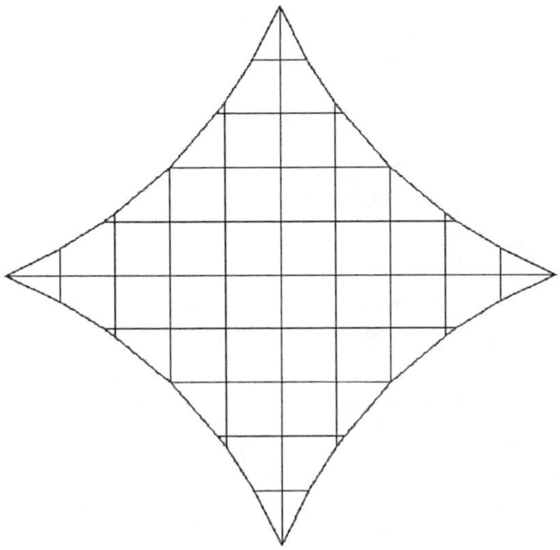

Figure 7.13 (a) Plan of the 'hyperbolic paraboloid' cable net.

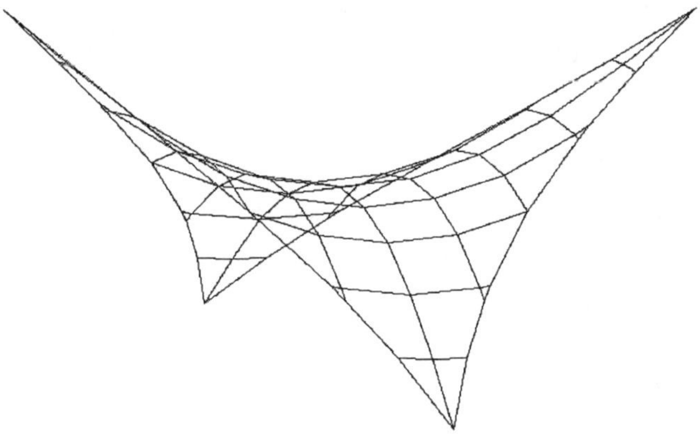

Figure 7.13 (b) Schematic of the hyperbolic paraboloid cable net.

force in cable 29 is $\sqrt{4000^2 + 8000^2} = 8944$ lbs

The grid method again gives a solution using HYPAR.DAT as the input file for LAYOUT.FOR following the previous example. One of the two output files FORT8 is also listed. In the final design (Fig. 7.15) the interior cable forces are of course not constant. Smoothing can be used to determine a shape with constant internal cable forces but in this case the real stiffness of the edge cable must be used if the plan shape of this structure is to be preserved.

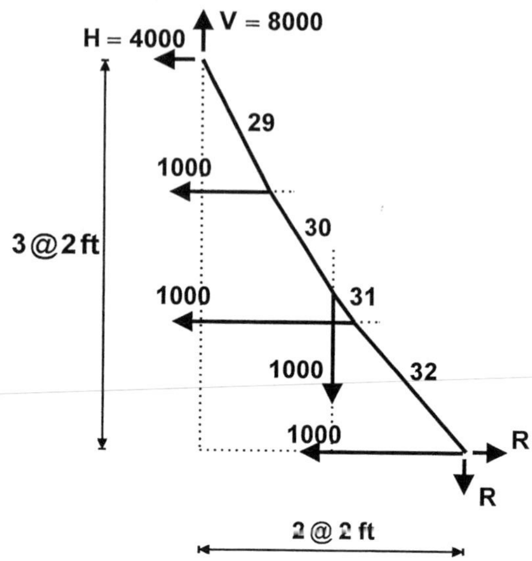

Figure 7.14 Edge cable for the hyperbolic paraboloid cable net.

Input file HYPAR.DAT

```
   36    23
1000.          1    7
1000.         07   13
1000.         13   17
1000.         17   20
1000.         20   23
1000.          2    8
1000.          8   14
1000.         14   18
1000.         18   22
1000.          3    9
1000.          9   15
1000.          4   10
1000.         10   16
1000.          5   12
1000.          2    1
1000.          8    7
1000.         14   13
1000.         18   17
1000.         21   20
1000.          3    2
1000.          9    8
1000.         15   14
1000.         19   18
1000.          4    3
1000.         10    9
1000.          5    4
1000.         11   10
1000.          6    5
8944.         21   23
9433.         22   21
8602.         19   22
9219.         15   19
9219.         16   15
```

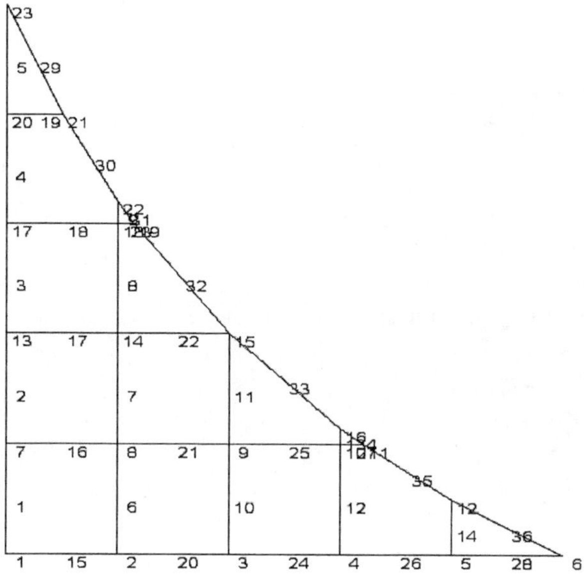

Figure 7.15 Node map for the hyperbolic paraboloid.

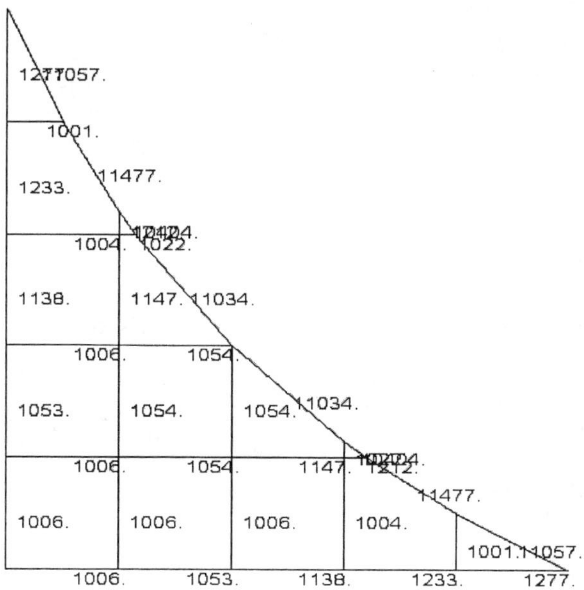

Figure 7.16 Cable forces in the hyperbolic paraboloid.

```
8602.          11   16
9433.          12   11
```

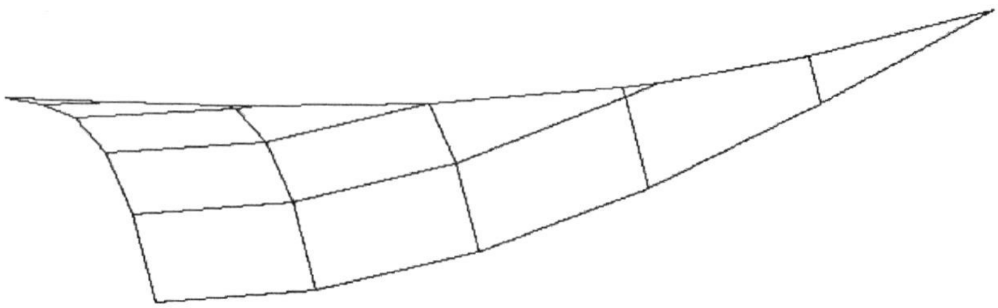

Figure 7.17 Equilibrium shape of 1/4 of the hyperbolic paraboloid.

```
8944.          6    12
0.          0.          0.             3     2     7     2     7
2.           0.0        0.0            1     8     3     8     1
4.           0.0        0.0            1     9     4     9     2
6.           0.0        0.0            1    10     5    10     3
8.           0.0        0.0            1    12     6    12     4
10.          0.0        10.          1 1
0.           2.         0.0            2     1     8    13     8
2.           2.         0.0                  2     9    14     7
4.           2.         0.0                  3    10    15     8
6.           2.         0.0                  4    11    16     9
6.40         2.         0.0                 10    12    16
8.           1.         0.                   5     6    11
0.           4.         0.             2     7    14    17    14
2.           4.         0.                   8    15    18    13
4.           4.         0.                   9    14    16    19
6.           2.285      0.                  10    11    15
0.           6.         0.             2    13    18    20    18
2.           6.         0.                  14    19    22    17
2.285        6.         0.                  18    15    22
0.           8.         0.             2    17    21    23    21
1.           8.         0.                  20    22    23
2.           6.40       0.                  18    19    21
0.          10.         0.           1 2
```

Output file FORT8 for the hyperbolic paraboloid

```
              COORDINATES                                    FORCE BALANCE
              X             Y             Z              X              Y              Z
 1 .00000E+00 .00000E+00 .50000E+01  .20000E+04  .20000E+04  .23817E-02
 2 .20000E+01 .00000E+00 .52205E+01  .26542E-04  .20000E+04  .11969E-02
 3 .40000E+01 .00000E+00 .58820E+01  .90543E-05  .20000E+04  .69918E-03
 4 .60000E+01 .00000E+00 .69699E+01  .13404E-04  .20000E+04  .19375E-02
 5 .80000E+01 .00000E+00 .84115E+01  .77365E-05  .20000E+04  .13206E-04
 6 .10000E+02 .00000E+00 .10000E+02  .00000E+00  .00000E+00  .00000E+00
 7 .00000E+00 .20000E+01 .47795E+01  .20000E+04 -.25352E-04  .14257E-02
 8 .20000E+01 .20000E+01 .50000E+01  .26542E-04 -.40700E-04  .16566E-02
 9 .40000E+01 .20000E+01 .56688E+01  .24420E-04 -.37987E-04  .73296E-03
10 .60000E+01 .20000E+01 .67931E+01  .17458E-04  .17434E-04  .38531E-02
11 .64000E+01 .20000E+01 .70671E+01 -.64792E+01 -.79566E+01  .90093E-02
12 .80000E+01 .10000E+01 .83747E+01  .58832E+00 -.39761E+00 -.19493E-03
13 .00000E+00 .40000E+01 .41180E+01  .20000E+04 -.53616E-04  .21760E-02
14 .20000E+01 .40000E+01 .43312E+01 -.15173E-04  .13650E-04  .84525E-03
15 .40000E+01 .40000E+01 .50000E+01 -.27346E+01 -.27344E+01  .63961E-02
16 .60000E+01 .22850E+01 .67329E+01  .72955E+01  .95622E+01  .17164E-01
17 .00000E+00 .60000E+01 .30301E+01  .20000E+04  .66451E-04  .99830E-03
```

```
18 .20000E+01 .60000E+01 .32069E+01  .23749E-04  .37939E-04  .46322E-02
19 .22850E+01 .60000E+01 .32671E+01  .95622E+01  .72954E+01  .18472E-01
20 .00000E+00 .80000E+01 .15885E+01  .20000E+04 -.37339E-05  .12033E-02
21 .10000E+01 .80000E+01 .16252E+01 -.39796E+00  5.8840E+00  .41335E-02
22 .20000E+01 .64000E+01 .29329E+01 -.79562E+01 -.64792E+01 -.88023E-03
23 .00000E+00 .10000E+02 .00000E+00  .00000E+00  .00000E+00  .00000E+00
```

```
        MEMBER OUTPUT
       +END  -END            FORCE
   1     1     7        0.10060590E+04
   2     7    13        0.10532775E+04
   3    13    17        0.11383743E+04
   4    17    20        0.12326866E+04
   5    20    23        0.12770435E+04
   6     2     8        0.10060590E+04
   7     8    14        0.10544263E+04
   8    14    18        0.11471840E+04
   9    18    22        0.12120830E+04
  10     3     9        0.10056666E+04
  11     9    15        0.10544264E+04
  12     4    10        0.10039003E+04
  13    10    16        0.10220560E+04
  14     5    12        0.10006747E+04
  15     2     1        0.10060591E+04
  16     8     7        0.10060591E+04
  17    14    13        0.10056667E+04
  18    18    17        0.10039001E+04
  19    21    20        0.10006747E+04
  20     3     2        0.10532782E+04
  21     9     8        0.10544270E+04
  22    15    14        0.10544270E+04
  23    19    18        0.10220565E+04
  24     4     3        0.11383762E+04
  25    10     9        0.11471854E+04
  26     5     4        0.12326910E+04
  27    11    10        0.12120878E+04
  28     6     5        0.12770487E+04
  29    21    23        0.11056922E+05
  30    22    21        0.11476976E+05
  31    19    22        0.10404198E+05
  32    15    19        0.11034478E+05
  33    16    15        0.11034482E+05
  34    11    16        0.10404225E+05
  35    12    11        0.11477012E+05
  36     6    12        0.11056961E+05
```

In this case of smoothing it is not possible to go directly from scratch to the final shape in one step using the analysis program. (The analysis does not converge.) But a two step procedure does work:

1) Fix the plane coordinates, specify the desired cable forces, set E=0 in all cables and iterate using the analysis program. The analysis program in this case acts like a modified form of the grid method except that the cable forces rather than their horizontal projections are specified. HYPARSM1.DAT is the input file used and TR3DSM.FOR compiled with a very small `ec=1`.

2) Release the artificial constraints of the above step and restore the real stiffness of the edge cable and iterate using the analysis program. When this step has been completed a proper shape has been found in which the

interior cables have constant forces.(See Figure 7.18.) The steps require that FORT600 be saved as HYPARSM2.DAT which is appropriately edited to become the input file for another run with a recompiled TR3DSM.FOR with `ec=25.D06`. Editing frees nodes 8, 9, 10, 11, 12, 14, 15, 18, 19, 21 and 22 from the horizontal constraints.

Listed below are the output file FORTEC1600 (FORT600 for the case of `ec=1`, step 1 above), the input file HYPARSM2.DAT (edited file FORT600, step 2 above) and output file FORTEC25600 (FORT600 for the case of `ec=25.D06`, step 2 above).

Input file HYPARSM1.DAT

```
   36    23
1000.          1     7
1000.         07    13
1000.         13    17
1000.         17    20
1000.         20    23
1000           2     0
1000.          8    14
1000.         14    18
1000.         18    22
1000.          3     9
1000.          9    15
1000.          4    10
1000.         10    16
1000.          5    12
1000.          2     1
1000.          8     7
1000.         14    13
1000.         18    17
1000.         21    20
1000.          3     2
1000.          9     8
1000.         15    14
1000.         19    18
1000.          4     3
1000.         10     9
1000.          5     4
1000.         11    10
1000.          6     5
8944.         21    23
9433.         22    21
8602.         19    22
9219.         15    19
9219.         16    15
8602.         11    16
9433.         12    11
8944.          6    12
0.          0.          0.          11
2.          0.0         0.0         11
4.          0.0         0.0         11
6.          0.0         0.0         11
8.          0.0         0.0         11
10.         0.0         10.         111
0.          2.          0.0         11
2.          2.          0.0         11
4.          2.          0.0         11
6.          2.          0.0         11
```

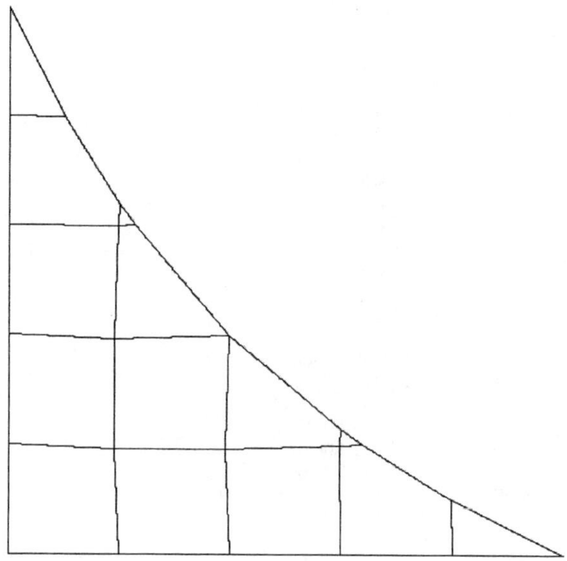

Figure 7.18 Hypar with constant cable forces.

```
6.40      2.      0.0     11
8.        1.      0.      11
0.        4.      0.      11
2.        4.      0.      11
4.        4.      0.      11
6.        2.285   0.      11
0.        6.      0.      11
2.        6.      0.      11
2.285     6.      0.      11
0.        8.      0.      11
1.        8.      0.      11
2.        6.40    0.      11
0.       10.      0.      111
```

Output file FORTEC1600

```
  36    23     0    1    10
 1000.000      1    7         .006
 1000.000      7   13         .006
 1000.000     13   17         .006
 1000.000     17   20         .006
 1000.000     20   23         .006
 1000.000      2    8         .006
 1000.000      8   14         .006
 1000.000     14   18         .006
 1000.000     18   22         .006
 1000.000      3    9         .006
 1000.000      9   15         .006
 1000.000      4   10         .006
 1000.000     10   16         .006
 1000.000      5   12         .006
 1000.000      2    1         .006
 1000.000      8    7         .006
```

```
 1000.000   14   13       .006
 1000.000   18   17       .006
 1000.000   21   20       .006
 1000.000    3    2       .006
 1000.000    9    8       .006
 1000.000   15   14       .006
 1000.000   19   18       .006
 1000.000    4    3       .006
 1000.000   10    9       .006
 1000.000    5    4       .006
 1000.000   11   10       .006
 1000.000    6    5       .006
 8944.275   21   23      1.000
 9433.247   22   21      1.000
 8602.237   19   22      1.000
 9219.220   15   19      1.000
 9219.220   16   15      1.000
 8602.236   11   16      1.000
 9433.246   12   11      1.000
 8943.412    6   12      1.000
     .000        .000    5.000110 0.  0.  0.
    2.000        .000    5.313110 0.  0.  0.
    4.000        .000    5.964110 0.  0.  0.
    6.000        .000    6.995110 0.  0.  0.
    8.000        .000    8.396110 0.  0.  0.
   10.000        .000   10.000111 0.  0.  0.
     .000       2.000    4.687110 0.  0.  0.
    2.000       2.000    5.000110 0.  0.  0.
    4.000       2.000    5.664110 0.  0.  0.
    6.000       2.000    6.762110 0.  0.  0.
    6.400       2.000    7.036110 0.  0.  0.
    8.000       1.000    8.344110 0.  0.  0.
     .000       4.000    4.035110 0.  0.  0.
    2.000       4.000    4.336110 0.  0.  0.
    4.000       4.000    5.000110 0.  0.  0.
    6.000       2.285    6.705110 0.  0.  0.
     .000       6.000    3.005110 0.  0.  0.
    2.000       6.000    3.237110 0.  0.  0.
    2.285       6.000    3.295110 0.  0.  0.
     .000       8.000    1.604110 0.  0.  0.
    1.000       8.000    1.656110 0.  0.  0.
    2.000       6.400    2.964110 0.  0.  0.
     .000      10.000     .000111 0.  0.  0.
```

Input file HYPARSM2.DAT

```
  36    23    0    1    10
 1000.000    1    7       .006
 1000.000    7   13       .006
 1000.000   13   17       .006
 1000.000   17   20       .006
 1000.000   20   23       .006
 1000.000    2    8       .006
 1000.000    8   14       .006
 1000.000   14   18       .006
 1000.000   18   22       .006
 1000.000    3    9       .006
 1000.000    9   15       .006
 1000.000    4   10       .006
 1000.000   10   16       .006
 1000.000    5   12       .006
 1000.000    2    1       .006
 1000.000    8    7       .006
```

```
1000.000   14   13      .006
1000.000   18   17      .006
1000.000   21   20      .006
1000.000    3    2      .006
1000.000    9    8      .006
1000.000   15   14      .006
1000.000   19   18      .006
1000.000    4    3      .006
1000.000   10    9      .006
1000.000    5    4      .006
1000.000   11   10      .006
1000.000    6    5      .006
8944.275   21   23     1.000
9433.247   22   21     1.000
8602.237   19   22     1.000
9219.220   15   19     1.000
9219.220   16   15     1.000
8602.236   11   16     1.000
9433.246   12   11     1.000
8943.412    6   12     1.000
    .000      .000     5.000110 0. 0. 0.
   2.000      .000     5.313110 0. 0. 0.
   4.000      .000     5.964110 0. 0. 0.
   6.000      .000     6.995110 0. 0. 0.
   8.000      .000     8.396110 0. 0. 0.
  10.000      .000    10.000111 0. 0. 0.
    .000     2.000     4.687110 0. 0. 0.
   2.000     2.000     5.000000 0. 0. 0.
   4.000     2.000     5.664000 0. 0. 0.
   6.000     2.000     6.762000 0. 0. 0.
   6.400     2.000     7.036000 0. 0. 0.
   8.000     1.000     8.344000 0. 0. 0.
    .000     4.000     4.035110 0. 0. 0.
   2.000     4.000     4.336000 0. 0. 0.
   4.000     4.000     5.000000 0. 0. 0.
   6.000     2.285     6.705000 0. 0. 0.
    .000     6.000     3.005110 0. 0. 0.
   2.000     6.000     3.237000 0. 0. 0.
   2.285     6.000     3.295000 0. 0. 0.
    .000     8.000     1.604110 0. 0. 0.
   1.000     8.000     1.656000 0. 0. 0.
   2.000     6.400     2.964000 0. 0. 0.
    .000    10.000      .000111 0. 0. 0.
```

Output file FORTEC25600

```
 36   23    0    1   10
1000.000    1    7      .006
1000.000    7   13      .006
1000.000   13   17      .006
1000.000   17   20      .006
1000.000   20   23      .006
1000.000    2    8      .006
1000.000    8   14      .006
1000.000   14   18      .006
1000.000   18   22      .006
1000.000    3    9      .006
1000.000    9   15      .006
1000.000    4   10      .006
1000.000   10   16      .006
1000.000    5   12      .006
1000.000    2    1      .006
1000.000    8    7      .006
```

```
  1000.000   14   13        .006
  1000.000   18   17        .006
  1000.000   21   20        .006
  1000.000    3    2        .006
  1000.000    9    8        .006
  1000.000   15   14        .006
  1000.000   19   18        .006
  1000.000    4    3        .006
  1000.000   10    9        .006
  1000.000    5    4        .006
  1000.000   11   10        .006
  1000.000    6    5        .006
 10588.110   21   23       1.000
 11025.270   22   21       1.000
 10161.760   19   22       1.000
 10747.010   15   19       1.000
 10747.010   16   15       1.000
 10161.760   11   16       1.000
 11025.270   12   11       1.000
 10588.110    6   12       1.000
      .000        .000     5.000110 0. 0. 0.
     2.000        .000     5.309110 0. 0. 0.
     4.000        .000     5.966110 0. 0. 0.
     6.000        .000     7.008110 0. 0. 0.
     8.000        .000     8.112110 0. 0. 0.
    10.000        .000    10.000111 0. 0. 0.
      .000       2.000     4.691110 0. 0. 0.
     1.916       1.916     5.000000 0. 0. 0.
     3.926       1.907     5.675000 0. 0. 0.
     5.973       1.982     6.784000 0. 0. 0.
     6.384       2.005     7.058000 0. 0. 0.
     7.984       1.002     8.364000 0. 0. 0.
      .000       4.000     4.034110 0. 0. 0.
     1.907       3.926     4.325000 0. 0. 0.
     3.993       3.993     5.000000 0. 0. 0.
     5.984       2.288     6.725000 0. 0. 0.
      .000       6.000     2.992110 0. 0. 0.
     1.982       5.973     3.216000 0. 0. 0.
     2.288       5.984     3.275000 0. 0. 0.
      .000       8.000     1.588110 0. 0. 0.
     1.002       7.984     1.636000 0. 0. 0.
     2.005       6.384     2.942000 0. 0. 0.
      .000      10.000      .000111 0. 0. 0.
```

One final comment on the case of edge cables. The above analysis does not seem to be sensitive to the edge cable forces specified initially. This allows the designer to start with a rough approximation of the edge cable forces (which of course correct themselves in the second step of the analysis) thus avoiding some tedious cable calculations.

The two steps required above in the use of the analysis program can be argued in terms of Newton's method where it is sometimes possible to improve convergence by taking smaller steps. In the example here, rather than trying to start from scratch and get shape in a single step, two steps are used.

7.6 Membrane Finite Element Model

If you prefer to use a membrane finite element rather than a cable element to represent fabric, that can be done as suggested above for cable nets. (See Figures 7.19-22 where the above two examples are redone using a finite element model.) These examples add little to the above discussion but it should be noted that the analysis seems to converge more slowly in these cases. Similarly, cases in which the geometry is "radial" rather than "flat" as it is here can be handled using appropriate variations of the above discussions.

Listed below are the data files MEMDOMSM.DAT and MEMHYPSM.DAT which are used with the membrane finite element program MEMSM.FOR to analyze the skylight dome and the hyperbolic paraboloid respectively. In terms of detail, FORT600 of step 1 for the DOME example is edited and saved as MEMDOMSM1.DAT which is used as input for step 2 whose FORT60 and FORT600 output files are saved as MEMDOMSM60 and MEMDOMSM600 respectively. Similarly MEMHYPSM1.DAT, MEMHYPSM60 and MEMHYPSM600 are obtained for the hyperbolic paraboloid example.

Input file MEMDOMSM.DAT

```
        15                10    16
0.        0.        0.            111
0.        21.5      0.            11
0.        43.       0.            11
0.        64.5                    11
0.        86.       30.           111
21.5                0.            111
21.5      21.5      0.            11
21.5      43.       0.            11
21.5      64.5      28.227        111
43.       0.        0.            111
43.       21.5      0.            11
43.       43.       22.832        111
64.5      0.        0.            111
64.5      21.5      13.578        111
86.        0.       0.            111
         1    6    2 1.                    46.51      46.51
         7    2    6 1.                    46.51      46.51
         2    7    3 1.                    46.51      46.51
         8    3    7 1.                    46.51      46.51
         3    8    4 1.                    46.51      46.51
         9    4    8 1.                    46.51      46.51
         4    9    5 1.                    46.51      46.51
         6   10    7 1.                    46.51      46.51
        11    7   10 1.                    46.51      46.51
         7   11    8 1.                    46.51      46.51
        12    8   11 1.                    46.51      46.51
         8   12    9 1.                    46.51      46.51
        10   13   11 1.                    46.51      46.51
        14   11   13 1.                    46.51      46.51
        11   14   12 1.                    46.51      46.51
        13   15   14 1.                    46.51      46.51
```

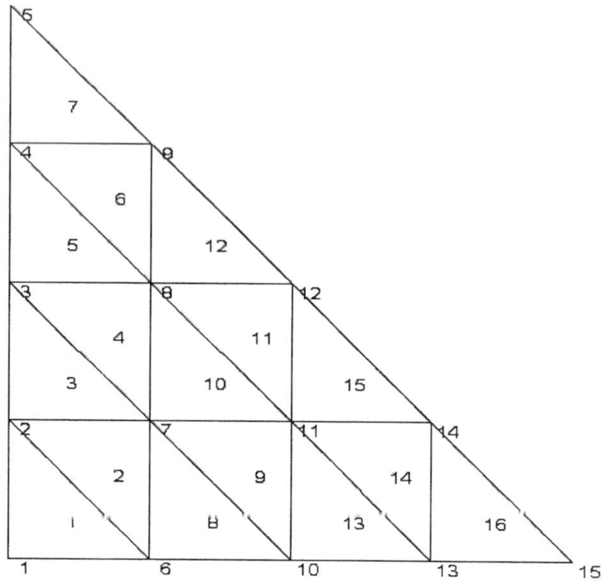

Figure 7.19 Finite element map of 1/8 of the skylight example.

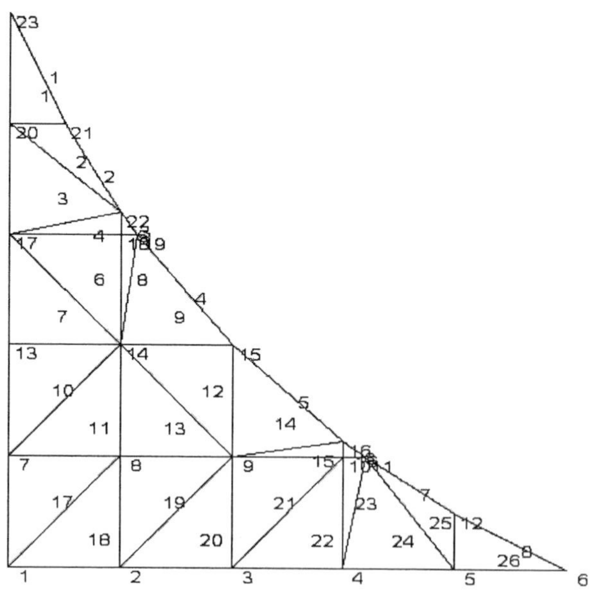

Figure 7.20 Finite element map of 1/4 of the hyperbolic paraboloid.

Input file MEMHYPSM.DAT

```
    8   23              10   26
0.        0.        0.        11
2.         0.0      0.0       11
4.         0.0      0.0       11
6.         0.0      0.0       11
8.         0.0      0.0       11
10.        0.0      10.       111
0.         2.       0.0       11
2.         2.       0.0       11
4.         2.       0.0       11
6.         2.       0.0       11
6.40       2.       0.0       11
8.         1.       0.        11
0.         4.       0.        11
2.         4.       0.        11
4.         4.       0.        11
6.         2.285    0.        11
0.         6.       0.        11
2.         6.       0.        11
2.285      6.       0.        11
0.         8.       0.        11
1.         8.       0.        11
2.         6.40     0.        11
0.        10.       0.        111
8944.        21   23 1.
9433.        22   21 1.
8602.        19   22 1.
9219.        15   19 1.
9219.        16   15 1.
8602.        11   16 1.
9433.        12   11 1.
8944.         6   12 1.
        20   21   23 1.                  500.       500.
        21   20   22 1.                  500.       500.
        20   17   22 1.                  500.       500.
        17   18   22 1.                  500.       500.
        18   19   22 1.                  500.       500.
        18   17   14 1.                  500.       500.
        13   14   17 1.                  500.       500.
        18   14   19 1.                  500.       500.
        14   15   19 1.                  500.       500.
        14   13    7 1.                  500.       500.
         7    8   14 1.                  500.       500.
        15   14    9 1.                  500.       500.
         8    9   14 1.                  500.       500.
        15    9   16 1.                  500.       500.
         9   10   16 1.                  500.       500.
        10   11   16 1.                  500.       500.
         7    1    8 1.                  500.       500.
         1    2    8 1.                  500.       500.
         8    2    9 1.                  500.       500.
         2    3    9 1.                  500.       500.
         9    3   10 1.                  500.       500.
         3    4   10 1.                  500.       500.
        11   10    4 1.                  500.       500.
         4    5   11 1.                  500.       500.
         5   12   11 1.                  500.       500.
         5    6   12 1.                  500.       500.
```

7.7 Patterning

The final step in the design of fabric structures is patterning. Fabric structures are usually made by joining strips of fabric which have been cut from rolls of cloth. How that is to be done is the subject of this section.
The object of the design process is to produce a structure which will assume a specified shape when it is prestressed. This implies that cables, for example, must be fabricated "too short" so that they will fit together when prestressed. Similarly, there is a step in patterning called "compensation" in which the stretch of the material under prestress is introduced into the patterning process. That step can be easily appended to the end of the patterning process and will not be discussed here.

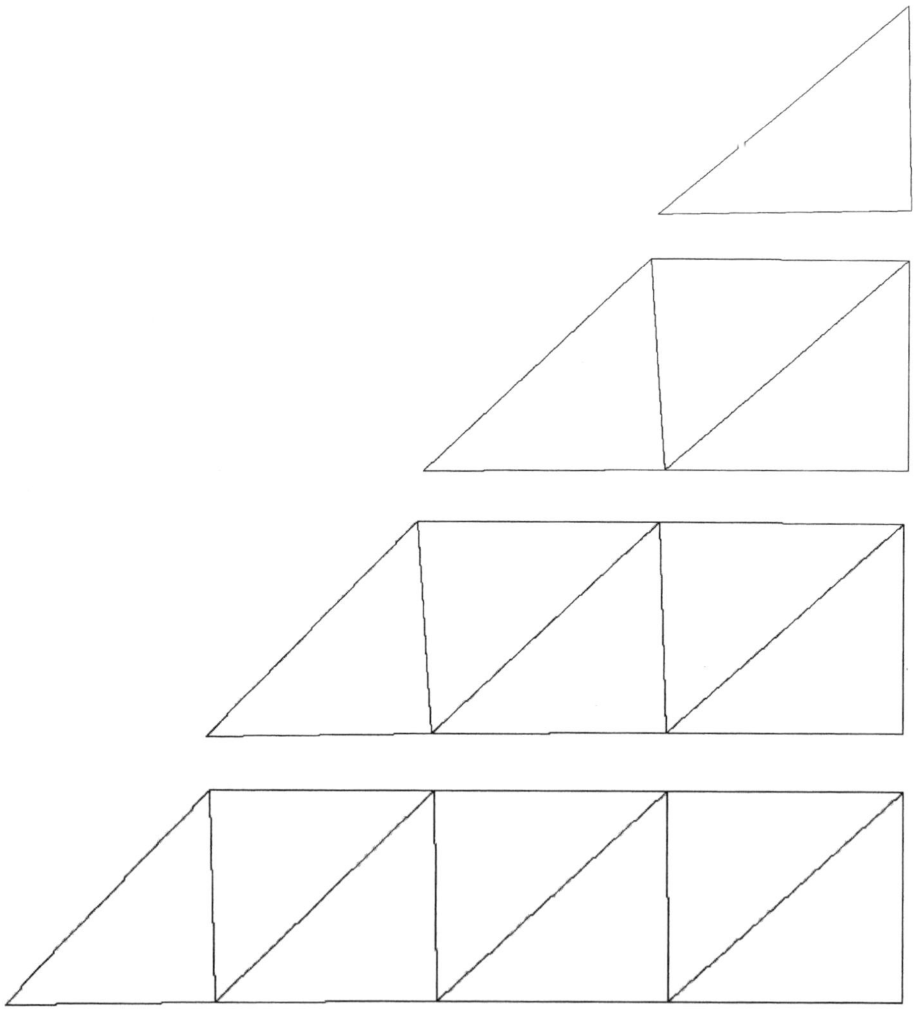

Figure 7.21 Patterning strips for the 1/8 skylight example.

The basic problem of patterning is the construction of a curved surface out of flat pieces of material. This can be done in the following manner. First, the surface must be "triangulated". This is the process by which the curved surface of the structure is approximated by a collection of flat triangular facets. It is usually done during the analysis phase of fabric structure design in order to deal with, for example, environmental loads. The next step is to select "strips" of triangles over the structure. These strips must

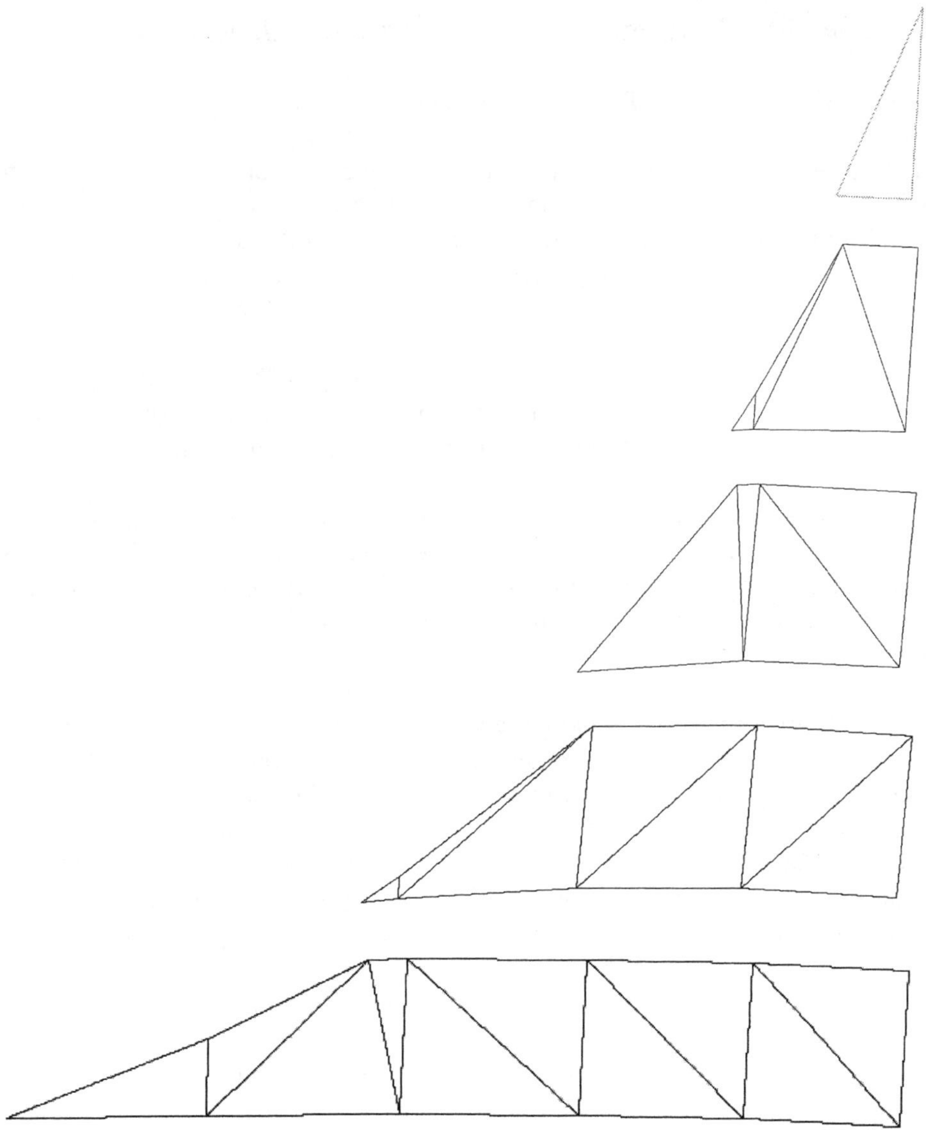

Figure 7.22 Patterning strips for 1/4 hyperbolic paraboloid.

cover the structure and must be SERIALLY connected. The point is that strips of serially connected triangles can be deformed into a flat sheet without stretching.

A patterning for the two examples discussed above is described in Figures 7.21-22. The patterning process will be described in more detail under the discussions of computer programs which follow.

7.8 Computer Programs For Cable Nets and Fabric Structures

7.8.1 LAYOUT.FOR (The grid method)

This program uses an iterative scheme (Gauss-Seidel) to compute the z coordinates over a grid of a cable net (finds shape) using the equations of equilibrium in the z direction. It might be noted that it would have been more efficient to solve these equations directly using for example Gaussian elimination since they are linear. At least, the iterative scheme used is simple and requires little computer storage.

With regard to the details of LAYOUT.FOR the example of Figure 7.1 is useful. In this case it is required to find the shape (compute the z coordinates) of a skylight which is formed using crossed circular arches with a rise of 30".

The data required by layout is included in the file DOME. In order to get started it is necessary to prepare a node map Figure 7.8 and to compute the z coordinates along the arch as described above in Example 7.2. Since this structure is 1/8 symmetric it is possible to work with the small piece of the structure shown.

The data file DOME first lists the number of bars and the number of nodes. It then lists the member description. (In this case the horizontal component of the member force and the two ends of each member.) Finally, node data is specified over the grid. At each node the z coordinates are set to be zero unless they are know to be fixed otherwise. Next, a "1" is used to indicate that the node is "fixed" in the z direction; then "4" is used to indicate terms along the diagonal (This is required later by the graphics routines.); node connectivity is then specified by listing adjacent nodes. (The repeated listing of nodes is used to cover the case of symmetry.)

With regard to the details of coding, following statement number 678, there is a primary DO loop set at 200 iterations. For each iteration a new set of z coordinates is computed keeping all but the coordinate which corresponds to diagonal term fixed. Figure 7.4 explains how the vertical force components in each bar are proportional to the z coordinates of the end of the bar. For each bar, the coefficients of these force terms are collected in

SUM and SUM1 which are subsequently used to compute new z coordinates. The final DO loop computes the residual error associated with the final coordinates.To run LAYOUT.FOR the DOS command is

LAYOUT DOME FORT8 FORT150

Here FORT8 is an output file and FORT150 is set up to pass information to subsequent programs. (DOME is, of course, the input file described above.) When the analysis program is used to find shape the DOS command is:

TR3DSM DOME FORT60 FORT600

7.8.2 LPLOT1.FOR

(This graphics routine constructs a 1/4 symmetric piece of a structure from a 1/8 symmetric piece.) As described for the example DOME above, it can be useful to construct the entire geometry of a structure even when less is required for reason of symmetry. This program handles the first step in so doing by reflecting the 1/8 solution about the 45 degree line. That is the function of the primary DO loop of this short program.

It can be run, immediately following LAYOUT.FOR, using the DOS commands

COPY FORT150 FORT50
DEL FORT150
LPLOT1 FORT50 FORT150

LPLOT1 at one point waits for keyboard input. The coordinate shift of the y-axis is then specified (Figure 7.7). In this case it is 86".

7.8.3 LAYOUTPLOT.FOR

(This graphics routine creates a full symmetric model of a structure given 1/4 of it.)The final graphics step takes the " DOME " from 1/4 symmetry to full symmetry. Generally, this program projects each node and bar in the 1/4 model into three new pieces to fill out the model. There are of course many cases in which, for reasons of symmetry, all three reflections are not required. It can be run, in this case immediately following LPLOT1.FOR, using the DOS commands

COPY FORT150 FORT50
DEL FORT150
LAYOUTPLOT FORT50 FORT150

7.8.4 PATTERN.FOR

(This program produces "patterns" from which fabric is to be cut. Note that patterning is done here using grid points from the shape finding algorithm. It could as well have been done using points interpolated over the shape.) This program creates the actual fabric dimensions so that material strips can be cut from rolls of fabric. It begins by reading output from the layout program (FORT150) from which it gets the three-dimensional shape of the structure. It then reads strip definitions. In the case of DOME these are shown in the file DOMEPAT.DAT which is listed below. It then produces coordinates and graphic output which describe how the fabric should be cut. It can be run with the following DOS commands after running LAYOUT.FOR:

COPY FORT150 FORT50
PATTERN FORT50 FORT60 FORT10 DOMEPAT.DAT

Here FORT60 is a scratch file. The cutting data appears in FORT10.

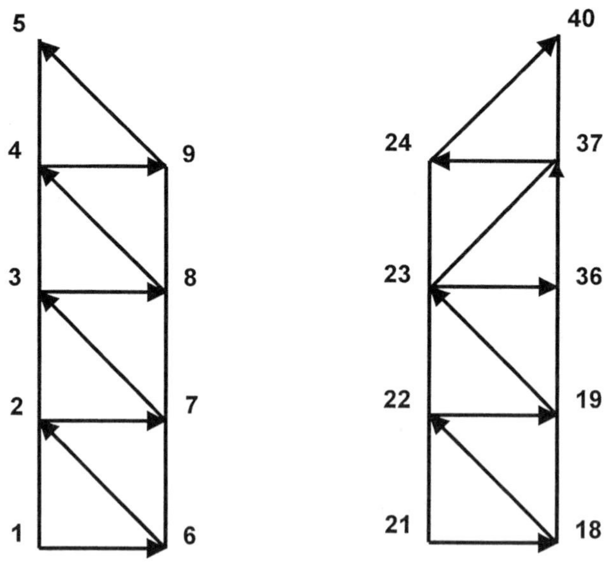

strip 1 from "DOME" example defined as 1 6 2 7 3 8 4 9 5

"fan" strip defined by sequence 21 18 22 19 23 36 37 -24 40

Figure 7.23 Numerical descriptions of strips.

Input file DOMEPAT.DAT

```
    4
    9
    1    6    2    7    3
    8    4    9    5
    7
    6   10    7   11    8
   12    9
    5
   10   13   11   14   12
    3
   13   15   14
```

Given a physical "strip" across a structural surface, this strip may be described by listing a sequence of nodes as indicated in Fig. 7.21. This sequence is understood by the computer as follows: Points 1 and 6 define a starting line; each additional node defines a triangle which is serially connected to the preceding one by using the line formed by the new node and the second node in the sequence back from it. The "fan strip" of Fig. 7.23 indicates a configuration in which this description does not work. In this case triangle 37-24-23 requires going back in the strip sequence 3 rather than 2 nodes. That fact is indicated by assigning a negative sign to node 24 in the strip sequence.

A three-dimensional description of a strip is now in place. The projection of this representation onto a plane can be done in the following manner. The orientation of the first edge (edge 1-6 from DOME above) of the strip on the plane is arbitrary. In a typical step, a new node (eg. node 2) describes a new triangle in space one side of which is the line from the last node to the new one. The real three-dimensional angle between the last two lines defined on the strip in then computed using the scalar product of unit vectors which describe the slopes of these lines. This angle allows the next point to be located on the plane projection of the strip. When all the points have been projected onto the plane, the strip is rotated so that it will lie within a piece of fabric of minimum width. Finally, suitable coordinates for the strip which are convenient for cutting are printed out.

With regard to programming details, PATTERN.FOR begins with some now familiar input and then produces a plot showing the length of lines in the surface to be patterned. This plot can be useful when checking patterns. Its primary do loop, (`DO 11 I=1, NSTR`), produces the patterning information for each strip. It begins by laying out arbitrarily the first line in the strip as described above. In the typical step, which begins with the statement `1200      IP=IABS(LIST(J+1)`, it constructs the unit vector which describes the slope of one side of the new triangle. It then computes the real three-dimensional angle between itself and the last triangle. This allows a new point to be projected onto the plane of the fabric.

After all points of the strip have been projected onto the fabric plane, the entire strip is rotated so that it can be contained within a piece of fabric of minimum width.

With regard to subroutines called by PATTERN.FOR:

- `ROTATE` which is used to "rotate" a strip into a piece of fabric of minimum width is constructed from the plotting routines which themselves determine the maximum dimensions of a rotated object.
- `CONVRT` creates a single array R given three arrays X, Y, and Z.
- `SPLOT` and `PLOT` are the usual plotting subroutines.

7.9 Problems

1. Set up the iterative scheme to determine the shape of a radially symmetric cable net. (Hint: The discussions of this chapter have been limited to relatively "flat" configurations; in the case of a radially symmetric configuration another approach may be used. Let ρ_i be the radial coordinate at level *i*. Write the radial equilibrium equation at level *i*, $f_i(\rho) = 0$. Solve these equations iteratively.

2. The square-based tents at the "Great Adventure Amusement Park" have a more complex radial geometry in which the cross sections of the structure are not circular. How would you find shape in cases like this.

CHAPTER 8

THREE-DIMENSIONAL BEAM-COLUMNS

8.1 Introduction

This chapter is concerned with the application of the perturbation method to the continuous problem of a three-dimensional beam. The reason for doing so is quite simple. It is now common to use the so-called beam-column equations to generate the element stiffness for plane frames. The question is then what should be done in the case of three-dimensional frames. This chapter suggests that if "beam-columns" are used in plane frames then "three-dimensional beam-columns" should be used in space frames. Just what is to be included in three-dimensional beam-columns is the subject of this chapter. The use of two-dimensional beam-columns has already been discussed above in Chapter 1 and Appendix 1. The derivation of the well-known beam-column equations using the methods of this section is discussed in Appendix 4.

Structural engineers have been well-served by the so-called "beam-columns" over the years. Roughly speaking, the beam-column equations include the effect of member displacement upon bending moment to the end that the bending stiffness becomes a function of the axial load (Gere, 1963). The beam-column equations, of course become singular at the flexural buckling load.

Typically, the term beam-column implies a two-dimensional beam in which there is bending about a single axis. This chapter is concerned with straight, three-dimensional prismatic beams in which there is bending about two axes and torsion. As will be seen, three-dimensional beams have the potential of many complex beam-column like interactions. A finite difference solution and a power series solution are proposed below which includes all possible interactions of initial stress resultants and can be used to generate the member stiffness in a three-dimensional frame.

Of the various interactions which can occur in three-dimensional beam problems, the best known is probably the phenomenon of torsional (lateral) buckling (Timoshenko, 1936). In the most simple case, the presence of bending moment about the strong axis of a beam weakens its effective stiffness about its weak axis (stress softening) leading eventually to lateral buckling. The presence of axial compression, of course, only makes matters worse (see below). Biot (1965) in his classic book discusses another case of interaction in which axial tension increases the torsional stiffness in a prismatic bar. More generally, it would not be a simple matter to characterize all possible types of interactions which can occur in three-dimensional beam-columns. Not knowing what to expect places an increasing importance on the problem formulation used. In fact, the interactions of stress resultants in three-dimensional beam-columns are not well understood toady. An indication of the possibilities is given in (MacBain *et a*l. 1999) using the using 3D graphical displays.

The fact is that we are comfortable when considering the softening effect of axial compression upon member stiffness in two-dimensional beam-columns and at least willing to think about the softening effect of bending moment in lateral buckling problems in three-dimensional beams. It will be seen below that the entire initial state of stress in a member can interact with the member stiffness matrix in a rather complex manner. Since any incremental analysis has initial states of stress associated with it which are not simply compression, there is every reason to use the appropriate three-dimensional "beam-column" described below.

This chapter describes a general approach which subsumes all the effects cited above. Numerical procedures for computing the member stiffness for three-dimensional beams are presented in which four coupled differential equations are solved numerically. In order to find the 6×6 member stiffness matrix for a three-dimensional beam, this system of equations must be solved 6 times. Several applications are discussed.

8.2 The Equations of Three-Dimensional Beam-Columns

The three-dimensional beam-column equations are now available in the literature (Spillers, 1990) and are presented in Table 1 below. This section reviews their derivation. For an alternative derivation see Chen *et al.* (1993).

First of all, a perturbation method (Green *et al.*, 1968) is used to describe the beam response. That is, a given straight beam, in equilibrium under given forces, is subjected to a small load perturbation. It will be seen that this method has the advantage of describing what are typically thought of as nonlinear effects without recourse to theories of large strain. It also produces the tangent stiffness now common in nonlinear analysis and describes buckling as the response to load perturbation becomes singular. The starting point is the equilibrium of a three-dimensional beam segment (Reissner, 1962) described by the force and moment equilibrium as

$$\begin{aligned} \mathbf{P}' + \mathbf{p} &= \mathbf{0} \\ \mathbf{M}' + \mathbf{m} + \mathbf{t} \times \mathbf{P} &= \mathbf{0} \end{aligned} \tag{8.1}$$

Here **P** and **M** are the usual force and moment stress-resultants with **p** and **m** the applied forces and moments and the prime symbol refers to the differentiation with respect to arc length. The vector **t** is of course the unit tangent vector which will sometimes be referred to as **i** in the applications cited below. In component form

$$\begin{aligned}\mathbf{P} &= P_x\mathbf{i} + P_y\mathbf{j} + P_z\mathbf{k} \\ \mathbf{M} &= M_x\mathbf{i} + M_y\mathbf{j} + M_z\mathbf{k}\end{aligned} \tag{8.2}$$

and the base vectors **i**, **j**, **k** selected so that P_x and M_x represent thrust and torque, respectively. Going from initial to the perturbed configuration, changes are indicated as

$$\begin{aligned}&\mathbf{P}^0 \to \mathbf{P}^1 = \mathbf{P}^0 + \varepsilon\overline{\mathbf{P}} \\ &\mathbf{M}^0 \to \mathbf{M}^1 = \mathbf{M}^0 + \varepsilon\overline{\mathbf{M}} \\ &\mathbf{i}^0 \to \mathbf{i}^1 = \mathbf{i}^0 + \varepsilon\bar{\mathbf{i}}; \mathbf{j}^0 \to \mathbf{j}^1 = \mathbf{j}^0 + \varepsilon\bar{\mathbf{j}}; \mathbf{k}^0 \to \mathbf{k}^1 = \mathbf{k}^0 + \varepsilon\overline{\mathbf{k}}\end{aligned} \tag{8.3}$$

where ε is the anticipated small parameter and the bar is used to distinguish the perturbation term. The perturbed terms of Eq. 8.2 are now inserted into the equilibrium equations of the perturbed configuration given in Eq. 8.1. In component form these equations are

$$P_x^1{}'\mathbf{i}^1 + P_x^1\mathbf{i}^1{}' + P_y^1{}'\mathbf{j}^1 + P_y^1\mathbf{j}^1{}' + P_z^1{}'\mathbf{k}^1 + P_z^1\mathbf{k}^1{}' + p_x^1\mathbf{i}^1 + p_y^1\mathbf{j}^1 + p_z^1\mathbf{k}^1 = \mathbf{0} \tag{8.4}$$

$$\begin{aligned}M_x^1{}'\mathbf{i}^1 + M_x^1\mathbf{i}^1{}' + M_y^1{}'\mathbf{j}^1 + M_y^1\mathbf{j}^1{}' + M_z^1{}'\mathbf{k}^1 + M_z^1\mathbf{k}^1{}' \\ + m_x^1\mathbf{i}^1 + m_y^1\mathbf{j}^1 + m_z^1\mathbf{k}^1 + \mathbf{t}^1 \times \mathbf{P}^1 = \mathbf{0}\end{aligned} \tag{8.5}$$

In order to complete the analysis, it is simply a matter of inserting the definition of the perturbations (Eq. 8.3) into Eqs. 8.4 and 8.5 and collecting terms in the coordinate directions. Before doing so it is convenient to introduce displacements in the following manner. Let **ω** represent the rotation vector associated with any beam element. If in the initial configuration the beam is straight and lies along the *x*-axis, for the case of small rotations **ω** can be written (from Fig. 5.2) as

$$\boldsymbol{\omega} = \theta_x\mathbf{i} - \delta_{z,x}\mathbf{j} + \delta_{y,x}\mathbf{k} \tag{8.6}$$

where θ_x is the torsional rotation, δ_z is the z-component of the beam displacement vector, δ_y is the y-component of the beam displacement can, of course, be used to determine the changes of the base vectors (from Eq. 5.18) as

$$\begin{aligned} \mathbf{i}^0 \to \mathbf{i}^1 &= \mathbf{i}^0 + \varepsilon\bar{\mathbf{i}} = \mathbf{i}^0 + \boldsymbol{\omega} \times \mathbf{i}^0 \\ \mathbf{j}^0 \to \mathbf{j}^1 &= \mathbf{j}^0 + \varepsilon\bar{\mathbf{j}} = \mathbf{j}^0 + \boldsymbol{\omega} \times \mathbf{j}^0 \\ \mathbf{k}^0 \to \mathbf{k}^1 &= \mathbf{k}^0 + \varepsilon\bar{\mathbf{k}} = \mathbf{k}^0 + \boldsymbol{\omega} \times \mathbf{k}^0 \end{aligned} \tag{8.7}$$

Clearly

$$\begin{aligned} \boldsymbol{\omega} \times \mathbf{i}^0 &= \delta_{y,x}\mathbf{j}^0 + \delta_{z,x}\mathbf{k}^0 \\ \boldsymbol{\omega} \times \mathbf{j}^0 &= -\delta_{y,x}\mathbf{i}^0 + \theta_x\mathbf{k}^0 \\ \boldsymbol{\omega} \times \mathbf{k}^0 &= -\delta_{z,x}\mathbf{i}^0 - \theta_x\mathbf{j}^0 \end{aligned} \tag{8.8}$$

Table 1 First-order equations.

	<u>Force Equilibrium</u>
$\mathbf{i}^0$ component:	$\bar{P}_x' + \bar{p}_x - P_y^0\delta_{y,xx} - P_z^0\delta_{z,xx} = 0$
$\mathbf{j}^0$ component:	$\bar{P}_y' + \bar{p}_y + P_x^0\delta_{y,xx} - P_z^0\theta_{x,x} = 0$
$\mathbf{k}^0$ component:	$\bar{P}_z' + \bar{p}_z + P_x^0\delta_{z,xx} + P_y^0\theta_{x,x} = 0$
	<u>Moment Equilibrium</u>
$\mathbf{i}^0$ component:	$\bar{M}_x' + \bar{m}_x - M_y^0\delta_{y,xx} - M_z^0\delta_{z,xx} = 0$
$\mathbf{j}^0$ component:	$\bar{M}_y' + \bar{m}_y + M_x^0\delta_{y,xx} - M_z^0\theta_{x,x} - \bar{P}_z = 0$
$\mathbf{k}^0$ component:	$\bar{M}_z' + \bar{m}_z + M_x^0\delta_{z,xx} + M_y^0\theta_{x,x} + \bar{P}_y = 0$

Table 2 Zero-order equations.

	<u>Force Equilibrium</u>
$\mathbf{i}^0$ component:	$P_x^0{}' + p_x^0 = 0$
$\mathbf{j}^0$ component:	$P_y^0{}' + p_y^0 = 0$
$\mathbf{k}^0$ component:	$P_z^0{}' + p_z^0 = 0$
	<u>Moment Equilibrium</u>
$\mathbf{i}^0$ component:	$M_x^0{}' + m_x^0 = 0$
$\mathbf{j}^0$ component:	$M_y^0{}' + m_y^0 - P_z^0 = 0$
$\mathbf{k}^0$ component:	$M_z^0{}' + m_z^0 + P_y^0 = 0$

The first-order equilibrium equations in the $\mathbf{i}^0$, $\mathbf{j}^0$, $\mathbf{k}^0$ directions are collected in Table 1. They are obtained by writing the equilibrium equations in the perturbed configuration and keeping terms which are linear in the small parameter ε. For convenience this parameter can then be set to one in which case the terms indicated by bars (Eq. 8.3) then represent the full perturbation. The zero-order equations are assumed to satisfy equilibrium in the $\mathbf{i}^0$, $\mathbf{j}^0$, $\mathbf{k}^0$ directions and in the initial configuration. They are a direct outcome of Eq. 8.1 in component form and presented in Table 2.

It should be noted that the assumptions of small displacement theory have been invoked above to allow arc length differentiation to be replaced by differentiation with respect to the space variable *x*.

8.3 The Member Stiffness Matrix

In the most simple case of elastic analysis, a straight beam segment with no cross-sectional warping can be described by four uncoupled events: axial deformation, torsional deformation, and bending deformation about 2 axes. If the two-dimensional beam-column equations are used to introduce nonlinear effects, there is a coupling between axial forces and bending. If the equations of Table 1 are used, they introduce a coupling between all the stress resultants present in the three-dimensional frame element. The resultant member stiffness matrix is a tangent stiffness matrix due to the use of the perturbation method.

First of all, from considerations of equilibrium it can be argued that the member stiffness matrix is a 6×6 matrix. Second, using Newton's method for nonlinear structural analysis, each step (iteration) of the nonlinear analysis becomes simply a linear analysis which uses the local tangent stiffness. Therefore the terms in the member stiffness matrix are "forces" due to unit "displacements". Third, if the member forces are chosen properly, the terms in the member stiffness matrix may be computed by introducing sequentially 6 discontinuities into the boundary conditions of the system of equations in Table 1:

- A unit axial discontinuity $\delta_x(L) = 1.0$, is introduced first (*i*=1).
- A unit torsional discontinuity $\theta_x(L) = 1.0$, is introduced next (*i*=2).
- Four flexural discontinuities are finally introduced, (*i*=3, 4, 5, 6):

$$\delta_{z,x}(L) = -1.0; \quad \delta_{y,x}(L) = 1.0;$$
$$\delta_{z,x}(0) = -1.0; \quad \delta_{y,x}(0) = 1.0;$$

Note that the four flexural discontinuities are those which are used in moment distribution: a unit rotation about an axis of flexure is applied at one end of a beam while the other end is held fixed. Two beam ends and *2* axes of flexure then imply four flexural discontinuities.

The next step will combine some of the equations in Table 1 to produce a system of four displacements $\delta_x, \delta_y, \delta_z$, and θ_x as unknowns. Quite generally and following linear beam theory, the second equation can be combined with the sixth equation eliminating the shear term $\overline{\mathrm{P}}_y$; similarly the third equation can be combined with the fifth equation in the table to eliminate the shear term $\overline{\mathrm{P}}_z$. Furthermore at this point all the member loads can be eliminated as not of interest,

$$\begin{aligned} m_x^0 = m_y^0 = m_z^0 = \overline{m}_x = \overline{m}_y = \overline{m}_z = 0 \\ p_x^0 = p_y^0 = p_z^0 = \overline{p}_x = \overline{p}_y = \overline{p}_z = 0 \end{aligned} \tag{8.9}$$

This implies that the initial axial thrust and torque, P_x^0, P_y^0, P_z^0 and M_x^0, must be constant and that the initial bending moments, M_y^0 and M_z^0, can at most be linear functions of x. The six equations of Table 1 then reduce to the following four equations:

$$\begin{aligned} \overline{P}_x{}' - P_y^0 \delta_{y,xx} - P_z^0 \delta_{z,xx} = 0 \\ \overline{M}_x{}' - M_y^0 \delta_{y,xx} - M_z^0 \delta_{z,xx} = 0 \\ \overline{M}_y{}'' - 2M_z^0{}' \theta_{x,x} + M_x^0 \delta_{y,xxx} - M_z^0 \theta_{x,xx} + P_x^0 \delta_{z,xx} = 0 \\ \overline{M}_z{}'' + 2M_y^0{}' \theta_{x,x} + M_x^0 \delta_{z,xxx} + M_y^0 \theta_{x,xx} - P_x^0 \delta_{y,xx} = 0 \end{aligned} \tag{8.10}$$

To complete the formulation, four constitutive equations are appended,

$$\overline{P}_x = k_x \delta_{x,x};\ \overline{M}_x = k_{\mathrm{T}} \theta_{x,x};\ \overline{M}_y = -k_y \delta_{z,xx};\ \overline{M}_z = k_z \delta_{y,xx}; \tag{8.11}$$

Here the k's are the usual spring constants from considerations of strength of materials. Some general comments on this system of equations can now be made:

- The last three equations are coupled and must be solved simultaneously; then the first equation can be integrated to complete the solution.

- The last two equations are fourth order in the beam displacements (like the linear elastic beam equations); the other two equations are second order.

• The equations themselves are linear in x since the initial moments M_y^0, M_z^0 may be linear in x. (Timoshenko (1936) remarks that equations of this type may be solved using Bessel functions.)

The rest of this section is concerned with some of the details required to construct the three-dimensional beam-column stiffness matrix which incidentally, is nonsymmetrical.

The 6×6 stiffness matrix is first obtained from the displacements that are computed for each discontinuity, i, from the following relations and then transformed into the perturbed coordinate system:

$$\begin{aligned}
\overline{P}_{xi}(L) &= \left(k_x \delta_{x,x}(L)\right)_i \\
\overline{M}_{xi}(L) &= \left(k_T \theta_{x,x}(L)\right)_i \\
\overline{M}_{yi}(L) &= \left(-k_y \delta_{z,xx}(L)\right)_i \\
\overline{M}_{zi}(L) &= \left(k_z \delta_{y,xx}(L)\right)_i \\
\overline{M}_{yi}(0) &= \left(-k_y \delta_{z,xx}(0)\right)_i \\
\overline{M}_{zi}(0) &= \left(k_z \delta_{y,xx}(0)\right)_i
\end{aligned} \tag{8.12}$$

The perturbed, stiffness matrix having elements, K_{ij}, becomes:

$$\begin{aligned}
&K_{1i} = \overline{P}_{xi}^1(L);\ K_{2i} = \overline{M}_{xi}^1(L);\ K_{3i} = \overline{M}_{yi}^1(L); \\
&K_{4i} = \overline{M}_{zi}^1(L);\ K_{5i} = \overline{M}_{yi}^1(0);\ K_{6i} = \overline{M}_{zi}^1(0);
\end{aligned} \tag{8.13}$$

The transformed components $\overline{P}_x^1(L)$, $\overline{M}_x^1(L)$, $\overline{M}_y^1(L)$, $\overline{M}_z^1(L)$, $\overline{M}_y^1(0)$, $\overline{M}_z^1(0)$, in the perturbed coordinate system are computed from the components of Eq. 12 by considering the change in the unit vectors due to a small rotation vector ω. Thus

$$\begin{aligned}
\overline{P}_x^1 \mathbf{i}^1 &= (\overline{P}_x - P_y^0 \delta_{y,x} - P_z^0 \delta_{z,x})\mathbf{i}^0 \\
\overline{P}_y^1 \mathbf{j}^1 &= (\overline{P}_y + P_x^0 \delta_{y,x} - P_z^0 \theta_x)\mathbf{j}^0 \\
\overline{P}_z^1 \mathbf{k}^1 &= (\overline{P}_z + P_y^0 \theta_x + P_x^0 \delta_{z,x})\mathbf{k}^0
\end{aligned} \tag{8.14}$$

$$
\begin{aligned}
\overline{M}_x^1 \mathbf{i}^1 &= (\overline{M}_x - M_y^0 \delta_{y,x} - M_z^0 \delta_{z,x})\mathbf{i}^0 \\
\overline{M}_y^1 \mathbf{j}^1 &= (\overline{M}_y + M_x^0 \delta_{y,x} - M_z^0 \theta_x)\mathbf{j}^0 \\
\overline{M}_z^1 \mathbf{k}^1 &= (\overline{M}_z + M_y^0 \theta_x + M_x^0 \delta_{z,x})\mathbf{k}^0
\end{aligned} \tag{8.15}
$$

The 3-D beam-column stiffness matrix may now be finalized as:

$$
\begin{bmatrix}
\overline{P}_{x1}(L) & \overline{P}_{x2}(L) & \overline{P}_{x3}(L) + P_z^0(L) & \overline{P}_{x4}(L) - P_y^0(L) & \overline{P}_{x5}(L) & \overline{P}_{x6}(L) \\
\overline{M}_{x1}(L) & \overline{M}_{x2}(L) & \overline{M}_{x3}(L) + M_z^0(L) & \overline{M}_{x4}(L) - M_y^0(L) & \overline{M}_{x5}(L) & \overline{M}_{x6}(L) \\
\overline{M}_{y1}(L) & \overline{M}_{y2}(L) - M_z^0(L) & \overline{M}_{y3}(L) & \overline{M}_{y4}(L) + M_x^0(L) & \overline{M}_{y5}(L) & \overline{M}_{y6}(L) \\
\overline{M}_{z1}(L) & \overline{M}_{z2}(L) + M_y^0(L) & \overline{M}_{z3}(L) - M_x^0(L) & \overline{M}_{z4}(L) & \overline{M}_{z5}(L) & \overline{M}_{z6}(L) \\
\overline{M}_{y1}(0) & \overline{M}_{y2}(0) & \overline{M}_{y3}(0) & \overline{M}_{y4}(0) & \overline{M}_{y5}(0) & \overline{M}_{y6}(0) + M_x^0(0) \\
\overline{M}_{z1}(0) & \overline{M}_{z2}(0) & \overline{M}_{z3}(0) & \overline{M}_{z4}(0) & \overline{M}_{z5}(0) - M_x^0(0) & \overline{M}_{z6}(0)
\end{bmatrix}
$$

8.4 Numerical Solutions

As a practical approach, each beam is divided into a number of segments and the solution of Eqs. 8.10 represented by a finite number of points using the method of finite differences. In the case of the solutions described below, twenty spaces have been taken along each beam. Using central differences and introducing fictitious points to handle the boundary conditions, seventy five simultaneous equations were solved on a PC for every case which involved the last three of Eqs. 8.10; once these equations have been solved together, the first of Eqs. 8.10 can be solved. The computed solutions are in agreement with the charts of Gere (1963) when those solutions were appropriate. Other numerical checks are described below. A FORTRAN subroutine, BMCOL3.FOR, which produces a 6×6 member stiffness matrix is included on the CD supplied with the text. A similar subroutine for three-dimensionanal beam elements of varying stiffness is described in Levy and Gal, (2002[A]).

Dividing members into short segments can be costly when thinking in terms of doing large, practical, three-dimensional frames. An alternative approach (Spillers and Rashidi, 1997) is to construct the member stiffness matrix using a power series approach. Eisenberger has used this approach for the solution of a rich variety of buckling and vibration problems over the last fifteen years. His elegant method is referred to, in the literature, as the *exact element method* (see for example Eisenberger, 1990). In it the four displacements are expanded in power series

$$\theta_x = \sum_{n=0}^{\infty} \alpha_n x^n \tag{8.16}$$

$$\delta_y = \sum_{n=0}^{\infty} \beta_n x^n \tag{8.17}$$

$$\delta_z = \sum_{n=0}^{\infty} \gamma_n x^n \tag{8.18}$$

$$\delta_x = \sum_{n=0}^{\infty} \lambda_n x^n \tag{8.19}$$

When these power series are inserted into the differential equations, recurrence relations occur between the coefficients. Setting the expression, which multiplies typically the term x^m, gives

$$k_x \lambda_m - P_y^0 \beta_m - P_z^0 \gamma_m = 0 \tag{8.20}$$

$$(m+2)\left[k_T \alpha_{m+2} + M_y^0 \beta_{m+2} + M_z^0 \gamma_{m+2}\right] + m\left[-P_z^0 \beta_{m+1} + P_y^0 \gamma_{m+1}\right] = 0 \tag{8.21}$$

$$-k_y (m+4)(m+3)(m+2)\gamma_{m+4} + M_x^0 (m+3)(m+2)\beta_{m+3} + 2P_y^0 \alpha_{m+1} + M_z^0 (m+2)\alpha_{m+2} + P_y^0 m\alpha_{m+1} + P_x^0 (m+2)\gamma_{m+2} = 0 \tag{8.22}$$

$$k_z (m+4)(m+3)(m+2)\beta_{m+4} + M_x^0 (m+3)(m+2)\gamma_{m+3} + 2P_z^0 \alpha_{m+1} - M_y^0 (m+2)\alpha_{m+2} + P_z^0 m\alpha_{m+1} - P_x^0 (m+2)\beta_{m+2} = 0 \tag{8.23}$$

Finally, a system of coupled, linear, algebraic equations in these coefficients results when the boundary conditions are applied.

By inspection of Eqs. 8.20 – 8.23, an elimination scheme for the coefficients can be constructed: the second of these equations can be used to eliminate α in terms of β and γ, the fourth equation can be used to eliminate β; and the third equation can be used to eliminate γ. Then in a secondary scheme, the first equation can be used to eliminate λ in terms of the coefficients already computed.

A FORTRAN subroutine BMCOLSR.FOR has been written that uses this series solution to compute the member stiffness matrix for 3D frames for the case of geometrically nonlinear structures. This routine replaces BMCOL3.FOR in the three-dimensional nonlinear frame analysis program that is described in Chapter 5 and FR3DNLSR.FOR is the resulting new program.

8.5 Special Cases

8.5.1 The Elastic Beam

When the initial forces are all zero, Eqs. 8.10 degenerate to the solution for an elastic beam in which axial compression, torsion, and y and z axis bending are all uncoupled:

$$\begin{aligned} k_x \delta_{x,xx} &= 0 \\ k_{\mathrm{T}} \theta_{x,xx} &= 0 \\ -k_y \delta_{z,xxxx} &= 0 \\ k_z \delta_{y,xxxx} &= 0 \end{aligned} \tag{8.24}$$

8.5.2 Two-Dimensional Beam-Columns

When only an initial axial load, P_x^0, is present, Eqs. 8.10 again uncouple as described above. In this case, the y and z axis bending is controlled by the familiar two-dimensional beam-column equations:

$$\begin{aligned} k_x \delta_{x,xx} &= 0 \\ k_{\mathrm{T}} \theta_{x,xx} &= 0 \\ -k_y \delta_{z,xxxx} + P_x^0 \delta_{z,xx} &= 0 \\ k_z \delta_{y,xxxx} - P_x^0 \delta_{y,xx} &= 0 \end{aligned} \tag{8.25}$$

This is the case which is most commonly used today in the nonlinear analysis of three-dimensional frames.

8.5.3. Lateral Buckling

If a constant initial moment M_z^0 is added to the problem, the member twist θ_x becomes coupled with δ_z while δ_x and δ_y remain uncoupled (Eq. 8.18 below). The third of these equations is particularly interesting and can be thought of as a beam-column equation in which the constant P_x^0 / k_y is replaced by $(P_x^0 - M_z^{0^2} / k_{\mathrm{T}}) / k_y$. With this correction Gere's beam-column charts can also be used in this case.

$$
\begin{aligned}
k_x \delta_{x,xx} &= 0 \\
k_{\mathrm{T}} \theta_{x,xx} - M_z^0 \delta_{z,xx} &= 0 \\
-k_y \delta_{z,xxxx} + (P_x^0 - \frac{M_z^{0^2}}{k_{\mathrm{T}}}) \delta_{z,xx} &= 0 \\
k_z \delta_{y,xxxx} - P_x^0 \delta_{y,xx} &= 0
\end{aligned}
\tag{8.18}
$$

8.5.4 A More Complex Case.

If an initial constant moment M_y^0 is now added, θ_x, δ_y, and δ_z, become fully coupled and it is no longer possible to invoke charts for well-known solutions. In this case the equations take on the form

$$
\begin{aligned}
k_x \delta_{x,xx} &= 0 \\
k_{\mathrm{T}} \theta_{x,xx} - M_y^0 \delta_{y,xx} - M_z^0 \delta_{z,xx} &= 0 \\
k_y \delta_{z,xxxx} + M_z^0 \theta_{x,xx} - P_x^0 \delta_{z,xx} &= 0 \\
k_z \delta_{y,xxxx} + M_y^0 \theta_{x,xx} - P_x^0 \delta_{y,xx} &= 0
\end{aligned}
\tag{8.19}
$$

The solution of Eqs. 8.19, the case of a beam subject to an initial state of axial load P_x^0 and constant biaxial bending M_y^0 and M_z^0 is now discussed. The first of Eqs. 8.19 is trivial. The remaining equations can be written in operator form as

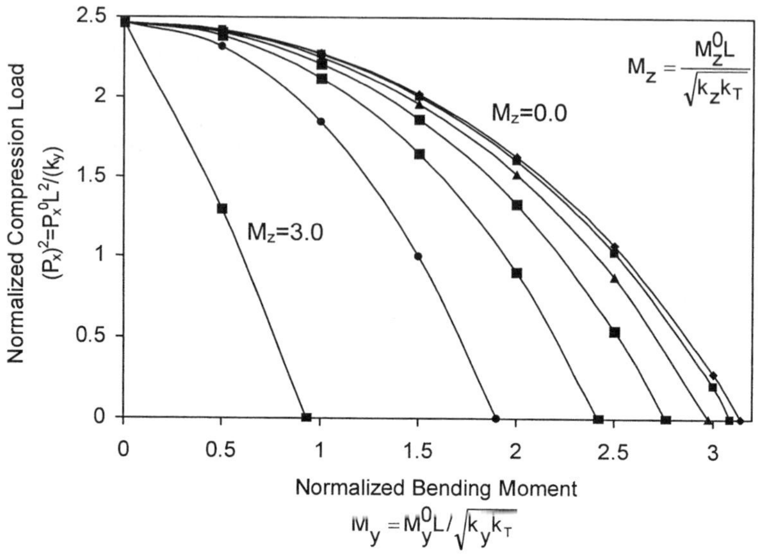

Figure 8.1 Buckling curves.

$$D^2 \begin{bmatrix} k_T & -M_z^0 & -M_y^0 \\ M_z^0 & k_y D^2 - P_x^0 & 0 \\ M_y^0 & 0 & k_z D^2 - P_x^0 \end{bmatrix} \begin{Bmatrix} \theta_x \\ \delta_z \\ \delta_y \end{Bmatrix} = \mathbf{0}$$

The determinantal equation for this system is

$$D^6 \left[k_T \left(k_y D^2 - P_x^0\right)\left(k_z D^2 - P_x^0\right) + M_z^{0\,2}\left(k_z D^2 - P_x^0\right) + M_y^{0\,2}\left(k_y D^2 - P_x^0\right)\right] = 0$$

which is basically quadratic in D^2 and thus should be regarded to be accessible. The solutions of this system generally have the form.

$$\theta_x = A_1 + A_2 x + A_3 x^2 + A_4 x^3 + A_5 x^4 + A_6 x^5 \\ + A_7 \sin k_1 x + A_8 \cos k_1 x + A_9 \sin k_2 x + A_{10} \cos k_2 x$$

$$\delta_z = B_1 + B_2 x + B_3 x^2 + B_4 x^3 + B_5 x^4 + B_6 x^5 \\ + B_7 \sin k_1 x + B_8 \cos k_1 x + B_9 \sin k_2 x + B_{10} \cos k_2 x$$

$$\delta_y = C_1 + C_2 x + C_3 x^2 + C_4 x^3 + C_5 x^4 + C_6 x^5$$
$$+ C_7 \sin k_1 x + C_8 \cos k_1 x + C_9 \sin k_2 x + C_{10} \cos k_2 x$$

where k_1 and k_2 are the roots of the quadratic described above. Coupling of these coefficients requires that the polynomial terms of order quadratic or higher must vanish and also provides eight relationships between the coefficients of the trigonometric terms. The ten boundary conditions then complete the problem statement.

For a particular case of a 3D cantilever under biaxial bending and axial compression, buckling loads are obtained using the finite difference method and presented in Figure 8.1. For $M_z^0 = M_y^0 = 0$ the result is in full agreement with Timoshenko's (1936) analytical solution. More cases, including added initial torsion are discussed in Levy and Gal (2002^{B}).

8.5.5 The Effect of Initial Torsion[1]

The discussion of special cases has thus far been motivated by practical applications and available solutions and therefore has been dominated by the effect of axial load. An alternative would have been to start with the case of initial torsion M_z^0. In this case the equations are

$$\begin{aligned} k_x \delta_{x,xx} &= 0 \\ k_{\mathrm{T}} \theta_{x,xx} &= 0 \\ -k_y \delta_{z,xxxx} + M_x^0 \delta_{y,xxx} &= 0 \\ k_z \delta_{y,xxxx} + M_x^0 \delta_{z,xxx} &= 0 \end{aligned} \tag{8.20}$$

Here only δ_y and δ_z are coupled and it is possible to solve the system by hand. For

$$\mathrm{D}^3 \begin{bmatrix} -k_y \mathrm{D} & M_x^0 \\ M_x^0 & k_z \mathrm{D} \end{bmatrix} \begin{Bmatrix} \delta_z \\ \delta_y \end{Bmatrix} = \mathbf{0}$$

[1]Professor Maciej Bieniek of Columbia University suggests that this case is of interest to and known to those working on the mechanics of drill shafts.

Here D represents the derivative d/dx. If the determinant of this system is set to zero it follows that

$$D^6\left(k_y k_z D^2 + M_x^{0^2}\right) = 0$$

and that δ_z and δ_y have the form:

$$\begin{aligned}\delta_z &= A_1 + A_2 x + A_3 x^2 + A_4 x^3 + A_5 x^4 + A_6 x^5 \\ &\qquad + A_7 \sin kx + A_8 \cos kx \\ \delta_y &= B_1 + B_2 x + B_3 x^2 + B_4 x^3 + B_5 x^4 + B_6 x^5 \\ &\qquad + B_7 \sin kx + B_8 \cos kx\end{aligned}$$

with $k^2 - M_x^{0^2}/(k_y k_z)$. Coupling of these coefficients implies that $A_4 = A_5 = A_6 = B_4 = B_5 = B_6 = 0$ and that

$$\begin{aligned}-k_y A_7 k + B_8 M_x^0 &= 0 \\ k_y A_8 k + B_7 M_x^0 &= 0\end{aligned}$$

Given the eight boundary conditions (flexure about two axes) of this system, the A's and B's can be computed explicitly by hand.

An interesting two-dimensional 'extension' to this case would be the effect of twist on the classical 2D beam-column. This case was solved in Levy and Gal (2002[B]) using the finite difference method and the results are presented in Figure 8.2 below.

There is a subset of Eqs. 8.20, in which torsion can produce flexural buckling which is accessible to hand calculation (Figure 8.3). In the case of Euler buckling a lateral displacement produces a bending moment $\overline{M}_z = P_x^0 \delta_y$; in the torsional buckling problem, lateral displacements generate bending moments from components of the applied torsion M_x^0. For this problem, δ_y and δ_z couple as

$$D\begin{bmatrix} -k_y D & M_x^0 \\ M_x^0 & k_z D \end{bmatrix}\begin{Bmatrix} \delta_z \\ \delta_y \end{Bmatrix} = \mathbf{0}$$

The determinantal equation in this case is

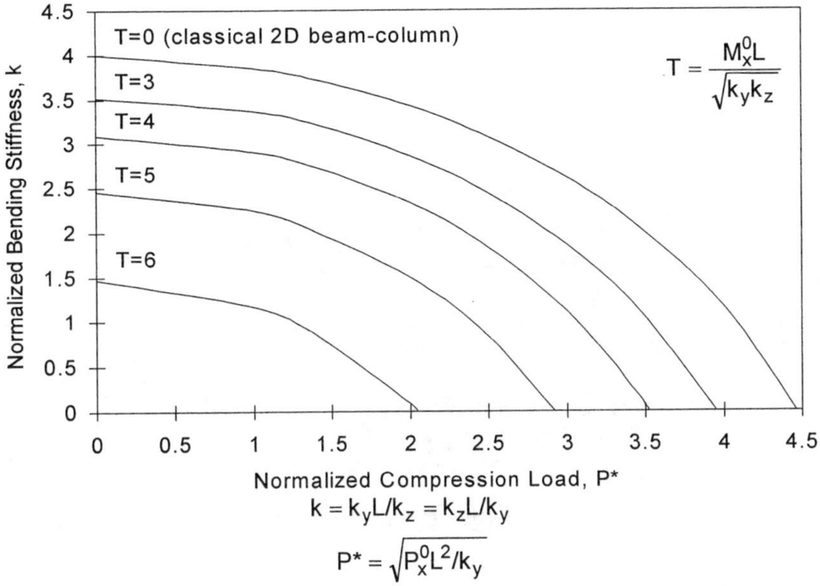

Figure 8.2 Effect of initial twist on classical beam-column interactions.

$$D^2\left(k_y k_z D^2 + M_x^{0^2}\right) = 0$$

which implies solutions of the form

$$\delta_z = A_1 + A_2 x + A_3 \sin kx + A_4 \cos kx$$
$$\delta_y = B_1 + B_2 x + B_3 \sin kx + B_4 \cos kx$$

with $k^2 = M_x^{0^2}/\left(k_y k_z\right)$. In this case coupling of the coefficients implies that $A_2 = B_2 = 0$ and that

$$-k_y A_3 k + B_4 M_x^0 = 0$$
$$k_y A_4 k + B_3 M_x^0 = 0$$

The problem statement is completed with the 4 boundary conditions $\delta_y = \delta_z = 0$ @ $x = 0$ and $x = l$. Setting the determinant of the matrix of the coefficients equal to zero in this case implies the characteristic equation

$$\cos kl = 1.0$$

The smallest nontrivial root is $kl = 2\pi$ or the critical buckling torque is

$$\left(M_x^0\right)_{\text{crit.}} = 2\pi \frac{\sqrt{k_y k_z}}{l}$$

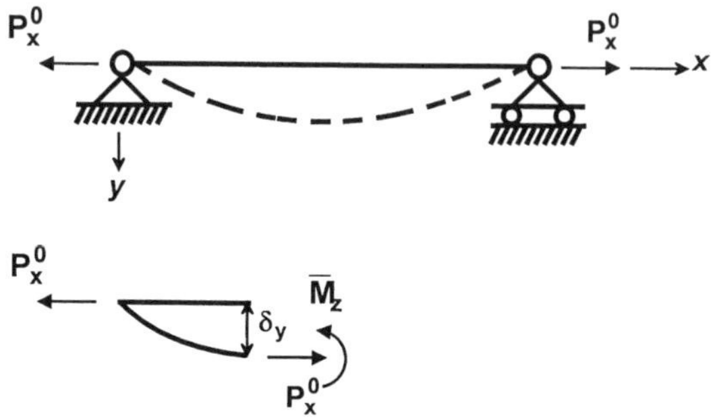

Euler Buckling

$$\overline{M}_z - P_x^0 \delta_y = 0 \text{ (equilibrium); } \overline{M}_z = k_z \delta_{y,xx} \text{ (constitutive equation)}$$

$$\delta_{y,xx} = \frac{P_x^0}{k_z} \delta_y$$

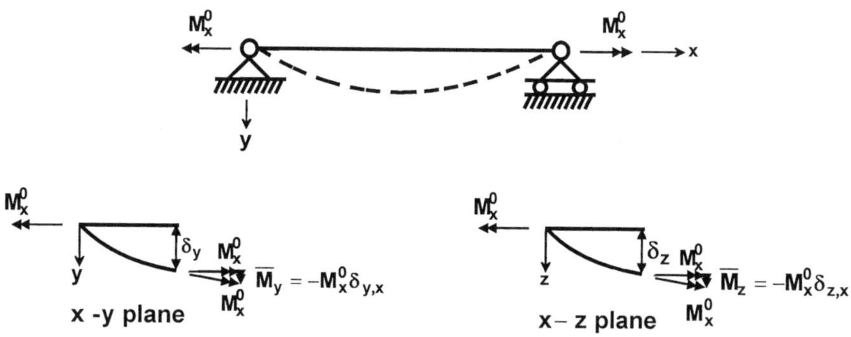

Torsional Buckling

$$\overline{M}_y + M_x^0 \delta_{y,x} = 0 \text{ (deformed equilibrium, x-y plane)}$$

$$\overline{M}_z + M_x^0 \delta_{z,x} = 0 \text{ (deformed equilibrium, x-z plane)}$$

$$\overline{M}_z = k_z \delta_{y,xx},\ \overline{M}_y = -k_y \delta_{z,xx} \text{ (constitutive equations)}$$

Figure 8.3 Simple buckling cases.

This chapter has explored several special cases of three-dimensional beam-column equations and suggests that the numerical procedures such as method of finite differences and the power series approach can be used when a solution of the general case is required. While not elegant, the method of finite differences has the added advantage of being unconcerned with the signs of the coefficients in the equations and thus includes both stress hardening and softening in a single algorithm.

One of the driving forces of this chapter is the need to construct a member stiffness matrix for the nonlinear analysis of three-dimensional frames. At the moment, many of the available computer programs and even theoretical discussions such as See and McConnell (1986) use two-dimensional beam-column theory to construct such a member stiffness matrix. Clearly such an approach leaves out the effect of lateral buckling. As pointed out here it also leaves out other possible three-dimensional couplings produced by initial loads.

8.6 Problems

1. The finite difference solution of the three-dimensional beam-columns is rather slow to execute. Reconstruct the alternative solution using the power series method.

2. Investigate the effect of member weight upon the buckling of a drill shaft.

3. Timoshenko claims that Bessel functions can be used to solve the lateral buckling problem when unequal end moments are used. Compare the use of Bessel functions with the use of the series method in this case.

4. Run the program BMCOL3.FOR using a single member cantilever and verify Timoshenko's lateral buckling results.

CHAPTER 9

NONLINEAR ANALYSIS OF SHELLS

9.1 Introduction

This chapter extends the material of Chapter 6, Nonlinear Analysis of Membranes, to include the effects of bending and the material of Chapter 5, Nonlinear Analysis of Space Frames, to remove rigid body rotations (Levy and Gal 2002^{C}). Chapter 6 discussed a “membrane shell”, using a finite element developed for fabric structures applications where the stress levels are almost constant. The approach to shell buckling of this chapter is quite general, building upon this work in Chapter 6.

The history of finite element applications to shell buckling is extensive going back to the work of Clough and Johnson (1968). The natural mode contribution of Argyris *et al.* (1977) was a major addition to shell theory. It was recently modified to include elastoplastic effects (Argyris *et al.* 2000). Horrigmoe and Bergan (1978) used the co-rotational method for nonlinear analysis and Bathe and Ho (1980), Hsiao (1987), Mohan and Kapania (1997), Peng and Crisfield (1992) improved element performance along those lines. The 3-D elasticity "degenerate" element of Ahmad *et al.* (1970) was followed by, among others, Bathe and Balourchi (1980), Hughes and Lui (1981), Dvorkin and Bathe (1984), and Buechter and Ramm (1992). In an excellent review, Ibrahimbegović (1997) addresses the various approaches and the complex issues involved.

Here the derivation of the geometric stiffness matrix is somewhat different but consistent with the approach used throughout this text. The linear equilibrium equations for a flat triangular shell element in its local coordinates system are first perturbed to yield the in-plane geometric stiffness matrix. Then out-of-plane considerations that involve the effect of rigid body rotations on member forces yield an out-of-plane geometric stiffness matrix. The shell element that was chosen for that purpose combines the constant stress triangle (CST) flat triangular membrane element (Zienkeiwicz (1977)) and of the discrete Kirchhoff theory (DKT) flat triangular plate element (Batoz *et al.* (1980)).

Finally a computer program, featuring incremental analysis and Newton’s method, geometric effects, pure deformation isolation, internal stress retrieval and updating of nodal forces and coordinates is presented and used to solve a number of problems that have appeared in the literature.

These problems include experiments, analytical results and numerical problems that were solved by the finite element method. They compare well with the results in the current literature. Some details in the derivation have been omitted. The interested reader can find them in Gal (2002).

9.2. The Geometric Stiffness Matrix of Triangular Element Shells

The local geometric stiffness matrix of the shell element is split into three distinct matrices:

$$\left[\mathbf{K}_{G}^{e}\right]_{TOTAL}^{shell} = \left[\mathbf{K}_{G}^{e}\right]_{IP}^{mem} + \left[\mathbf{K}_{G}^{e}\right]_{IP}^{plate} + \left[\mathbf{K}_{G}^{e}\right]_{OP}^{shell} \tag{9.1}$$

where the first, second and third terms on the R.H.S. of Eq. 9.1 represent the in-plane geometric stiffness matrix of the membrane, the in-plane geometric stiffness matrix of the plate and the out-of-plane geometric stiffness matrix of the shell element respectively. The total, 'tangential' stiffness matrix for use in nonlinear analysis will include, in addition, the linear elastic stiffness matrices of a plane stress triangular element (membrane) and that of a triangular plate element.

The geometrically nonlinear triangular shell element has eighteen local degrees of freedom (DOF's): 3 displacements and 3 rotations at each node. The membrane element contributes to nine displacement DOF's only. The basic three noded constant stress triangular flat element has only six local displacement DOF's that are shown in Figure 9.1. The out-of-plane contribution (the normal stiffness) of the membrane element to the basic local shell element is a displacement DOF in the direction normal to the plane of the element. This nonzero contribution was encountered earlier in prestressed trusses as the ability to carry load perpendicular to the bars.

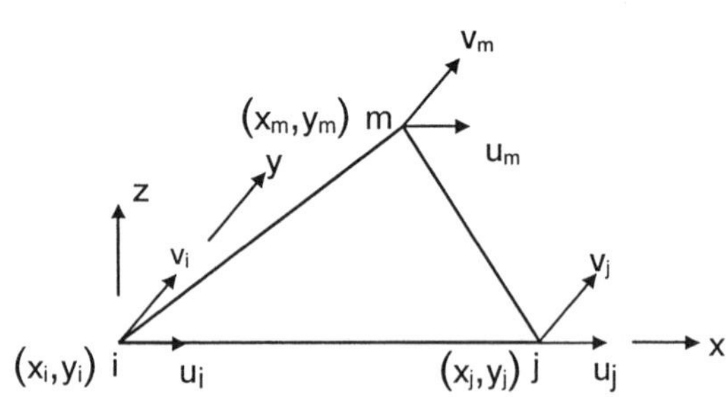

Figure 9.1 Triangular membrane element.

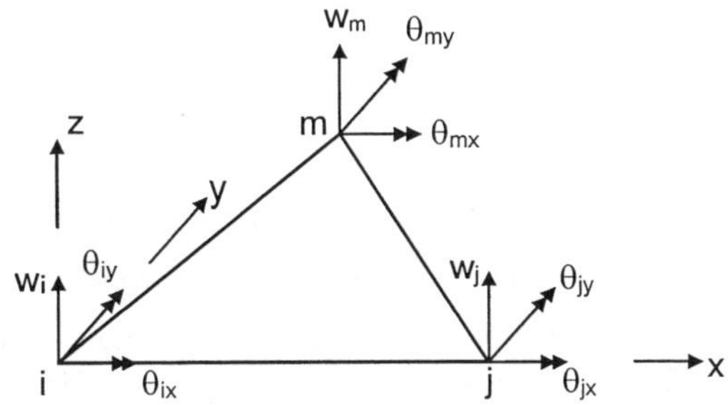

Figure 9.2 Triangular plate element.

The flexural element contributes to eighteen local DOF's. The basic three noded plate triangular flat element has only nine DOF's that are shown in Figure 9.2. The in-plane contribution to the basic local element adds two displacement DOF's in the plane of the element. The out-of-plane contribution adds a rotation DOF in the direction normal to the plane of the element.

9.2.1 In-Plane Contribution of the Triangular Membrane Element

The elastic stiffness matrix in the local coordinate system of the membrane element has the general form

$$\left[\mathbf{K}_{\mathrm{E}}^{e}\right]^{\mathrm{mem}} = \int_A \mathbf{B}_{\mathrm{mem}}^{\mathrm{T}} \mathbf{D} \mathbf{B}_{\mathrm{mem}} \, t dx dy \tag{9.2}$$

and is explicitly presented in Eqs. 6.17 and 6.18.

The local in-plane contribution of the membrane element to the geometric stiffness matrix is the gradient of the nodal force vector (Eq. 6.21)

$$\left[\mathbf{K}_{\mathrm{G}}^{e}\right]_{\mathrm{IP}}^{\mathrm{mem}} = \nabla \mathbf{F}_{\mathrm{mem}}^{\mathrm{e}} \tag{9.3}$$

and is explicitly given in Eqs.6.22 and 6.23.

9.2.2 In-Plane Contribution of the Triangular Plate Bending Element

The general form of the elastic stiffness matrix of the DKT triangular plate element may be written as:

$$\left[\mathbf{K}_{\mathrm{E}}^{e}\right]^{\mathrm{plate}} = 2A\int_{0}^{1}\int_{0}^{1-\eta} \mathbf{B}_{\mathrm{plate}}^{\mathrm{T}}\left(\frac{t^{3}}{12}\mathbf{D}\right)\mathbf{B}_{\mathrm{plate}}\,d\xi\,d\eta \tag{9.4}$$

where ξ and η are the usual area coordinates and $\mathbf{B}_{\mathrm{plate}}$, which appears explicitly in Batoz *et al.* (1980) as a function of ξ and η is slightly more complex than the $\mathbf{B}_{\mathrm{mem}}$ of the membrane. Here again the local in-plane contribution of the plate element to the geometric stiffness matrix will be derived as the gradient for fixed **M**, of the element nodal force vector which is given as:

$$\begin{aligned}\mathbf{F}_{\mathrm{plate}}^{e} &= \left(2A\int_{0}^{1}\int_{0}^{1-\eta}\frac{t^{3}}{12}\mathbf{B}_{\mathrm{plate}}^{\mathrm{T}}\mathbf{D}\mathbf{B}_{\mathrm{plate}}\,d\xi\,d\eta\right)\mathbf{q}^{e} \\ &= 2A\int_{0}^{1}\int_{0}^{1-\eta}\mathbf{B}_{\mathrm{plate}}^{\mathrm{T}}\left(\frac{t^{3}}{12}\mathbf{D}\mathbf{B}_{\mathrm{plate}}\mathbf{q}^{e}\right)d\xi\,d\eta = 2A\int_{0}^{1}\int_{0}^{1-\eta}\mathbf{B}_{\mathrm{plate}}^{\mathrm{T}}\mathbf{M}\,d\xi\,d\eta\end{aligned} \tag{9.5}$$

where

$$\mathbf{M} = \left\{M_{xx}, M_{yy}, M_{xy}\right\}^{\mathrm{T}} \tag{9.6}$$

$$\mathbf{F}_{\mathrm{plate}}^{e} = \left\{\mathbf{F}_{i}^{e}, \mathbf{F}_{j}^{e}, \mathbf{F}_{m}^{e}\right\}^{\mathrm{T}};\ \mathbf{F}_{r}^{e} = \left\{F_{rz}, M_{rx}, M_{ry}\right\}^{\mathrm{T}} \quad r = i, j, m \tag{9.7}$$

$$\mathbf{q}_{e} = \left\{w_{1}\quad \theta_{x1}\quad \theta_{y1}\quad w_{2}\quad \theta_{x2}\quad \theta_{y2}\quad w_{3}\quad \theta_{x3}\quad \theta_{y3}\right\} \tag{9.8}$$

The in-plane geometric stiffness matrix may be written symbolically as:

$$\left[\mathbf{K}_{\mathrm{G}}^{e}\right]_{\mathrm{IP}}^{\mathrm{plate}} = \nabla\mathbf{F}_{\mathrm{plate}}^{e} = \begin{bmatrix} (\mathbf{A}_{ii})_{\mathrm{IP}}^{\mathrm{plate}} & (\mathbf{A}_{ij})_{\mathrm{IP}}^{\mathrm{plate}} & (\mathbf{A}_{im})_{\mathrm{IP}}^{\mathrm{plate}} \\ (\mathbf{A}_{ji})_{\mathrm{IP}}^{\mathrm{plate}} & (\mathbf{A}_{jj})_{\mathrm{IP}}^{\mathrm{plate}} & (\mathbf{A}_{jm})_{\mathrm{IP}}^{\mathrm{plate}} \\ (\mathbf{A}_{mi})_{\mathrm{IP}}^{\mathrm{plate}} & (\mathbf{A}_{mj})_{\mathrm{IP}}^{\mathrm{plate}} & (\mathbf{A}_{mm})_{\mathrm{IP}}^{\mathrm{plate}} \end{bmatrix} \tag{9.9}$$

where

$$\left(\mathbf{A}_{rs}\right)_{\mathrm{IP}}^{\mathrm{plate}} = \begin{bmatrix} \dfrac{\partial F_{rz}}{\partial x_s} & \dfrac{\partial F_{rz}}{\partial y_s} & 0 \\ \dfrac{\partial M_{rx}}{\partial x_s} & \dfrac{\partial M_{rx}}{\partial y_s} & 0 \\ \dfrac{\partial M_{ry}}{\partial x_s} & \dfrac{\partial M_{ry}}{\partial y_s} & 0 \end{bmatrix} \tag{9.10}$$

The expressions for the individual terms of the geometric stiffness matrix were obtained in closed form using symbolic algebra.

The above follows directly from the discussion, in Chapter 3 on the derivation of the geometric stiffness matrix for trusses. When applied to the plate, chain rule differentiation with respect to the coordinates yields

$$\begin{aligned} \mathbf{dF}_{\mathrm{plate}}^{e} = & \sum_{r=i,j,m} \left[\frac{\partial}{\partial x_r} \left(2A \int_0^1 \int_0^{1-\eta} \left(\mathbf{B}_{\mathrm{plate}}^{\mathrm{T}}\right)_{\mathrm{fixed}} \mathbf{M} \mathrm{d}\xi \mathrm{d}\eta \right) \mathrm{d}x_r \right. \\ & \left. + \frac{\partial}{\partial y_r} \left(2A \int_0^1 \int_0^{1-\eta} \left(\mathbf{B}_{\mathrm{plate}}^{\mathrm{T}}\right)_{\mathrm{fixed}} \mathbf{M} \mathrm{d}\xi \mathrm{d}\eta \right) \mathrm{d}y_r \right] \\ & + \sum_{r=i,j,m} \left[\frac{\partial}{\partial x_r} \left(2A \int_0^1 \int_0^{1-\eta} \mathbf{B}_{\mathrm{plate}}^{\mathrm{T}} \mathbf{M}_{\mathrm{fixed}} \mathrm{d}\xi \mathrm{d}\eta \right) \mathrm{d}x_r \right. \\ & \left. + \frac{\partial}{\partial y_r} \left(2A \int_0^1 \int_0^{1-\eta} \mathbf{B}_{\mathrm{plate}}^{\mathrm{T}} \mathbf{M}_{\mathrm{fixed}} \mathrm{d}\xi \mathrm{d}\eta \right) \mathrm{d}y_r \right] \end{aligned} \tag{9.11}$$

where $\mathrm{d}x_r = u_r; \quad \mathrm{d}y_r = v_r; \quad r = i, j, m$.

The first expression on the R.H.S. of Eq. 9.11 returns the elastic stiffness matrix whereas the second expression becomes the geometric stiffness matrix. It is left to clarify what is meant by $\mathbf{M}_{\mathrm{fixed}}$ in Eq. 9.11. The moment vector **M**, of Eq. 9.6 contains three components that are functions of ξ and η and defined as:

$$M_{xx}(\xi,\eta) = \left(M_{xx}^{j} - M_{xx}^{i}\right)\xi + \left(M_{xx}^{m} - M_{xx}^{i}\right)\eta + M_{xx}^{i} \tag{9.12}$$

$$M_{yy}(\xi,\eta) = \left(M_{yy}^{j} - M_{yy}^{i}\right)\xi + \left(M_{yy}^{m} - M_{yy}^{i}\right)\eta + M_{yy}^{i} \tag{9.13}$$

$$M_{xy}(\xi,\eta) = \left(M_{xy}^{j} - M_{xy}^{i}\right)\xi + \left(M_{xy}^{m} - M_{xy}^{i}\right)\eta + M_{xy}^{i} \qquad (9.14$$

where

$$\{M_{xx}^{i} \quad M_{yy}^{i} \quad M_{xy}^{i}\}^{\mathrm{T}} = \mathbf{M}(\xi = 0, \eta = 0) \tag{9.15}$$

$$\{M_{xx}^{j} \quad M_{yy}^{j} \quad M_{xy}^{j}\}^{\mathrm{T}} = \mathbf{M}(\xi = 1, \eta = 0) \tag{9.16}$$

$$\{M_{xx}^{m} \quad M_{yy}^{m} \quad M_{xy}^{m}\}^{\mathrm{T}} = \mathbf{M}(\xi = 0, \eta = 1) \tag{9.17}$$

When the components of **M** (Eqs. 9.12-9.14) are inserted into Eq. 9.11 it is the values of these components at the nodes (Eqs. 9.15-9.17) that are held fixed.

9.2.3 Out-of-Plane Contribution to the Shell Geometric Stiffness Matrix

The out-of-plane stiffness matrix for the membrane was discussed in section 6.3. This section will derive the out-of-plane geometric stiffness matrix for the plate along the same lines and subsequently present a combined out-of-plane contribution. The derivation starts with the change in a vector G, due to a small rotation that is given by Goldstein (1950) as

$$\mathbf{dG} = \boldsymbol{\omega} \times \mathbf{G} \tag{9.18}$$

where $\boldsymbol{\omega}$ is the rigid body rotation vector due to changes in the geometry. In terms of joint displacements with respect to the local coordinate system, two components θ and $-\phi$, of the rigid body rotation are obtained from Figure 6.2. The third component is chosen for the plate arbitrarily, as the local z-rotation of node i. Recall that for the membrane this component is included in the in-plane contribution.

$$\begin{aligned}
\omega_{x'} &= \frac{c-e}{ea}(\boldsymbol{\delta}_i)_{z'} - \frac{c}{ea}(\boldsymbol{\delta}_j)_{z'} + \frac{1}{a}(\boldsymbol{\delta}_m)_{z'} \\
\omega_{y'} &= \frac{a}{ea}(\boldsymbol{\delta}_i)_{z'} - \frac{a}{ea}(\boldsymbol{\delta}_j)_{z'} \\
\omega_{z'} &= \theta_{iz}
\end{aligned} \tag{9.19}$$

where $(\boldsymbol{\delta}_r)_{z'}$ is the displacement in the local (primed) z–direction of node r.

At each node the forces, moments and rotations may be written as

$$\mathbf{F}_r = F_{rz}\mathbf{k}' ;\ \mathbf{M}_r = M_{rx}\mathbf{i}' + M_{ry}\mathbf{j}' + M_{rz}\mathbf{k}' ;\ \boldsymbol{\omega} = \omega_{x'}\mathbf{i}' + \omega_{y'}\mathbf{j}' + \omega_{z'}\mathbf{k}' ;$$

for $r = i, j, m$ and the changes in the force and moment vectors are given in more detail as

$$\mathrm{d}\mathbf{F}_r = -\mathbf{F}_r \times \boldsymbol{\omega} = \begin{bmatrix} 0 & F_{rz} & 0 \\ -F_{rz} & 0 & 0 \\ 0 & 0 & 0 \end{bmatrix} \begin{Bmatrix} \omega_{x'} \\ \omega_{y'} \\ \omega_{z'} \end{Bmatrix} \tag{9.20}$$

$$\mathrm{d}\mathbf{M}_r = -\mathbf{M}_r \times \boldsymbol{\omega} = \begin{bmatrix} 0 & M_{rz} & -M_{ry} \\ -M_{rz} & 0 & M_{rx} \\ M_{ry} & -M_{rx} & 0 \end{bmatrix} \begin{Bmatrix} \omega_{x'} \\ \omega_{y'} \\ \omega_{z'} \end{Bmatrix} \tag{9.21}$$

The out-of-plane stiffness contribution of the plate is now obtained as:

$$(\mathrm{d}\mathbf{F}^e)_{\mathrm{plate}} = \left[\mathbf{K}_{\mathrm{G}}\right]_{\mathrm{OP}}^{\mathrm{plate}} \cdot \boldsymbol{\delta}^e = -\mathbf{F}^e \times \boldsymbol{\omega} = \begin{bmatrix} \mathbf{F}^i \\ \mathbf{F}^j \\ \mathbf{F}^m \end{bmatrix} \left[\mathbf{A}_i \quad \mathbf{A}_j \quad \mathbf{A}_m\right] \begin{Bmatrix} \mathbf{d}_i \\ \mathbf{d}_j \\ \mathbf{d}_m \end{Bmatrix} \tag{9.22}$$

where

$$\mathbf{F}^e = \begin{Bmatrix} \mathbf{F}_i \\ \mathbf{M}_i \\ \mathbf{F}_j \\ \mathbf{M}_j \\ \mathbf{F}_m \\ \mathbf{M}_m \end{Bmatrix} ;\ \mathbf{F}^r = \begin{bmatrix} 0 & F_{rz} & 0 \\ -F_{rz} & 0 & 0 \\ 0 & 0 & 0 \\ 0 & M_{rz} & -M_{ry} \\ -M_{rz} & 0 & M_{rx} \\ M_{ry} & -M_{rx} & 0 \end{bmatrix} \tag{9.23}$$

Now $\boldsymbol{\omega}$ can be described in terms of the displacement vector as:

$$\boldsymbol{\omega} = \mathbf{A}\boldsymbol{\delta}^e = \left[\mathbf{A}_i \quad \mathbf{A}_j \quad \mathbf{A}_m\right] \begin{Bmatrix} \mathbf{d}_i \\ \mathbf{d}_j \\ \mathbf{d}_m \end{Bmatrix} \tag{9.24}$$

where

$$(\mathbf{A}_i)=\begin{bmatrix} 0 & 0 & -\dfrac{e-c}{ae} & 0 & 0 & 0 \\ 0 & 0 & 1/e & 0 & 0 & 0 \\ 0 & 0 & 0 & 0 & 0 & 1 \end{bmatrix} \tag{9.25}$$

$$(\mathbf{A}_j)=\begin{bmatrix} 0 & 0 & -\dfrac{c}{ae} & 0 & 0 & 0 \\ 0 & 0 & -1/e & 0 & 0 & 0 \\ 0 & 0 & 0 & 0 & 0 & 0 \end{bmatrix} \tag{9.26}$$

$$(\mathbf{A}_m)=\begin{bmatrix} 0 & 0 & 1/a & 0 & 0 & 0 \\ 0 & 0 & 0 & 0 & 0 & 0 \\ 0 & 0 & 0 & 0 & 0 & 0 \end{bmatrix} \tag{9.27}$$

and

$$\{\mathbf{d}_i\}=\begin{Bmatrix} (\delta_i)_{x'} \\ (\delta_i)_{y'} \\ (\delta_i)_{z'} \\ (\theta_i)_{x'} \\ (\theta_i)_{y'} \\ (\theta_i)_{z'} \end{Bmatrix};\ \{\mathbf{d}_j\}=\begin{Bmatrix} (\delta_j)_{x'} \\ (\delta_j)_{y'} \\ (\delta_j)_{z'} \\ (\theta_j)_{x'} \\ (\theta_j)_{y'} \\ (\theta_j)_{z'} \end{Bmatrix};\ \{\mathbf{d}_m\}=\begin{Bmatrix} (\delta_m)_{x'} \\ (\delta_m)_{y'} \\ (\delta_m)_{z'} \\ (\theta_m)_{x'} \\ (\theta_m)_{y'} \\ (\theta_m)_{z'} \end{Bmatrix} \tag{9.28}$$

Carrying out Eq. 9.22 for the plate and adding Eq. 6.38 (adjusted to 18 DOF's) for the membrane, results in the following out-of-plane shell geometric stiffness matrix with respect to the local coordinate system:

$$\left[\mathbf{K}_{\mathrm{G}}^{e}\right]_{\mathrm{OP}}^{\mathrm{shell}}=\begin{bmatrix} (\mathbf{A}_i)_1 & (\mathbf{A}_i)_2 & (\mathbf{A}_i)_3 \\ (\mathbf{A}_j)_1 & (\mathbf{A}_j)_2 & (\mathbf{A}_j)_3 \\ (\mathbf{A}_m)_1 & (\mathbf{A}_m)_2 & (\mathbf{A}_m)_3 \end{bmatrix} \tag{9.29}$$

where

$$(\mathbf{A}_r)_1 = \begin{bmatrix} 0 & 0 & a_r & 0 & 0 & 0 \\ 0 & 0 & b_r & 0 & 0 & 0 \\ 0 & 0 & \alpha_r & 0 & 0 & 0 \\ 0 & 0 & c_r & 0 & 0 & -M_{ry} \\ 0 & 0 & d_r & 0 & 0 & M_{rx} \\ 0 & 0 & e_r & 0 & 0 & 0 \end{bmatrix} \tag{9.30}$$

$$(\mathbf{A}_r)_2 = \begin{bmatrix} 0 & -0 & f_r & 0 & 0 & 0 \\ 0 & -0 & g_r & 0 & 0 & 0 \\ 0 & 0 & \beta_r & 0 & 0 & 0 \\ 0 & -0 & h_r & 0 & 0 & 0 \\ 0 & -0 & i_r & 0 & 0 & 0 \\ 0 & 0 & j_r & 0 & 0 & 0 \end{bmatrix} \tag{9.31}$$

$$(\mathbf{A}_r)_3 = \begin{bmatrix} 0 & 0 & 0 & 0 & 0 & 0 \\ 0 & 0 & k_r & 0 & 0 & 0 \\ 0 & 0 & \lambda_r & 0 & 0 & 0 \\ 0 & 0 & 0 & 0 & 0 & 0 \\ 0 & 0 & l_r & 0 & 0 & 0 \\ 0 & 0 & m_r & 0 & 0 & 0 \end{bmatrix} \tag{9.32}$$

$$\alpha_r = -(\mathbf{F}_r)_{y'} \frac{e-c}{ae} - \frac{(\mathbf{F}_r)_{x'}}{e};\ \beta_r = -(\mathbf{F}_r)_{y'} \frac{c}{ae} + \frac{(\mathbf{F}_r)_{x'}}{e};\ \lambda_r = \frac{(\mathbf{F}_r)_y}{a}; a_r = \frac{F_{rz}}{e};$$

$$b_r = F_{rz} \frac{e-c}{ae}; c_r = \frac{M_{rz}}{e}; d_r = M_{rz} \frac{e-c}{ae}; e_r = -M_{ry} \frac{e-c}{ae} - \frac{M_{rx}}{e};$$

$$f_r = -\frac{F_{rz}}{e};\ g_r = F_{rz} \frac{c}{ae};\ h_r = -\frac{M_{rz}}{e};\ i_r = M_{rz} \frac{c}{ae};\ j_r = -M_{ry} \frac{c}{ae} + \frac{M_{rx}}{e};$$

$$k_r = -\frac{F_{rz}}{a}; l_r = -\frac{M_{rz}}{a}; m_r = \frac{M_{ry}}{a};.$$

9.3 Element Pure Deformational Rotations and Translations

For a given set of nodal displacements it is necessary to isolate the pure elastic deformations from the rigid body ones so as to proceed with the determination of the internal stresses and the evaluation of the element nodal forces. The internal stresses are needed to construct the geometric stiffness matrix and the nodal forces are used to calculate the unbalanced load as the analysis progresses. It is the local shell element pure translations and rotations that are of interest and will be discussed below. The translations are related to the membrane element whereas the rotations to the plate element. Internal stresses follow from the linear constitutive relations: 1) $\boldsymbol{\sigma} = \mathbf{D}\boldsymbol{\varepsilon}$ for the membrane, where $\boldsymbol{\sigma}$ comprises the stress components σ_x, σ_y and τ, and $\boldsymbol{\varepsilon}$ comprises the strain components ε_x, ε_y and γ_{xy} respectively and 2) $\mathbf{M} = \dfrac{t^3}{12}\mathbf{D}\mathbf{B}_{\text{plate}}\mathbf{q}_{\text{e}}$ for the plate, where $\mathbf{q}_{\text{e}}$ is the pure local deformation vector that includes one lateral displacement and two in-plane rotations at each node. Since lateral displacements are trivially zero in the local coordinates system only rotations are discussed.

9.3.1 Stress Retrieval in the Membrane Finite Element

Figure 9.3 describes the membrane element exact geometric size in its initial and current configurations. The membrane strains are obtained directly as:

$$\varepsilon_x = \frac{\mathrm{AC'} - \mathrm{AC}}{\mathrm{AC}} \tag{9.33}$$

$$\varepsilon_y = \frac{\mathrm{B'D'} - \mathrm{BD}}{\mathrm{BD}} \tag{9.34}$$

$$\gamma_{xy} = \angle\mathrm{DBD'} - \angle\mathrm{DBD''} = \frac{\mathrm{AD'} - \mathrm{AD}}{\mathrm{BD}} - \frac{(\mathrm{AC'} - \mathrm{AC})\dfrac{\mathrm{AD}}{\mathrm{AC}}}{\mathrm{BD}} \tag{9.35}$$

The second term of Eq. 9.33 is argued as follows. Each point along AC will be stretched by a fraction of the total stretch CC′. That fraction is determined by the relative position of the point along AC. Point D stretches to DD″ and therefore the length DD″ will be equal to CC′ multiplied by the fraction AD/AC.

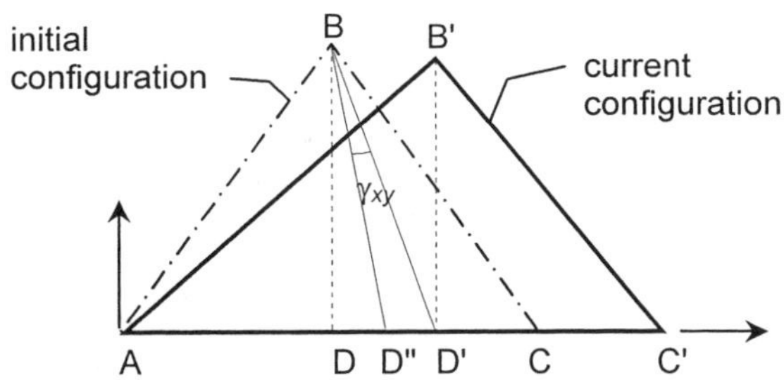

Figure 9.3 Membrane pure strains of the shell element.

9.3.2 Stress Retrieval in the Plate Finite Element

Figure 9.4 describes two positions of a given element at two consecutive iterations k and $k+1$ with normal unit vectors n and n′ respectively. Moreover the unit vectors n_i, n_j, n_m, which are normal to tangent planes at the nodes of the elastically deformed element define the pure nodal rotations relative to n′. Note that the rigid body motion of the element is not fully determined by the unit vector n′ since rotation about the element plane is still possible given n'.

At any given iteration, analysis will provide for each node, three rotation components θ_X, θ_Y, θ_Z (in addition to three displacements). These rotations are given with respect to the global coordinate system. Euler's theorem implies that any finite rotation can be described as a simple rotation of magnitude α about some fixed axis described by a unit vector n*. Let

$$\boldsymbol{\theta} = \mathbf{n}^* \alpha \tag{9.36}$$

describe this rotation. Given the rotation components: $\boldsymbol{\theta} = \theta_X \mathbf{i} + \theta_Y \mathbf{j} + \theta_Z \mathbf{k}$ and $\alpha = \sqrt{\theta_X^2 + \theta_Y^2 + \theta_Z^2}$ and the unit vector $\mathbf{n}^*$ is computed from $\mathbf{n}^* = (\theta_X \mathbf{i} + \theta_Y \mathbf{j} + \theta_Z \mathbf{k})/\alpha$. Here **i**, **j**, **k**, belong to the global coordinates X, Y, Z. Following Noble (1969), a rotation matrix **R** that transforms the components of a vector from global to local coordinates is related to Euler's rigid body motion through

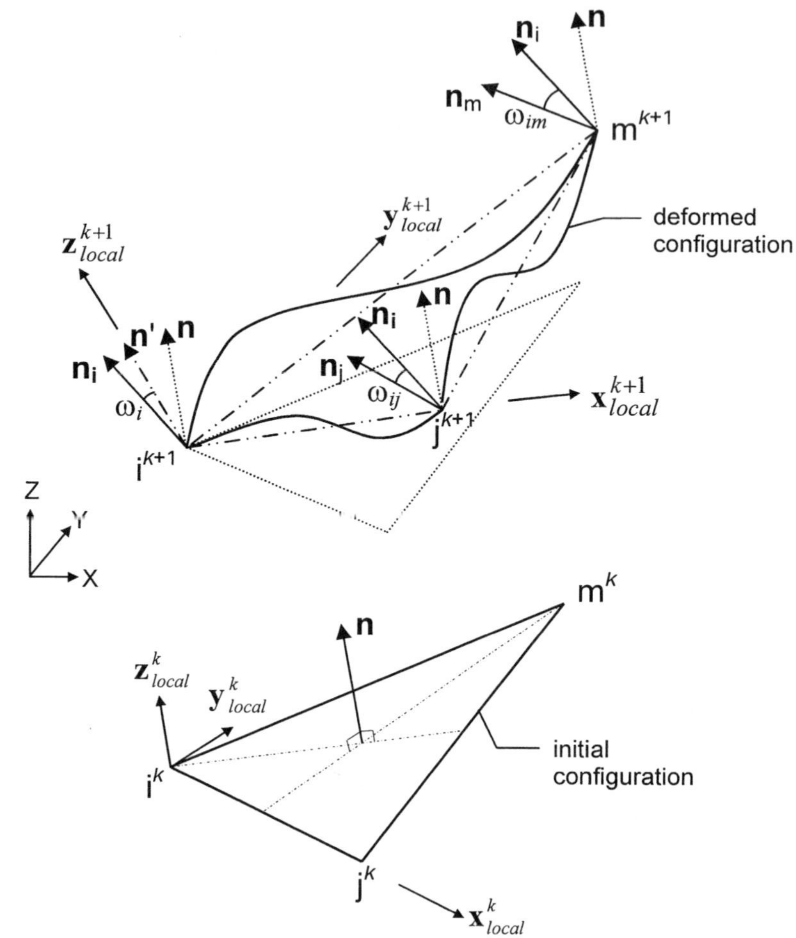

Figure 9.4 Pure flexural rotations of the shell element.

$$
\mathbf{R}^{\mathrm{T}} = \begin{bmatrix} \cos\alpha + c\overset{*}{n}_x^{2} & c\overset{*}{n}_x\overset{*}{n}_y - \overset{*}{n}_z\sin\alpha & c\overset{*}{n}_x\overset{*}{n}_z + \overset{*}{n}_y\sin\alpha \\ c\overset{*}{n}_x\overset{*}{n}_y + \overset{*}{n}_z\sin\alpha & \cos\alpha + c\overset{*}{n}_y^{2} & c\overset{*}{n}_z\overset{*}{n}_y - \overset{*}{n}_x\sin\alpha \\ c\overset{*}{n}_x\overset{*}{n}_z - \overset{*}{n}_y\sin\alpha & c\overset{*}{n}_z\overset{*}{n}_y + \overset{*}{n}_x\sin\alpha & \cos\alpha + c\overset{*}{n}_z^{2} \end{bmatrix} \tag{9.37}
$$

where $c = 1 - \cos\alpha$. Actually any rigid body motion may be described equivalently using either a rotation matrix $\mathbf{R}$ or the above representation which involves $\mathbf{n}^*$ and α. Let the small deformational rotation vector $\boldsymbol{\omega}_i$ (Figure 9.4) describe the rotation between the vector $\mathbf{n}'$ and $\mathbf{n}_i$ with the properties

$$\boldsymbol{\omega}_i \propto \mathbf{n}_i \times \mathbf{n}' ; \; \|\boldsymbol{\omega}_i\| = \cos^{-1}(\mathbf{n}' \cdot \mathbf{n}_i) \tag{9.38}$$

or equivalently

$$\boldsymbol{\omega}_i = \sin^{-1}(\|\mathbf{n}_i \times \mathbf{n}'\|) \cdot \frac{\mathbf{n}_i \times \mathbf{n}'}{\|\mathbf{n}_i \times \mathbf{n}'\|} \tag{9.39}$$

where $\boldsymbol{\omega}_i$ is the pure rotation at node i in global coordinates. The pure rotations at the other elemental nodes, j and m, may be computed as:

$$\boldsymbol{\omega}_r = \boldsymbol{\omega}_i + \boldsymbol{\omega}_{ri} \; ; \; r = j,m \tag{9.40}$$

Here $\boldsymbol{\omega}_{ri}$ describes the relative rotation of joint r with respect to joint i and is the vector form (Euler's rigid body motion) of the compound rotations $\mathbf{R}_j^{\mathrm{T}}\mathbf{R}_i$. Let this compound rotations matrix be called $\mathbf{R}_{ij}$ so that $\mathbf{R}_{ij} = \mathbf{R}_j^{\mathrm{T}}\mathbf{R}_i$. Then the unit vector describing the rigid body rotation (Eq. 9.34) is given (Noble, 1969) as:

$$\mathbf{n}^*_{ji} = \frac{\overline{\boldsymbol{\omega}}_{ji}}{\|\overline{\boldsymbol{\omega}}_{ji}\|} \tag{9.41}$$

where

$$\overline{\boldsymbol{\omega}}_{ji} = \left(R^{\mathrm{T}}_{ij,23} - R^{\mathrm{T}}_{ij,32}\right)\mathbf{i} + \left(R^{\mathrm{T}}_{ij,31} - R^{\mathrm{T}}_{ij,13}\right)\mathbf{j} + \left(R^{\mathrm{T}}_{ij,21} - R^{\mathrm{T}}_{ij,12}\right)\mathbf{k} \tag{9.42}$$

and the magnitude of rotation as:

$$\alpha_{ji} = \cos^{-1}\left(\frac{R^{\mathrm{T}}_{ij,11} + R^{\mathrm{T}}_{ij,22} + R^{\mathrm{T}}_{ij,33} - 1}{2}\right) \tag{9.43}$$

Finally the relative rotation vector becomes

$$\boldsymbol{\omega}_{ji} = \mathbf{n}^*_{ji} \, \alpha_{ji} \tag{9.44}$$

These pure rotations vector has components in the global coordinate system. In order to complete the computation of the vector $\mathbf{q}_e$ it is necessary to transform this vector to the present local coordinate system.

9.4 Computer Program P16-SHELLNL.FOR

This program performs nonlinear analysis of thin shells in one of two distinct ways – force control (constant force increments) or generalized displacement control (the cylindrical arc-length method). The total stiffness matrix `C(N,N)` is assembled at every iteration from the individual contributions which are explicit. Subroutine `RASM` constructs this stiffness matrix, `C(N,N)`, (size 18×18) in the global coordinates system given the five individual stiffness matrices in local coordinates. This five stiffness matrices are: the linear elastic stiffness matrix of the membrane element `PLSTR`; the linear elastic stiffness matrix of the shell, element `KEPLATE1+KEPLATE2`; the in-plane geometric stiffness matrix of the membrane element `MGSTIF`; the in-plane geometric stiffness matrix of the shell element `KGPLATE` and `KGOUTPLANE`, the out of plane geometric stiffness matrix for both the membrane and shell elements.

The unbalanced load vector is stored in in the array `P(N)` and its norm stored in `ERROR`. Convergence at any iteration n is achieved when $\mathrm{ERROR}^{n+1} < 10^{-5} * \mathrm{ERROR}^{n}$. The system of equations is solved using `SOLV`. Membrane stresses are calculated in `UPDATE`. For element number `L` the membrane stresses σ_x, σ_y, and τ_{xy} are stored in the 1st, 2nd and 3rd columns of `PSIG` respectively. The internal moments for the plate element are calculated in `UPDATE1` and stored in `PSIG` as follows: The moments M_{xx}, M_{yy} and M_{xy} of node i are stored in the 4th, 5th and 6th columns respectively and those of nodes j and k are stored in the 7th, 8th, 9th, 10th, 11th and 12th columns respectively. `UPDATE1` uses the pure deformation vector in local coordinates `(DP(9))` that is obtained in `RESET` for `ISTRESS=0` according to the method of section 9.3.2 and linear stress evaluation (section 9.3). The option `ISTRESS=1` works with full nodal deformations rather than pure deformations. These (array `DS`) are transformed to element coordinates using `TRANS` and stored in array `DS1`. `PSDP` then extracts `DP` for linear stress evaluation. One must be aware that this option is load step dependent for convergence and in many cases, having converged, might result in a wrong answer.

As for the input the program uses formatted read statements. General data is read form an input file using:

```
      READ(15,150)
NN,NS,NCONS,NIT,NFE,NSTEP,E,ANU,IARC,ID,ISTRESS
  150 FORMAT(6I5,2D10.2/,3I5)
```

where `NN, NS, NCONS, NIT, NFE` and `NSTEP` are the number of nodes, the number of fixed supports, the number of additional constraints, the number of iterations, the number of finite elements and the number of load steps respectively. `E` is Young's modulus, `ANU` is Poisson's ratio, `IARC` is the index that will determine force control analysis (`IARC=0`) or displacement controlled analysis (`IARC=1`), `ID` is the number of iterations to be used for the arc-length method and `ISTRESS` will determine which of two methods are to be used for determining internal moments.

Data pertaining to nodes is read using:

```
      READ(15,156)(R(3*K-2),R(3*K-
1),R(3*K),PSAVE(6*K-5),PSAVE(6*K-4)
     1 ,PSAVE(6*K-3),PSAVE(6*K-2),PSAVE(6*K-
1),PSAVE(6*K),K=1,NN)
  156 FORMAT (9D10.2)
```

where the arrays `R` and `PSAVE` store nodal coordinates and external loads respectively.

Data pertaining to the finite elements is read using:

```
      IF(NFE.GT.0)
READ(15,1158)((IJM(L,I),I=1,3),TH(L),(PSIG(L,I)
     1 ,I=1,12),L=1,NFE)
 1158 FORMAT(5X,3I5,13D10.2)
```

Here `IJM` is the element connectivity matrix. For element number `L` the *i* node number is stored in the 1st column of `IJM` and the *j* and *m* node numbers are stored in the 2nd and 3rd columns respectively. `TH` is the element thickness and initial prestress is read into `PSIG`.

Finally it is left to the reader to study the main program and its subroutines before producing any meaningful results from the output that is generated. Care has to be taken with the data parameters and generated output when invoking the arc-length method that is coded in the subroutine `DISPLACEMENTCONTROL.FOR`.

9.5 Examples

This section describes the analysis of seven examples of diverse nature and shows the general usefulness of the shell element in identifying snap through buckling, reverse snap, bifurcation and large rotation elastica like problems of rigid nature. All the examples were run using the same SHELLNL.FOR. Each example was run as a project bearing its name. Input and output data files are identified by the example they represent.

9.5.1 Example 9.1. Bending of a Cantilever Plate

An initially flat cantilever plate (Figure 9.5) is subjected to a bending moment at its tip. The analytical solution to this classical elastica problem corresponds to a beam "rolled-up" into a circular arc of radius r and is given by $1/r = M/EI$ where M is the applied end moment. A complete circle is achieved at $M = 2\pi EI/L$ where L is the length of the cantilever. This example was run using twenty elements with the following physical and geometrical properties: Young's modulus E=1.2 $\times 10^6$ kN/m^2, Poisson's ratio of 0.0, L=10.0 m, b=1.0 m and h=0.1m.

SHELLNL91.EXE activates the datafile SHELLNL91.DAT to produce an outputfile SHELLNL91.OUT. Actually the program runs with a.dat and produces a.out. The files have been renamed accordingly.

Data file SHELLNL91.DAT

```
   22     2     0    10    20   100 120000.00           0.0
    0     0     1
10.000000  .000000 .000000        .00        .00        .00        0.3d1
10.000000 1.000000 .000000        .00        .00        .00        0.3d1
 9.000000  .000000 .000000        .00        .00        .00
 9.000000 1.000000 .000000        .00        .00        .00
 8.000000  .000000 .000000        .00        .00        .00
 8.000000 1.000000 .000000        .00        .00        .00
 7.000000  .000000 .000000        .00        .00        .00
 7.000000 1.000000 .000000        .00        .00        .00
 6.000000  .000000 .000000        .00        .00        .00
6.000000 1.000000 .000000        .00        .00        .00
 5.000000  .000000 .000000        .00        .00        .00
 5.000000 1.000000 .000000        .00        .00        .00
4.000000  .000000 .000000        .00        .00        .00
 4.000000 1.000000 .000000        .00        .00        .00
 3.000000  .000000 .000000        .00        .00        .00
 3.000000 1.000000 .000000        .00        .00        .00
 2.000000  .000000 .000000        .00        .00        .00
 2.000000 1.000000 .000000        .00        .00        .00
 1.000000  .000000 .000000        .00        .00        .00
 1.000000 1.000000 .000000        .00        .00        .00
 0.000000  .000000 .000000        .00        .00        .00
 0.000000 1.000000 .000000        .00        .00        .00
        1     2     4        0.1
        1     3     4        0.1
        3     4     6        0.1
        3     5     6        0.1
        5     6     8        0.1
        5     7     8        0.1
        7     8    10        0.1
        7     9    10        0.1
        9    10    12        0.1
        9    11    12        0.1
       11    12    14        0.1
       11    13    14        0.1
       13    14    16        0.1
       13    15    16        0.1
       15    16    18        0.1
       15    17    18        0.1
```

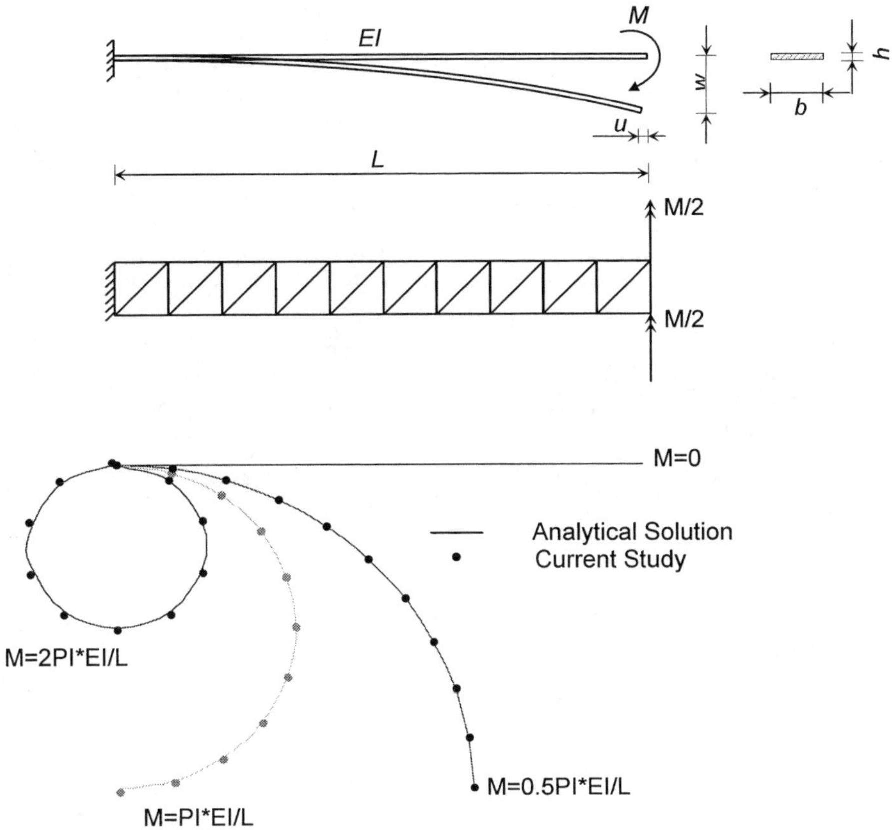

Figure 9.5 Bending of a cantilever plate.

```
17    18    20         0.1
17    19    20         0.1
19    20    22         0.1
19    21    22         0.1
```

Output file SHELLNL91.OUT

```
NUMBER OF NODES                        22
NUMBER OF SUPPORTS                      2
NUMBER OF CONSTRAINTS                   0
NUMBER OF ITERATIONS                   10
NUMBER OF FINITE ELEMENTS              20
NUMBER OF LOAD STEPS                  100

E=    120000.000000000
NU=   0.000000000000000E+000
IARC= 0       Id= 0     ISTRESS= 1

       COORDINATES                LOADS
```

```
          X       Y       Z        PX       PY       PZ       MX       MY       MZ
 1 10.000 0.0000 0.0000  0.0000  0.0000  0.0000  0.0000  0.0300  0.0000
 2 10.000 1.0000 0.0000  0.0000  0.0000  0.0000  0.0000  0.0300  0.0000
:

22 0.0000 1.0000 0.0000  0.0000  0.0000  0.0000  0.0000  0.0000  0.0000

FINITE ELEMENT DATA
NO.NODES TH  PRESTRESS
   I J M     SIGX SIGY TAU MXXI MYYI MXYI MXXJ MYYJ MXYJ MXXM MYYM MXYM

 1 1 2 4 0.1 0.00 0.00 .00 0.00 0.00 0.00 0.00 0.00 0.00 0.00 0.00 0.00
 2 1 3 4 0.1 0.00 0.00 .00 0.00 0.00 0.00 0.00 0.00 0.00 0.00 0.00 0.00
:

20192122 0.1 0.00 0.00 .00 0.00 0.00 0.00 0.00 0.00 0.00 0.00 0.00 0.00

****ITERATION NUMBER   1
     LOAD STEP          1
        COORDINATES           LOADS
          X       Y       Z        PX       PY       PZ       MX       MY       MZ
 1 10.000 0.0000 0.0000  0.0000  0.0000  0.0000  0.0000  0.0300  0.0000
 2 10.000 1.0000 0.0000  0.0000  0.0000  0.0000  0.0000  0.0300  0.0000
:

22 0.0000 1.0000 0.0000  0.0000  0.0000  0.0000  0.0000  0.0000  0.0000

  ERROR =         0.42426406D-01

DISPLACEMENTS
              X            Y            Z            THX          THY         THZ
 1 .00000D+00 .00000D+00-.29999D+00 -.27749D-14  .59999D-01  .00000D+00
 2 .00000D+00 .00000D+00-.29999D+00 -.26976D-14  .59999D-01  .000000+00
:

20 0.00000D+00 .00000D+00-.29999D-02-.36180D-15  .59999D-02  .00000D+00

 ****ITERATION NUMBER   2
     LOAD STEP          1

COORDINATES
              X            Y            Z
 1   0.10000D+02 0.00000D+00-0.30000D+00
 2   0.10000D+02 0.10000D+01-0.30000D+00
:

22    0.00000D+00 0.10000D+01 0.00000D+00

UNBALANCED LOADS
          PX           PY           PZ            MX           MY          MZ
 1-.97233D+01-.66613D-12 .55423D+00 -.37484D-15  .19360D-14  .10301D-15
 2-.97233D+01-.33199D-12 .55423D+00  .84504D-15  .20123D-15  .33434D-16
:

22 .00000D+00 .00000D+00 .00000D+00  .00000D+00  .00000D+00  .00000D+00

  ERROR =         0.14705609D+02

 ****ITERATION NUMBER   3
     LOAD STEP        100

COORDINATES
              X            Y            Z
 1 -0.48448D+00-0.95310D-12-0.64903D-01
 2 -0.48448D+00 0.10000D+01-0.64903D-01
```

```
⋮
22 0.00000D+00 0.10000D+01 0.00000D+00


UNBALANCED LOADS
            PX          PY          PZ           MX           MY           MZ

 1 .76428D-12-.22160D-12-.17364D-12  .17039D-11 -.13767D-13 -.25241D-11
 2-.15034D-11 .14142D-13-.15389D-12  .16934D-11  .44409D-15 -.25312D-11
⋮
22 .00000D+00 .00000D+00 .00000D+00  .00000D+00  .00000D+00  .00000D+00

  ERROR =         0.10658200D-10
```

9.5.2 Example 9.2. Simply Supported Plate

A square plate is subjected to a uniform pressure q as shown in Figure 9.6. One quarter of the plate was analyzed using eighteen elements. The response is compared in Figure 9.7 to the analytical solution reported by Levy (1942). This example was run using Young's modulus E=10^7 psi, Poisson's ratio $\nu = \sqrt{0.1}$ and h=0.12in and a=24in. Following is a truncated listing of SHELLNL92.DAT (full geometry) and SHELLNL92.OUT that have been invoked by SHELLNL92.EXE.

Data file SHELLNL92.DAT

```
   49     0    72    10    72    2010000000.0.31622777
    0     6     1
     .0000      .0000       .0000
     .0000     4.0000       .0000
```

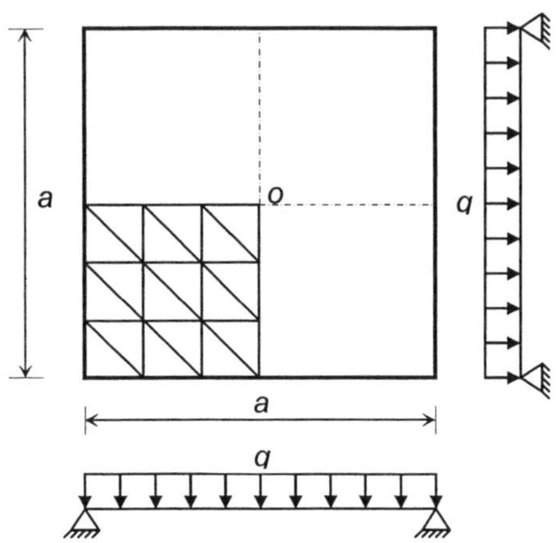

Figure 9.6 Simply supported uniformly loaded plate.

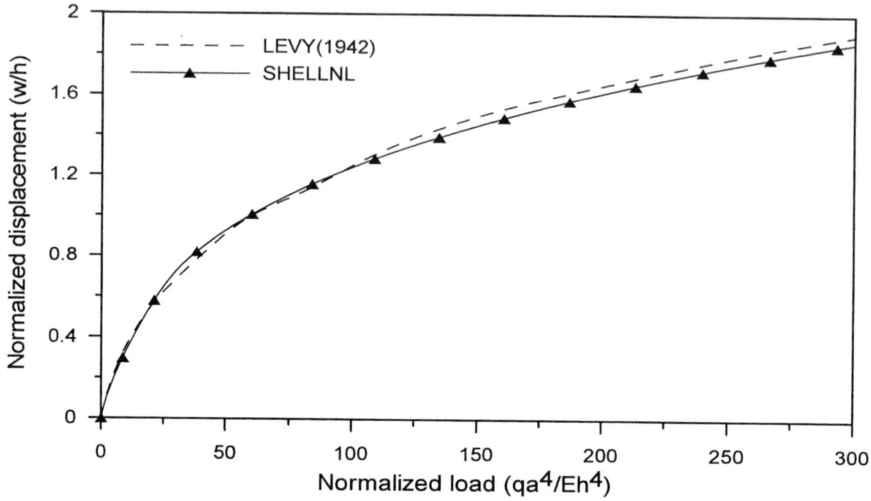

Figure 9.7 Displacement at Point O in Simply Supported Plate.

```
  .0000    8.0000     .0000
  .0000   12.0000     .0000
  .0000   16.0000     .0000
  .0000   20.0000     .0000
  .0000   24.0000     .0000
 4.0000     .0000     .0000
 4.0000    4.0000     .0000                         -30.0
 4.0000    8.0000     .0000                         -30.0
 4.0000   12.0000     .0000                         -30.0
 4.0000   16.0000     .0000                         -30.0
    :
20.0000   16.0000     .0000                         -30.0
20.0000   20.0000     .0000                         -30.0
20.0000   24.0000     .0000
24.0000     .0000     .0000
24.0000    4.0000     .0000
24.0000    8.0000     .0000
24.0000   12.0000     .0000
24.0000   16.0000     .0000
24.0000   20.0000     .0000
24.0000   24.0000     .0000

 1    1
 1    2
 1    3
 2    1
 2    2
 2    3
    :
48    1
48    2
48    3
49    1
49    2
49    3
      1    2    8     .1200
      2    3    9     .1200
```

```
        3    4   10      .1200
        :
       46   47   40      .1200
       47   48   41      .1200
       48   49   42      .1200
```

Data file SHELLNL92.OUT

```
 NUMBER OF NODES                    49
 NUMBER OF SUPPORTS                  0
 NUMBER OF CONSTRAINTS              72
 NUMBER OF ITERATIONS               10
 NUMBER OF FINITE ELEMENTS          72
 NUMBER OF LOAD STEPS               20
 E=   10000000.0000000
 NU=  0.316227770000000
 IARC=          0       Id=           6    ISTRESS=          1

       COORDINATES            LOADS
       X       Y       Z       PX       PY       PZ       MX       MY       MZ
 1 .00000 .00000 .00000  .00000  .00000  .00000  .00000  .03000  .00000
 2 .00000 4.0000 .00000  .00000  .00000  .00000  .00000  .03000  .00000
:
49 24.000 24.000 .00000  .00000  .00000  .00000  .00000  .00000  .00000

CONSTRAINTS DATA
NO.  NODE NO DOF NO
  1        1      1
  2        1      2
  3        1      3
:
72       49      3

FINITE ELEMENT DATA
NO.NODES TH PRESTRESS
   I J M     SIGX SIGY TAU MXXI MYYI MXYI MXXJ MYYJ MXYJ MXXM MYYM MXYM
 1 1 2 8 .12 0.00 0.00 .00 0.00 0.00 0.00 0.00 0.00 0.00 0.00 0.00 0.00
 2 2 3 9 .12 0.00 0.00 .00 0.00 0.00 0.00 0.00 0.00 0.00 0.00 0.00 0.00
:
72484942 .12 0.00 0.00 .00 0.00 0.00 0.00 0.00 0.00 0.00 0.00 0.00 0.00

 ****ITERATION NUMBER   1
     LOAD STEP          1

        COORDINATES           LOADS
       X       Y       Z       PX       PY       PZ       MX       MY       MZ
 1 .00000 .00000 .00000  .00000  .00000  .00000  .00000  .03000  .00000
 2 .00000 4.0000 .00000  .00000  .00000  .00000  .00000  .03000  .00000
:
49 24.000 24.000 .00000  .00000  .00000  .00000  .00000  .00000  .00000

  ERROR =        0.75000001D+01

DISPLACEMENTS
           X           Y           Z          THX          THY          THZ
 1 .00000D+00 .00000D+00 .00000D+00 -.15193D-04  .15193D-04  .00000D+00
 2 .00000D+00 .00000D+00 .00000D+00  .94171D-04  .53708D-02  .000000+00
:
49 .00000D+00 .00000D+00-.29999D-02  .15193D-04 -.15193D-04  .00000D+00
```

```
****ITERATION NUMBER   2
     LOAD STEP           1

COORDINATES
               X            Y            Z
  1 0.00000D+00 0.00000D+00 0.00000D+00
  2 0.00000D+00 0.40000D+01 0.00000D+00
:
49 0.24000D+02 0.24000D+02 0.00000D+00

UNBALANCED LOADS
          PX          PY          PZ          MX          MY          MZ
 1 .00000D+00 .00000D+00 .27713D+01 -.97145D-15 -.13878D-16  .00000D+00
 2 .82918D+02 .35790D+02-.23393D+01 -.16772D-03 -.16952D-05  .23976D-03
:
49 .00000D+00 .00000D+00 .27713D+01  .26645D-14  .14988D-14  .00000D+00
:
  ERROR =          0.62699235D+03

 ****ITERATION NUMBER  10
     LOAD STEP          20

COORDINATES
               X            Y            Z

  1 0.00000D+00 0.00000D+00 0.00000D+00
  2 0.00000D+00 0.40000D+01 0.00000D+00
:
25 0.12000D+02 0.12000D+02-0.22360D+00
:
49 0.24000D+02 0.24000D+02 0.00000D+00

UNBALANCED LOADS
          PX          PY          PZ          MX          MY          MZ

 1 .00000D+00 .00000D+00 .99917D+01  .11102D-15  .88818D-15  .00000D+00
 2 .47620D+03 .15675D+03-.22866D+02  .19789D-07  .39397D-07  .30931D-06
:
25 .11033D-08 .40631D-09 .76922D-10  .18616D-10 -.45734D-13  .21851D-09
:
49 .00000D+00 .00000D+00 .99917D+01  .17764D-14  .55511D-15  .00000D+00
  ERROR =          0.80893953D-03
```

9.5.3 Example 9.3. Analysis of a Shallow Cylindrical Shell

A cylindrical shell of dimensions L=254mm and h=12.7mm (Figure 9.8) that is pinned along its longitudinal edges and free at its curved edges is loaded at its center point. Analysis is performed for two cases of thickness: t=12.7mm and t=6.35mm. Young's modulus was taken as E=3.10275 kN/mm^2 and Poisson's ratio as 0.3. Using symmetry one quarter of the shell was analyzed with a 10×10 grid. Figure 9.9 shows a comparison to the results reported by

Hsiao (1987) with the phenomenon of snap-back clearly seen for the case of t=6.35mm. Following is a truncated listing of SHELLNL93.DAT (for the case t=12.7mm) that have been invoked by SHELLNL93.EXE. Only highlights of the enormous output file SHELLNL93.OUT are presented. It starts with the usual echo of the data and then goes through the iterative process. Following loadstep 1 note that at "`****ITERATION NUMBER 3`" the error was "`ERROR = 0.21105481D-03`". Error convergence criteria are met at iteration 4. Therefore, iterations cease and the next load step takes over. Two more parameters that relate to the arc-length method are printed. These are the length squared "`DL2= 29.3281303301849`" and the load factor "`FACT= 9.314575418727726E-002`". From `FACT` and the updated coordinates at this stage one point on each curve of Fig. 9.8 (t=12.7) has been generated.(In this case by looking at nodes `1` and `111`.) As the analysis proceeds a "`NEGATIVE TERM ON THE DIAGONAL AT 724`" is encountered. This is expected since critical points are bypassed by the arc-length method (loadsteps 5 to 7).

Data file SHELLNL93.DAT

```
121     0    97    10   200    10    3.10275          0.3
  1     6     1
    .0000      .0000     12.6895
    .0000    25.3996     12.5625
    :
```

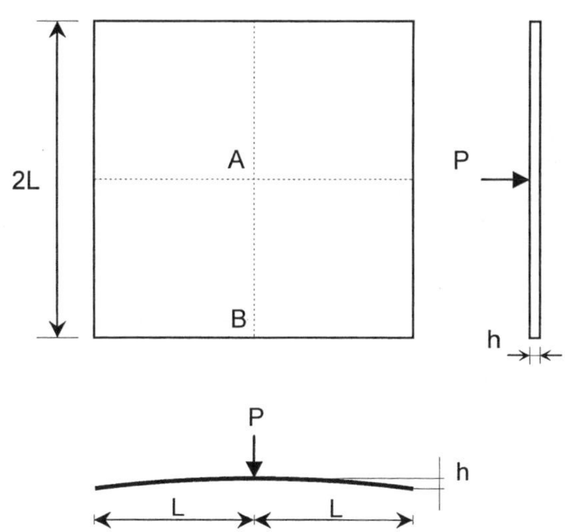

Figure 9.8 Longitudinally simply supported shallow cylindrical shell.

```
     .0000  228.2915     2.4094
     .0000  253.5769      .0000
   25.4000     .0000    12.6895
:

   25.4000  253.5769      .0000
   50.8000     .0000    12.6895
:

   50.8000  253.5769      .0000
   76.2000     .0000    12.6895
:

   76.2000  253.5769      .0000
```

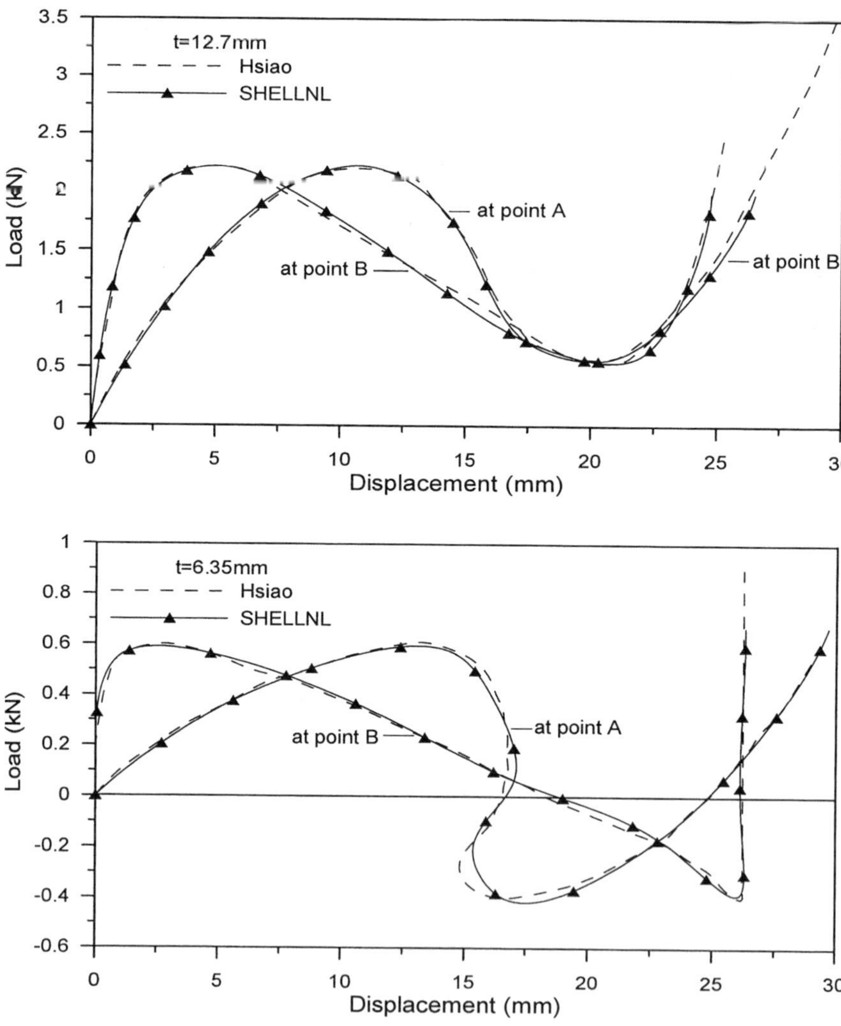

Figure 9.9 Displacements at points a and b of the shallow cylindrical shell.

```
  101.6000      .0000   12.6895
:
  101.6000   253.5769      .0000
  127.0000      .0000   12.6895
:
  127.0000   253.5769      .0000
  152.4000      .0000   12.6895
:
  152.4000   253.5769      .0000
  177.8000      .0000   12.6895
:
  177.8000   253.5769      .0000
  203.2000      .0000   12.6895
:
  203.2000   253.5769      .0000
  228.6000      .0000   12.6895
:
  228.6000   253.5769      .0000
  254.0000      .0000   12.6895                       -1.2
:
  254.0000   253.5769      .0000
     1    2
     1    4
     1    6
    11    1
    11    2
    11    3
    22    1
    22    2
    22    3
:
   121    1
   121    2
   121    3
   121    5
   121    6
   120    1
   120    5
   120    6
:
   112    1
   112    5
   112    6
   111    1
   111    2
   111    4
   111    5
   111    6
   100    2
   100    4
   100    6
:
    12    2
    12    4
    12    6
          1    2    12    12.7000
          2    3    13    12.7000
```

```
⋮
     10    11    21    12.7000
     12    13    23    12.7000
⋮
     21    22    32    12.7000
     23    24    34    12.7000
⋮
     32    33    43    12.7000
     34    35    45    12.7000
⋮
     43    44    54    12.7000
     45    46    56    12.7000
⋮
     54    55    65    12.7000
     56    57    67    12.7000
⋮
     65    66    76    12.7000
     67    68    78    12.7000
⋮
     76    77    87    12.7000
     78    79    89    12.7000
⋮
     87    88    98    12.7000
     89    90   100    12.7000
⋮
     98    99   109    12.7000
    100   101   111    12.7000
⋮
    109   110   120    12.7000
     12    13     2    12.7000
⋮
     21    22    11    12.7000
     23    24    13    12.7000
⋮
     32    33    22    12.7000
     34    35    24    12.7000
⋮
     43    44    33    12.7000
     45    46    35    12.7000
⋮
     54    55    44    12.7000
     56    57    46    12.7000
⋮
     65    66    55    12.7000
     67    68    57    12.7000
⋮
     76    77    66    12.7000
     78    79    68    12.7000
⋮
     87    88    77    12.7000
     89    90    79    12.7000
```

```
:
       98   99   88    12.7000
      100  101   90    12.7000
:
      109  110   99    12.7000
      111  112  101    12.7000
:
      120  121  110    12.7000
```

Data file SHELLNL93.OUT

```
 NUMBER OF NODES                    121
 NUMBER OF SUPPORTS                   0
 NUMBER OF CONSTRAINTS               97
 NUMBER OF ITERATIONS                10
 NUMBER OF FINITE ELEMENTS          200
 NUMBER OF LOAD STEPS                10
 E=    3.10275000000000
 NU=  0.300000000000000
 IARC=              1       Id=            6    ISTRESS=            1


       COORDINATES              LOADS
           X        Y        Z        PX       PY       PZ       MX       MY       MZ
  1 .00000 .00000 12.690 .00000  .00000  .00000  .00000  .00000  .00000
  2 .00000 25.400 12.562 .00000  .00000  .00000  .00000  .00000  .00000
:
111 254.00 .00000 12.690 .00000  .00000  .00000  .00000  .00000  .00000
:
121 254.00 253.58 .00000 .00000  .00000  .00000  .00000  .00000  .00000

CONSTRAINTS DATA
 NO. AT NODES DOF NO

  1          1      2
  2          1      4
  3          1      6
:
 66        111      1
 67        111      2
 68        111      4
 69        111      5
 70        111      6
:
 95         12      2
 96         12      4
 97         12      6

FINITE ELEMENT DATA
NO NODES TH    PRESTRESS
  I J M       SIGX SIGY TAU MXXI MYYI MXYI MXXJ MYYJ MXYJ MXXM MYYM MXYM
1 1 212 12.7 0.00 0.00 .00 0.00 0.00 0.00 0.00 0.00 0.00 0.00 0.00 0.00
2 2 313 12.7 0.00 0.00 .00 0.00 0.00 0.00 0.00 0.00 0.00 0.00 0.00 0.00
:
 ****ITERATION NUMBER    1
     LOAD STEP           1
  ERROR =        0.12000000D+00
```

```
DL2=   29.3281303301849        FACT=  0.100000000000000
****ITERATION NUMBER   2
    LOAD STEP           1
 ERROR =         0.24398085D-01
DL2=   29.3281303301849        FACT=  9.999999999999987E-002
****ITERATION NUMBER   3
    LOAD STEP           1
 ERROR =         0.21105481D-03
DL2=   29.3281303301849        FACT=  9.314575418727726E-002
****ITERATION NUMBER   4
    LOAD STEP           1
 ERROR =         0.21182225D-04

****ITERATION NUMBER   1
    LOAD STEP           2
 ERROR =         0.14683566D+00
DL2=   58.6562606603699        FACT=  0.215506995139774

****ITERATION NUMBER   3
    LOAD STEP           2
 ERROR =         0.93162090D-03
DL2=   58.6562606603699        FACT=  0.203092997001754
****ITERATION NUMBER   4
    LOAD STEP           2
 ERROR =         0.75035584D-04
DL2=   234.625042641479        FACT=  0.421975519876304

****ITERATION NUMBER   4
    LOAD STEP           4
 ERROR =         0.14925216D-03

****ITERATION NUMBER   1
    LOAD STEP           5
 ERROR =         0.11689717D+00
DL2=   469.250085282959        FACT=  0.519347694081416
****ITERATION NUMBER   2
    LOAD STEP           5
 ERROR =         0.20282994D+00
NEGATIVE TERM ON THE DIAGONAL AT   724
NEGATIVE TERM ON THE DIAGONAL AT   724
DL2=   469.250085282959        FACT=  0.519338641344191
****ITERATION NUMBER   3
    LOAD STEP           5
 ERROR =         0.25778760D-01
NEGATIVE TERM ON THE DIAGONAL AT   724
NEGATIVE TERM ON THE DIAGONAL AT   724
DL2=   469.250085282959        FACT=  0.463308867908064
****ITERATION NUMBER   4
 ERROR =         0.60527602D-03
NEGATIVE TERM ON THE DIAGONAL AT   724
NEGATIVE TERM ON THE DIAGONAL AT   724
DL2=   469.250085282959        FACT=  0.463216649218310
****ITERATION NUMBER   5
    LOAD STEP           5
 ERROR =         0.81589471D-04
****ITERATION NUMBER   1
    LOAD STEP           6
 ERROR =         0.13534852D-01
NEGATIVE TERM ON THE DIAGONAL AT   724
NEGATIVE TERM ON THE DIAGONAL AT   724
DL2=   703.875127924438        FACT=  0.474493978037810
****ITERATION NUMBER   2
    LOAD STEP           6
 ERROR =         0.40802921D-04
```

```
NEGATIVE TERM ON THE DIAGONAL AT    712
NEGATIVE TERM ON THE DIAGONAL AT    712
DL2=    703.875127924438        FACT=   0.392607019349170
****ITERATION NUMBER    8
    LOAD STEP           10
 ERROR =         0.17222246D-03
DL2=    1140.27770723759        FACT=   0.414133102394352
****ITERATION NUMBER    9
    LOAD STEP           10
 ERROR =         0.14731258D-03
DL2=    1140.27770723759        FACT=   0.414103471180779
****ITERATION NUMBER   10
    LOAD STEP           10
 ERROR =         0.12847919D-03
DL2=    1140.27770723759        FACT=   0.414076522020857
```

9.5.4 Example 9.4. Leicester's Shallow Spherical Shell

A spherical cap having a square base whose side is 61.8034in originates from a sphere of radius R=100in. It is subjected to a concentrated load at the crown with all edges hinged and immovable. This example was run using a 10×10 grid on a quarter of the cap with the following physical and geometrical constants: Young's modulus E=10^3 ksi, Poisson's ratio $\nu = 0.3$ and a thickness of 3.9154in. Figure 9.10 show how well results compare to those reported by Leicester (1968) who used Fourier transforms to solve his nonlinear differential equations.

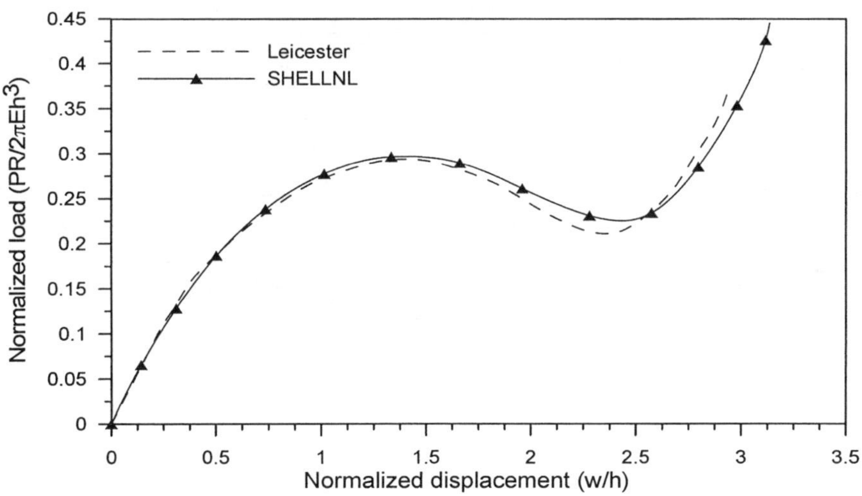

Figure 9.10 Displacement at crown of spherical shell.

9.5.5 Example 9.5. Mescall's Shallow Spherical Shell

Mescall (1965) reports analytic results for large displacement analysis of a spherical shallow shell subjected to a concentrated load at the crown. Two cases of support conditions are considered: rigid supports and vertically restrained supports. Figure 9.11 shows how well results compare for the rigidly supported shell that strain hardens and the vertically restrained shell that snaps. The geometry of the shallow shell is formed as the intersection of a cone whose tip lies at the center of a shell of radius 4.76in and having a tip angle of 21.8°. The thickness of the shell is taken as t=0.01576in, Young's modulus E=10^7psi and Poisson's ratio $\nu = 0.3$.

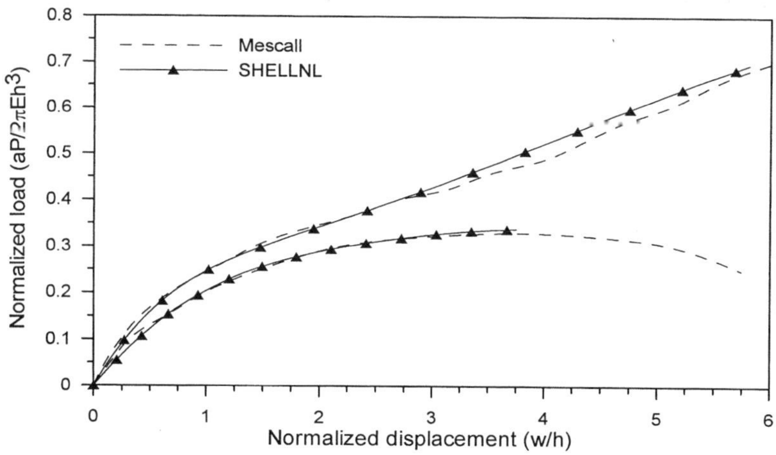

Figure 9.11 Displacements of Mescall's shell.

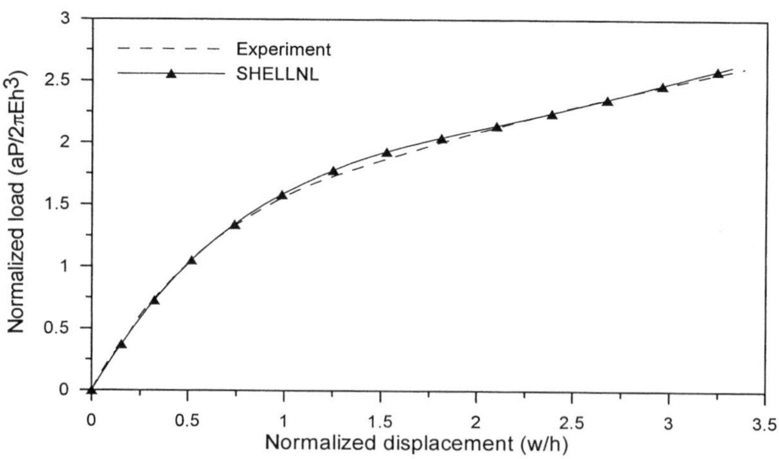

Figure 9.12 Displacements of Penning's experiment.

Penning (1966) performed an experiment on a shallow spherical shell with a circular base of 4.0in that is cut from a sphere of radius 19.8in. The thickness of the shell is taken as t=0.01576in, Young's modulus $E=10^7$psi and Poisson's ratio $\nu = 0.3$. Results are displayed in Figure 9.12.

9.5.6 Example 9.6. Open Hemispherical Shell

A hemispherical shell (Figure 9.13) with an 18° opening at the top is loaded by two orthogonal pairs of diametrically opposite equal forces. One pair is directed inwards whereas the other pair is directed outwards. Using symmetry, only one quarter of the shell was loaded and modeled using 16 latitudes by sixteen longitudes. In this case $E=6.825*10^6$ milliN/mm^2; t=0.04mm; Poisson's ratio of 0.3. Results are compared in Figure 9.14 with those of Buechter and Ramm (1992) and Simo *et al.* (1990).

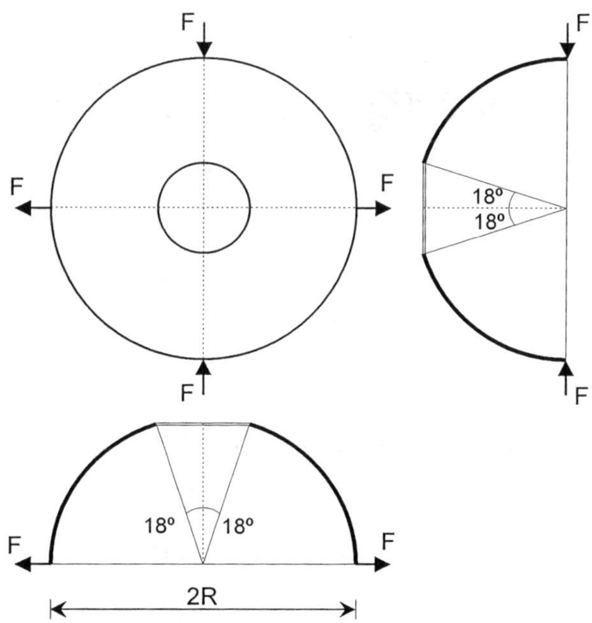

Figure 9.13 Open hemispherical shell.

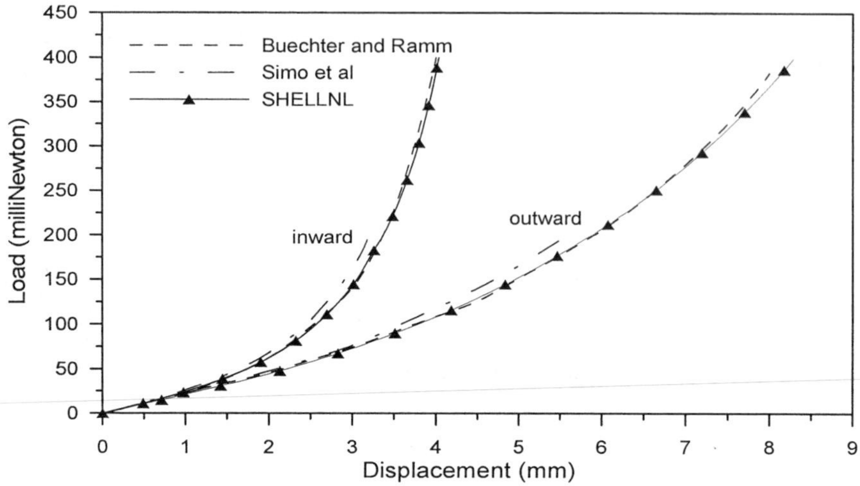

Figure 9.14 Open hemispherical shell.

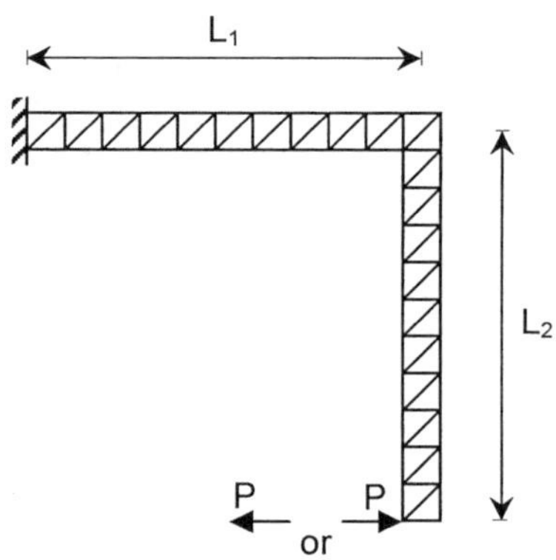

Figure 9.15 Lateral buckling of an L-shape frame.

9.5.7 Example 9.7. Lateral Buckling of an L-Frame

An L-shaped frame that is clamped on one end and subjected to an in-plane load at the other end (Figure 9.15) experiences lateral buckling. Interestingly, this problem exhibits two different bifurcation loads, depending on the load direction. Buckling was reached at between 1.1N-1.2N for the outward loading and between 0.7N-0.8N for the inward. Excellent agreement was found with the results of Simo *et al.* (1990) for both bifurcation loads and with that of Battini and Pacoste (2001) for the outward direction. Here L_1=L_2=240 mm; h=30.0 mm; t=0.6 mm; E=71240

N/mm^2; Poisson's ratio of 0.3.

An experiment for the outward loading case was carried out by Spillers *et al.* (1993). They reported a buckling load of 25 lbs. Here the buckling load was attained at between 24-25 lbs for L_1=10.64 in; L_2=10.5 in; h=1.0 in; t=0.12 in; $E=1.08 \times 10^7$ psi; Poisson's ratio of 0.3.

9.6 Some Remarks

A somewhat more complex approach for the derivation of the geometric stiffness matrix for triangular shell elements has been presented in this chapter. It separates the in-plane and the out-of-plane geometric effects when taking the gradient of the nodal forces that defines the geometric stiffness matrix. This approach is complete in the sense that all contributions to the response of the same order of magnitude are included.

Whereas this approach depends on an a priori chosen linear elastic finite element it is independent of large strain formulations that are needed otherwise. The nonlinear element is as good as the linear element it originates from. This approach has its rigor and its success is ensured in incremental analysis since equilibrium is satisfied in the deformed geometry.

Moreover the new technique is presented for the isolation of the pure deformations that allows stress retrieval via linear constitutive relations proved successful with respect to convergence and accuracy.

The presented shell element is overall efficient, reliable, accurate and stable not failing in any of the diverse examples on which it was tested.

9.7. Problems

1. Modify the computer program to solve an orthotropic thin shell.

2. Modify the computer program so as to inflate a rubber balloon.

3. Modify the computer program to inflate a rubber balloon that is strengthened by a cablenet.

REFERENCES

Ahmad, S., Irons, B. M., and Zienkiewicz, O. C., "Analysis of Thick and Thin Shell Structures by Curved Finite Elements", *Int. J. Numer. Methods Eng.*, Vol. 2, pp. 419-451, 1970.

Aitken, A.C., Determinants and Matrices, Oliver and Boyd, NY City, 1958.

ANSYS, Engineering Analysis Systems, Swanson Analysis Systems Inc., Box 65, Houston, PA, 1987.

Argyris, J. H., Angelopoulos, T., and Bichat, B., "A General Method for the Shape Finding of Lightweight Tension Structures", *Computer Methods in Applied Mechanics and Engineering*, 3, 135-149, 1974.

Argyris, J. H., Dunne, P. C., Malejannakis, G., and Schelkle, E., "A Simple Triangular Facet Shell Element With Applications to Linear and Non-Linear Equilibrium and Elastic Stability Problems", *Comp. Meth. Appl. Mech. & Engng.*, Vol. 11, 1977, pp. 97-131. and Vol. 10, 1977, pp.371-403.

Argyris, J. H., Papadrakakis, M., and Karapitta, L., "Elastoplastic Analysis of Shells With The Triangular Element TRIC", *The Fourth International Colloquium on Computational of Shell & Spatial Structures*, Chania-Crete, Greece, 2000.

Bathe, K. J., and Ho, L. W., "A Simple and Effective Element for Analysis of General Shell Structures", *Computers & Structures*, Vol. 13, pp. 673-681,1980.

Bathe, K. J., and Bolourchi, S., "A Geometric and Material Nonlinear Plate and Shell Element", *Computers & Structures*, Vol. 11, pp. 23-48, 1980.

Batoz, J. L., Bathe, K. J., and Ho, L. W., "A Study of Three Noded Triangular Plate Bending Elements", *Int. J. Num. Meth. Engng.*, Vol. 15, pp. 1771-1812, 1980.

Battini, J.-M., and Pacoste, C., "Co-rotational Beam Elements With Warping Effects in Instability Problems", *Comput. Methods Appl. Mech. Engrg.*, Vol. 191, pp. 1755-1789, 2001.

Baumbauer, J. C., and Tiersten, H. F., "Nonlinear Electrostatic Equations for Small Fields Superimposed on a Bias", *Journal of the Acoustic Society of America*, 54, 4, 1017-1034, 1973.

Biot, M. A., Mechanics of Incremental Deformation, John Wiley & Sons, New York, 1965.

Buechter, N., and Ramm, E., "Shell Theory Versus Degeneration – A Comparison in Large Rotation Finite Element Analysis", *Int. J. for Num. Meth Eng.*, Vol. 34, pp. 39-59, 1992.

Chan, S. L and Zhou, Z. H, "Pointwise Equilibrating Polynomial for Nonlinear Analysis of Frames", *Journal of Structural Engineering,* ASCE, 120, 6 1703-1717, 1994.

Chen, W. F., and Lui, E. M., Stability Design of Steel Frames, CRC Press, Boca Raton, FL, 1993.

Clough, R., W., and Johnson, C., J., "A Finite Element Approximation for the Analysis of Thin Shells", *Int. J. Num. Meth. Eng.*, Vol. 4, pp. 43-60, 1968.

Crisfield, M. A., "A Fast Incremental / Iterative Solution Procedure That Handles "snap-through"", *Comput. Struct.*, Vol. 13, pp. 55-62, 1981.

Dvorkin, E., N., and Bathe, K., J., "A Continuum Mechanics Based Four-Node Shell Element for General Non-Linear Analysis" *Eng. Comp.*, Vol. 1, 1984, pp. 77-88.

Eisenberger, M., ""Exact static and dynamic stiffness matrices for generalvariable cross section member", *AIAA Journal.*, Vol. 28, pp. 1105-1109, 1990.

Gal, E., "Triangular Shell Element for Geometrically Nonlinear Structural Analysis", Ph.D. Thesis, Technion, (2002).

Gere, James M., Moment Distribution, D. Van Nostrand Co., Princeton NJ, 1963.

Goldstein, Herbert, Classical Mechanics, Addison-Wesley Publishing Company, Reading, MA, 1950.

Green, A.E., Knops, R.J.,and Laws, N., "Large Deformations, Superimposed Small Deformations, and the Stability of Elastic Rods", *Inter. Journal Solids and Structures*, 4, 555-577, 1968.

Haber, R.B., and Abel, J.F., "Initial Equilibrium Solution Method for Cable Reinforced Membranes. Part 1 - Formulation", *Computer Methods in Applied Mechanics and Engineering*, 30, 263-284, 1982.

Horrigmoe, G., and Bergan, P. G., "Nonlinear Analysis of free-form Shells by Flat Finite Elements", *Comp. Meth. Appl. Mech. & Engng.*, Vol. 16, pp. 11-35, 1978.

Hsiao, K.M., "Nonlinear Analysis of General Shell Structures by Flat triangular Shell Element", *Computers & Structures*, Vol. 13, pp. 665-675, 1987.

Hughes, T.J.R., and Lui, W.K., "Nonlinear Finite Element Analysis of Shells: Part 1 Three-Dimensional Shells", *Comp. Meth. Appl. Mech. & Engng.*, Vol. 26, pp. 331-362, 1981.

Isaacson, Eugene, and Keller, Herbert Bishop, Analysis of Numerical Methods, John Wiley Co., New York City, 1966.

Ibrahimbegović, A., "Stress Resultant Geometrically Exact Shell Theory for Finite Rotations and its Finite Element Implementation", Appl. Mech. Rev., Vol. 50, no. 4, 1997, pp. 199-226.

Izzuddin, B. A., and Elnashai, A. S. "Eulerian Formulation for Large-Displacement Analysis of Space Frames", *Journal of Engineering Mechanics,* ASCE, 119, 3, 549-569, 1993.

Krysl, Petr, "Computer Stiffness Matrices for Buckling Analysis of Frames", *Journal of Engineering Mechanics,* ASCE, 119, 2, 225-237, 1993.

Kuo, Shyh-Rong, Yang, Yeong-Bin, and Chou, Jung-Hua, "Nonlinear Analysis of Space Frames with Finite Rotations", *Journal of Structural Engineering,* ASCE, 119, 1, 1-15, 1993.

Leicester, R. H., "Finite Deformations of Shallow Shells", *Proc. ASCE,* Vol. 94 (EM6, pp. 1409-1423), 1968.

Leonard, John William, Tension Structures, McGraw Hill, N Y City, 1988.

Levy, R., "Linear Programming in Prestressed Trusses", *Computers and Structures,* Vol. 33, No.2, pp. 421-425, 1989.

Levy, R., and Brill, D.R., "Optimization for Exact Stability of 2-Bar Trusses", Computer Utilization in Structural Engineering, pp. 520-529, (edited by James K. Nelson, Jr.), American Society of Civil Engineers, New York, 1989.

Levy, R., and Gal, E., "Three-Dimensional Non-Prismatic Beam-Columns", International Journal of Structural Stability and Dynamics, Vol. 2, No. 3, pp. 395-408, 2002[A].

Levy, R., and Gal, E., "Buckling and Stress Softening of Beam-Columns under complex Three-Dimensional Loading", International Journal of Structural Stability and Dynamics, Vol. 2, No. 4, pp. 487-498, 2002[B].

Levy, R. and Gal, E., "Three Noded Shell Finite Element for Large Rotations", Proceedings of WCCM V, Fifth World Congress on Computational Mechanics, Vienna, July 7-12, 2002[C], paper no. 80327.

Levy, R., and Hanaor, A., "Optimal Design of Prestressed Trusses", Shells, Membranes and Space Frames, Proceedings of the IASS Symposium on Membrane Structures and Space Frames, Osaka, September 15th-19th, 1986, Vol. 3, pp. 207-214, (edited by K. Heki), Elsevier Science Publishers B.V., Amsterdam, 1986.

Levy, R., Vilnay, O., and Acheampong, K.B., "Exact Geometry Considerations in Buckling Analysis of Trusses", *Computers and Structures,* Vol. 4-1, No. 6, pp. 1241-1248, 1991.

Levy, S., "Bending of Rectangular Plates With Large Deflections", *Tech. Notes*, NACA, No. 846, 1942.

MacBain, K., Saadeghvaziri, A., and Spillers, W. R., "Examples of Stress Hardening and Softening in Three-Dimensional Beams", *Journal of Structural Engineering,* ASCE, 129, 9, 1073-1075, 1999.

Mescall, J. F., "Large Deflections of Spherical Shells Under Concentrated Loads", *ASME Applied Mechanics Division*, pp. 936-938, 1965.

Mohan, P., and Kapania, R. K., "Geometrically Nonlinear Analysis of Composite Plates and Shells Using a Flat Triangular Shell Element", *38th AIAA/ASME/ASCE/AHS/ASC Structures., Struct. Dyn. and Mat. Conf.,* Kissimmee, FL, pp. 2347-2361 (AIAA Paper 97-1233), 1997.

Noble, B., Applied Linear Algebra, Prentice-Hall, New Jersey, 1969.

Otto, Frei (editor), Tensile Structures, MIT Press, Cambridge, MA 1973.

Peng, X., and Crisfield, M. A., "A Consistent Co-rotational Formulation for Shell Using The Constant Stress/ Constant Moment Triangle" *Int. J. Num. Meth. Engng.*, Vol. 35, pp. 1829-1847, 1992.

Penning, F. A., "Experimental Buckling Modes of Clamped Shallow Shells Under Concentrated Load", *ASME Applied Mechanics Division*, pp. 247-304, 1966.

Reissner, Eric, "Variational Considerations for Elastic Beams and Shells", *J. Engr. Mech. Division, Proc.* ASCE, 88, EM1, Feb. 1962, pp. 23-57.

Schek, H.J., "The force density method for form finding and computation of general networks", *Computer Methods in Applied Mechanics and Engineering*, 3, 1974, pp 115-134.

See, Thomas, and McConnell, Richard E., "Large Displacement Elastic Buckling of Space Structures", *Journal of Structural Engineering,* Vol. 112, no. 5, May 1986, pp. 1052-1069.

Siev, A., and Eidelman, J.,"Stress Analysis of Prestressed Suspended Roofs", *Proc. ASCE*, 90, ST4, pp. 103-121, August 1964.

Simo, J. C., Fox, D. D., and Rifai, M. S., "On A Stress Resultant Geometrically Exact Shell Model. Part III: Computational Aspects of The Nonlinear Theory", *Comput. Methods Appl. Mech. Engrg.*, Vol. 79, pp. 21-70, 1990.

Spillers, W. R., Automated Structural Analysis: An Introduction, Pergamon Press, New York, 1972. (This out of print text has been re-released in soft cover by Franklin Book Company, Elkins Park, PA, 1992)

Spillers, W. R., "Geometric Stiffness Matrix for Space Frames", *Computers and Structures*, 36, 1, 1990, pp 29-37.

Spillers, W. R., "A Note on Some Qualitative Results for Beams and Rods", *Quarterly Applied Mathematics*, XLVIII, 3 September 1990, pp 575-580.

Spillers, W. R., Saadeghvaziri, A., and Luke A., "An Example of Three Dimensional Frame Buckling", *Computers and Structures*, Vol. 47, No. 3, pp 483-486, 1993.

Spillers, W. R., Schlogel, Mark, and Pilla, Dominick, "A Simple Membrane Finite Element", *Computers and Structures*, 45, 1, 181-183, 1993.

Spillers, W. R., and Rashidi, S., "Member Stiffness for Three-Dimensional Beam-Columns", *Journal of Structural Engineering,* ASCE, 123, 7, 971-972, 1997.

Stubbs, N., and Fluss, H., "Modelling Fabric Reinforced Composites", Proc. 7th Canadian Cong. Appl. Mech., Sherbrooke, Quebec, pp. 113-114, 1979.

System/360 Scientific Subroutine Package, IBM Corp., White Plains,New York, 1970.

Timoshenko, S., Theory of Elastic Stability, McGraw Hill Book Co.,New York, 1936.

Timoshenko, S., and Goodier, J. N., Theory of Elasticity, McGraw Hill, NY 1951.

Zienkiewicz, O. C., The Finite Element Method, 3 rd edition, McGraw-Hill, NY, 1977.

APPENDIX 1

MEMBER STIFFNESS WHEN BEAM-COLUMN EFFECTS ARE INCLUDED

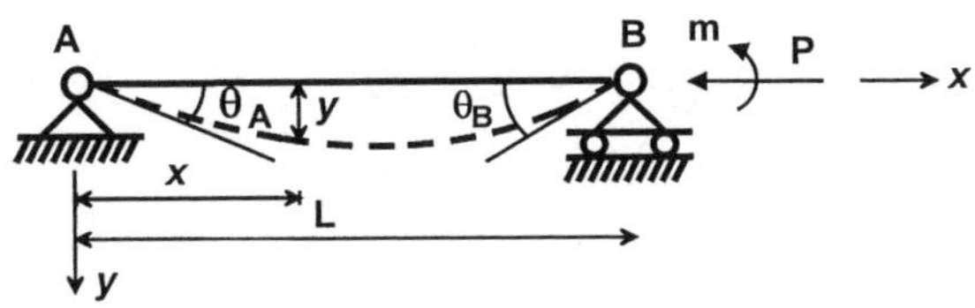

Figure A1.1 Simply supported beam.

This appendix derives the (moment distribution stiffness) of a member when beam-column effects are included for two cases: the pinned-fixed and the fixed-fixed beams. Consider the statically determinate beam of Figure A1.1 which is subjected to a compressive load P and an end moment m. In this case the internal bending moment at any point x can be written as

$$M = \frac{mx}{L} + Py \tag{A1.1}$$

If the elastic constitutive equation, $M = -EIy''$, is used it follows that

$$\frac{mx}{L} + Py = -EIy'' \tag{A.1.2}$$

In this case y is simply

$$y = A\sin kx + B\cos kx - \frac{mx}{PL}$$
$$k^2 = \frac{P}{EI} \tag{A1.3}$$

Using the boundary conditions:

y=0	@	x=0	=>	B=0
y=0	@	x=L	=>	A=m/(P sin kL)

The slope at $x=L$ is then

$$y'_{x=L} = \frac{m}{P}\left(k \cot kL - \frac{1}{L}\right) \tag{A1.4}$$

and, therefore

$$\theta_{\mathrm{B}} = -\frac{m}{p}\left(k \cot kL - \frac{1}{L}\right) \tag{A1.5}$$

or

$$\theta_{\mathrm{B}} = -\frac{mL}{3EI}\alpha \tag{A1.6}$$

where

$$\alpha = \frac{3}{kL}\left(\frac{1}{kL} - \cot kL\right) \tag{A1.7}$$

The member stiffness which is the moment per unit rotation and is denoted for this pinned case $K_{\mathrm{AB}}^{\mathrm{P}}$, becomes

$$K_{\mathrm{AB}}^{\mathrm{P}} = \frac{3EI}{L} \cdot \frac{1}{\alpha} \tag{A1.8}$$

This stiffness was used in Chapter 1 and is described in Figure 1.11 by the curve marked "Compression. Far End Simply Supported".

The member stiffness for the fixed-fixed case will be derived via superposition by supplying a moment at A to eliminate the rotation θ_{A} and then adding the effect of this moment on the rotation at B. Figure A1.2 illustrates the process. θ_{A} is derived from $(y')_{x=0}$ as

$$\theta_{\mathrm{A}} = \frac{ml}{6EI}\beta \tag{A1.9}$$

where

$$\beta = \frac{6}{kl}\left(\frac{1}{\sin kL} - \frac{1}{kL}\right) \tag{A1.10}$$

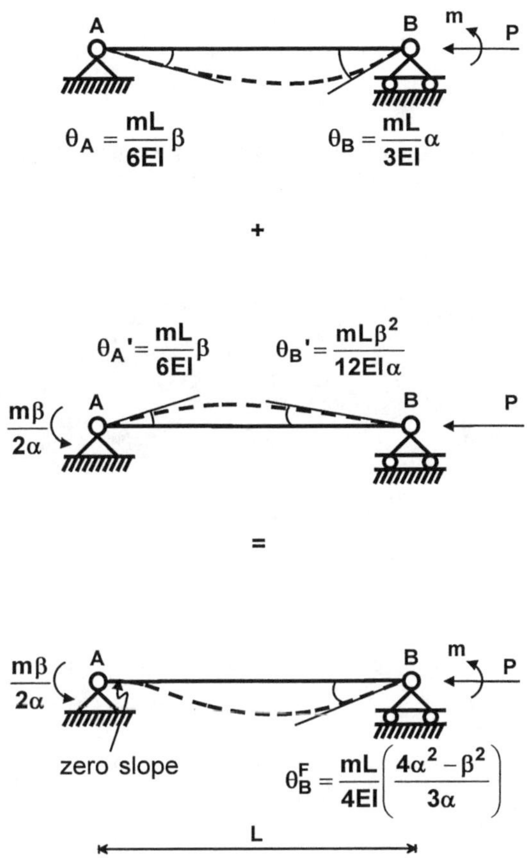

Figure A1.2 Rotation via superposition.

The moment required to produce a zero slope at A is $m\beta/(2\alpha)$ with an additional rotation θ_B', at B of

$$\theta_B' = -\left(\frac{m\beta}{2\alpha}\right)\frac{L}{6EI}\beta \tag{A1.11}$$

The total rotation for the fixed case is thus

$$\theta_B^F = -\frac{mL\beta^2}{12EI\alpha} + \frac{mL\alpha}{3EI} = \frac{mL}{4EI}\left(\frac{4\alpha^2 - \beta^2}{3\alpha}\right) \tag{A1.12}$$

and the member stiffness becomes,

$$K_{\mathrm{AB}}^{\mathrm{F}} = \frac{4EI}{L}\left(\frac{3\alpha}{4\alpha^2 - \beta^2}\right) \tag{A1.13}$$

This stiffness is described in Figure 1.11 by the curve marked "Compression. Far End Fixed".

APPENDIX 2

DETERMINANTS

A structure has buckled when its system matrix has become singular or $|\mathbf{K}_E + \mathbf{K}_G| = \mathbf{0}$. When that happens a zero will appear on the diagonal of the system matrix during the process of Gaussian elimination. Since it is unlikely to hit upon the exact buckling load when the applied load is incremented in steps, it is most common to go past the buckling load which will generally result in the determinant becoming negative. The computer programs discussed in these notes monitor the diagonal terms of the system matrix during the solution process looking for terms which are not positive. The logic of doing so is the subject of this appendix.

Cramer's rule has the inverse of a matrix **A** as the transpose of the matrix of the cofactors divided by the determinant. Similarly, the determinant of the matrix **A** can be described in terms of its first row elements a_{1j} and their cofactors $|\mathbf{A}_{1j}|$ as

$$|\mathbf{A}| = a_{11}|\mathbf{A}_{11}| + a_{12}|\mathbf{A}_{12}| + ... + a_{1n}|\mathbf{A}_{1n}|$$

(See Aitkin, 1958.) This allows the determinant of an $n \times n$ matrix to be defined inductively in terms of the determinant of a smaller matrix and eventually only having to deal with a scalar.

There are two rules of determinants which are important here:

1. A determinant is unaltered in value when any row or column is added to a constantmultiple of any other row or column.

2. The determinant of a triangular matrix is the product of its diagonal elements.

It is clear from the definition of a determinant that if you divide any row of a matrix by a constant, you change its determinant by this constant. The point is that rule 1 implies that Gaussian elimination can be performed in such a manner that the value of the determinant of the system matrix does not change. Rule 2 implies that when the value of the determinant goes to zero, a zero must appear along the diagonal at the end of the elimination process.

APPENDIX 3

THE ROTATION MATRIX

Geometric modeling, including the modeling of structures, typically involves the concept of global and local coordinate systems (Figure A3.1). The global coordinate system is of course the coordinate system in which the equations of equilibrium are written and is fixed in space; the local coordinate system is fixed in the member and describes the orientation of the member. Since these two coordinate systems differ only by a rotation (if they are both orthogonal) the rotation matrix becomes an important tool of structural analysis.

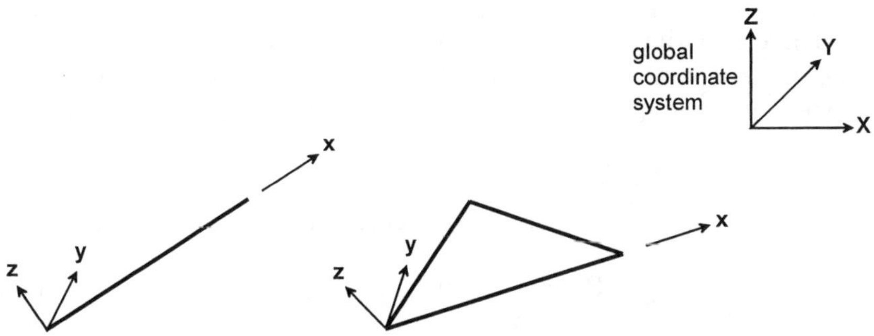

Figure A3.1 Global and local coordinate systems.

It is most easy to deal with the two-dimensional rotation matrix (Figure A3.2). From simple geometric arguments it follows that

$$A_{x'} = A_x \cos\theta + A_y \sin\theta$$

$$A_{y'} = -A_x \sin\theta + A_y \cos\theta$$

or in matrix form

$$\mathbf{A}' = \mathbf{R}\mathbf{A}$$

where

$$\mathbf{A}' = \begin{Bmatrix} A_{x'} \\ A_{y'} \end{Bmatrix}; \ \mathbf{R} = \begin{bmatrix} \cos\theta & \sin\theta \\ -\sin\theta & \cos\theta \end{bmatrix}; \ \mathbf{A} = \begin{Bmatrix} A_x \\ A_y \end{Bmatrix}$$

By direct multiplication it follows that $\mathbf{R}^{\mathrm{T}}\mathbf{R} = \mathbf{R}\mathbf{R}^{\mathrm{T}} = \mathbf{I}$.

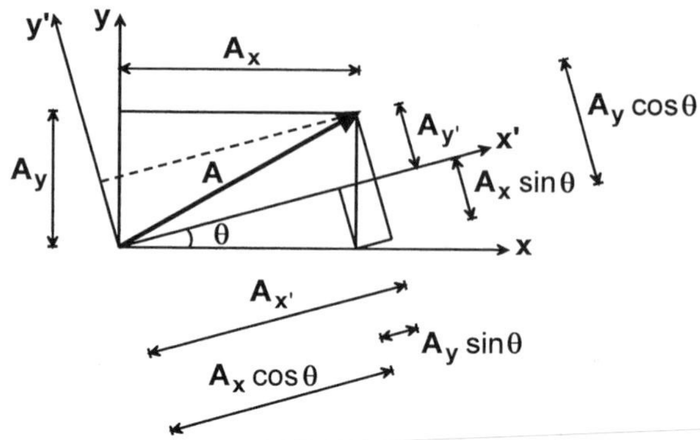

Figure A3.2 Two-dimensional rotation

There is a simple transition from two-dimensional rotation matrices to three-dimensional rotation matrices. First of all, two-dimensional rotations can be thought of as (three-dimensional) rotations about the z-axis. Then

$$\mathbf{R}_z = \begin{bmatrix} \cos\theta & \sin\theta & 0 \\ -\sin\theta & \cos\theta & 0 \\ 0 & 0 & 1 \end{bmatrix} \tag{A3.1}$$

By permutations of the labels of the coordinate axes it follows that rotations about the x and y axes can be described as

$$\mathbf{R}_x = \begin{bmatrix} 1 & 0 & 0 \\ 0 & \sin\theta & \cos\theta \\ 0 & \cos\theta & -\sin\theta \end{bmatrix}; \mathbf{R}_y = \begin{bmatrix} \cos\theta & 0 & -\sin\theta \\ 0 & 1 & 0 \\ \sin\theta & 0 & \cos\theta \end{bmatrix}$$

One way to describe three-dimensional rotations is through the use of compound rotations. A compound rotation is simply a sequence of rotations. This description is particularly useful when the rotations in the sequence are rotations about coordinate axes (which are essentially two-dimensional rotations). In this case the use of compound rotations has the effect of reducing three-dimensional rotations to two-dimensional rotations. When rotations are performed in sequence, the composite rotation matrix is the product of the individual rotation matrices used. For example, let a rotation be described physically as a sequence in which rotation $1 \Rightarrow \mathbf{R}_1$, rotation $2 \Rightarrow \mathbf{R}_2$, rotation $3 \Rightarrow \mathbf{R}_3$ then

$$\mathbf{x}' = \mathbf{R}_1\mathbf{x} \qquad \mathbf{x}'' = \mathbf{R}_2\mathbf{x}' \qquad \mathbf{x}''' = \mathbf{R}_3\mathbf{x}''$$

and

$$\mathbf{x}''' = \mathbf{R}_3\mathbf{x}'' = \mathbf{R}_3\mathbf{R}_2\mathbf{x}' = \mathbf{R}_3\mathbf{R}_2\mathbf{R}_1\mathbf{x}_1$$

which implies that

$$\mathbf{R} = \mathbf{R}_3\mathbf{R}_2\mathbf{R}_1$$

This gives a way to construct the rotation matrix using rotations about coordinate axes: That is, describe the physical rotation as a sequence of rotations about coordinate axes and then form the rotation matrix as the product of the rotation matrices described above.

Another approach to the rotation matrix comes from the fact that

"*The rows of the rotation matrix are the base vectors (coordinate unit vectors) of the local coordinate system*"

This result requires the fact that

"*The transpose of the rotation matrix is its inverse, i.e.* $\mathbf{R}^{\mathrm{T}} = \mathbf{R}^{-1}$"

To show the latter is an easy matter. By definition length does not change under rotation so neither does the square of the length. If **x'=Rx** then

$$\left|\mathbf{x}^2\right| = \mathbf{x}^{\mathrm{T}}\mathbf{x} = \mathbf{x}'^{\mathrm{T}}\,\mathbf{x}' = \mathbf{x}^{\mathrm{T}}\mathbf{R}^{\mathrm{T}}\mathbf{R}\mathbf{x}$$

or

$$\mathbf{x}^{\mathrm{T}}\left(\mathbf{I} - \mathbf{R}^{\mathrm{T}}\mathbf{R}\right)\mathbf{x} = \mathbf{0}$$

Since this is true for arbitrary **x** it follows that

$$\mathbf{I} - \mathbf{R}^{\mathrm{T}}\mathbf{R} = \mathbf{0}$$

or

$$\mathbf{R}^{\mathrm{T}}\mathbf{R} = \mathbf{I}$$

It is now possible to show that the rows of **R** are the base vectors of the local coordinate system. Clearly

$$\mathbf{x}' = \mathbf{R}\mathbf{x}$$

and

$$\mathbf{x} = \mathbf{R}^{\mathrm{T}}\mathbf{x}'$$

Let

$$\mathbf{x}' = \begin{Bmatrix} 1 \\ 0 \\ 0 \end{Bmatrix} = \mathbf{i}'$$

Since $\mathbf{x} = \mathbf{R}^{\mathrm{T}}\mathbf{x}'$ it follows that

$$\begin{Bmatrix} R_{11} \\ R_{12} \\ R_{13} \end{Bmatrix} = \begin{Bmatrix} (\mathbf{i}')_x \\ (\mathbf{i}')_y \\ (\mathbf{i}')_z \end{Bmatrix}$$

Repeating this procedure for the other two coordinate axes it follows that

$$\mathbf{R} = \begin{bmatrix} (\mathbf{i}')_x & (\mathbf{i}')_y & (\mathbf{i}')_z \\ (\mathbf{j}')_x & (\mathbf{j}')_y & (\mathbf{j}')_z \\ (\mathbf{k}')_x & (\mathbf{k}')_y & (\mathbf{k}')_z \end{bmatrix} = \begin{Bmatrix} \mathbf{i}' \\ \mathbf{j}' \\ \mathbf{k}' \end{Bmatrix} \tag{A3.2}$$

Finally some topics which will be returned to when three-dimensional frames are discussed. First of all, it is clear that finite rotations do not behave as vectors since the order in which rotations are applied makes a difference, (see Figure A3.3). Clearly to describe the finite rotation of an object it is necessary to use a rotation matrix rather than a rotation vector. What then is the rotation *vector* used in linear elastic analysis? It will be shown now that the rotation matrix is equivalent to a "vector-like" quantity. It is then the

Taylor series representation of this quantity which gives rise to the small rotation vector of linear mechanics.

First, note that the rotation matrix which is used to describe the relationship between the local and the global coordinate systems can be used in kinematics to describe rigid body rotation. To do so, the local coordinate system can be regarded as fixed in a rotating rigid body. Let $\mathbf{x}_{\text{old}}$ and $\mathbf{x}_{\text{new}}$ refer to the coordinates of a point before and after rotation, (see Figure A3.4). It follows that Euler's theorem of rigid body motion implies that any finite rotation (or sequence of rotations) can be described as a simple rotation of magnitude α about some fixed axis described by a unit vector

$$\mathbf{n}^* = \left((\mathbf{n}^*)_x, (\mathbf{n}^*)_y, (\mathbf{n}^*)_z\right)$$

Following Noble, (Noble, 1969, p. 421),

This implies that the rotation matrix **R** may be written in terms of α and the components of **n** as

$$\mathbf{x}_{\text{new}} = \mathbf{R}^{\mathrm{T}}\mathbf{x}_{\text{old}} \tag{A3.3}$$

$$\mathbf{x}_{\text{new}} = \mathbf{x}_{\text{old}} - (1-\cos\alpha)\{\mathbf{x}_{\text{old}} - (\mathbf{x}_{\text{old}}\cdot\mathbf{n}^*)\mathbf{n}^*\} + \sin\alpha(\mathbf{n}^*\times\mathbf{x}_{\text{old}})$$

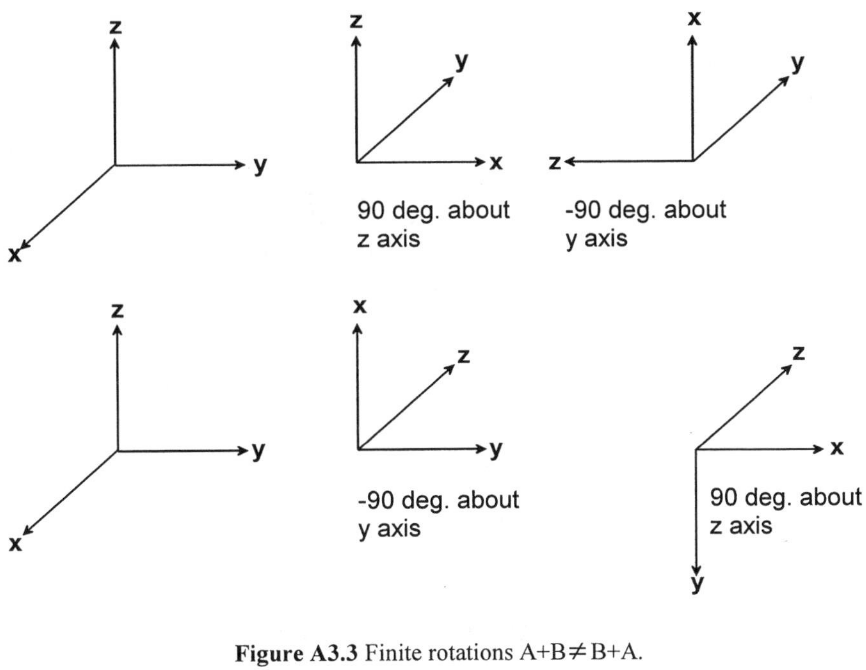

Figure A3.3 Finite rotations A+B≠B+A.

$$\mathbf{R}^{\mathrm{T}} = \begin{bmatrix} \cos\alpha + c n_x^{*2} & c n_x^* n_y^* - n_z^* \sin\alpha & c n_x^* n_z^* + n_y^* \sin\alpha \\ c n_x^* n_y^* + n_z^* \sin\alpha & \cos\alpha + c n_y^{*2} & c n_y^* n_z^* - n_x^* \sin\alpha \\ c n_z^* n_x^* - n_y^* \sin\alpha & c n_z^* n_y^* + n_x^* \sin\alpha & \cos\alpha + c n_z^{*2} \end{bmatrix}$$

where

$$c = 1 - \cos\alpha$$

Furthermore as Noble indicates, given **R** it is a simple matter to reconstruct **n*** and α since

$$\cos\alpha = \left(R_{11}^{\mathrm{T}} + R_{22}^{\mathrm{T}} + R_{33}^{\mathrm{T}} - 1\right)/2 \tag{A3.4}$$

and

$$n_x^* : n_y^* : n_z^* = R_{23}^{\mathrm{T}} - R_{32}^{\mathrm{T}} : R_{31}^{\mathrm{T}} - R_{13}^{\mathrm{T}} : R_{12}^{\mathrm{T}} - R_{21}^{\mathrm{T}}$$

It follows that any rigid body motion may be described equivalently using either a rotation matrix **R** or the above representation which involves **n*** and α.

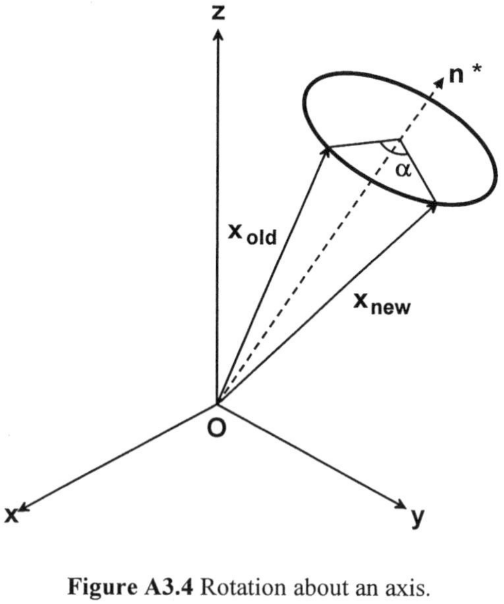

Figure A3.4 Rotation about an axis.

Let

$$\boldsymbol{\theta} = \mathbf{n} * \alpha \tag{A3.5}$$

be called a rotation. If $\boldsymbol{\theta}(\mathbf{x})$ possess a Taylor series expansion, then $d\boldsymbol{\theta} = \nabla\boldsymbol{\theta} \cdot d\mathbf{x}$. It is this $d\boldsymbol{\theta}$ which is referred to as small rotation vector of linear structural analysis.

APPENDIX 4

PERTURBATION METHODS APPLIED TO PLANE BEAMS

This appendix examines the special case of pertrubation methods applied to plane beams. The more general case of three-dimensional beams is addressed as a separate chapter. The results obtained here are simply the beam-column equations commonly found in discussions of strength of materials. This appendix is included in an attempt to show how the perturbation method works for a simple case of continuous systems.

The starting point is the vector form of the equations of equilibrium for a three-dimensional beam

$$\begin{gathered}\mathbf{P}'+\mathbf{p}=\mathbf{0}\\ \mathbf{M}'+\mathbf{m}+\mathbf{t}\times\mathbf{P}=\mathbf{0}\end{gathered} \tag{A4.1}$$

Here **P**, **M** = force and moment stress resultant vectors respectively; **p**, **m** = force and moment applied loads respectively. For plane beams these equations take on the form

$$P_x{}'+p_x=0; \quad P_y{}'+p_y=0; \quad M_z{}'+m_z+(\mathbf{t}\times\mathbf{P})_z=0; \tag{A4.2}$$

For an initially straight beam, $\mathbf{t}=\mathbf{i}^0$. This reduces Eq. A4.2 to

$$P_x^0{}'+p_x^0=0; \quad P_y^0{}'+p_y{}^0=0; \quad M_z^0{}'+m_z^0+P_y^0=0; \tag{A4.3}$$

When written in their perturbed form, where $\mathbf{i}^1=\mathbf{i}^0+\boldsymbol{\omega}\times\mathbf{i}^0$ and $\mathbf{j}^1=\mathbf{j}^0+\boldsymbol{\omega}\times\mathbf{j}^0$ and $\boldsymbol{\omega}=y'\mathbf{k}^0$, Eq. A4.1 appear as

$$\begin{gathered}\frac{\mathrm{d}}{\mathrm{d}x}\left[\left(P_x^0+\varepsilon\bar{P}_x\right)(\mathbf{i}^0+y'\mathbf{j}^0)+(P_y^0+\varepsilon\bar{P}_y)(\mathbf{j}^0-y'\mathbf{i})\right]\\ +(p_x^0+\varepsilon\bar{p}_x)(\mathbf{i}^0+y'\mathbf{j}^0)+(p_y^0+\varepsilon\bar{p}_y)(\mathbf{j}^0-y'\mathbf{i})=\mathbf{0}\end{gathered} \tag{A4.4}$$

$$\frac{\mathrm{d}}{\mathrm{d}s}\left[\left(M_z^0+\varepsilon\overline{M}_z\right)\right]+m_z^0+\varepsilon\overline{m}_z+\left((\mathbf{i}^0+\varepsilon\bar{\mathbf{i}})\times(\mathbf{P}^0+\varepsilon\overline{\mathbf{P}})\right)_z=0 \tag{A4.5}$$

where the perturbed quantities such as the thrust appear as the zero or initial solution plus the perturbation written as ε times a term with a bar over it i.e.

$$\begin{aligned}&\mathbf{P}^0 \rightarrow \mathbf{P}^1 = \mathbf{P}^0 + \varepsilon\overline{\mathbf{P}}\\&\mathbf{M}^0 \rightarrow \mathbf{M}^1 = \mathbf{M}^0 + \varepsilon\overline{\mathbf{M}}\\&\mathbf{i}^0 \rightarrow \mathbf{i}^1 = \mathbf{i}^0 + \varepsilon\overline{\mathbf{i}};\ \mathbf{j}^0 \rightarrow \mathbf{j}^1 = \mathbf{j}^0 + \varepsilon\overline{\mathbf{j}};\mathbf{k}^0 \rightarrow \mathbf{k}^1 = \mathbf{k}^0 + \varepsilon\overline{\mathbf{k}}\end{aligned} \tag{A4.6}$$

Setting $p_x^0 = p_y^0 = m_z^0 = \overline{p}_x = \overline{p}_y = \overline{m}_z = 0$ as of no interest and ε=1, and collecting first order terms it follows that

$$\begin{aligned}&\left(\overline{P}_x - P_y^0 y'\right)' = 0\\&\left(\overline{P}_y + P_x^0 y'\right)' = 0\\&\overline{M}_z' + \overline{P}_y = 0\end{aligned} \tag{A4.7}$$

The second and third of these equations can be combined (with the perturbation parameter set to 1) to give the well-known beam column equations

$$\overline{M}_z'' - P_x^0 y'' = 0 \tag{A4.8}$$

The first equation which describes the perturbed thrust is uncoupled from the other two.

APPENDIX 5

INTRODUCTION TO THE COMPUTER PROGRAMS

A5.1 Introduction

It is the contention of the authors that the best use of this book will be made by those who become actively involved with the computer programs included in it. Actively involved can mean many things. At the best level, the reader will use the work here as a springboard into other things of his or her interest, perhaps even finding something that will help with an ongoing project. On a more casual level, the reader my simply come to understand nonlinear effects more clearly.

The way we learn is a complex matter. Can you learn something by simply reading it once? How many hours must be spent to comprehend a page of difficult material? What IS difficult material? How much of a concept is it necessary to understand in order to be able to apply the concept? If canned computer programs exist for structural analysis, how much do you need to know in order to be able to run them? What **should** a structural engineer know? Without answering any of these questions, it is our contention that a structural engineer can learn much from simple programming even if he or she has no intention of programming for real applications. This is particularly true in the case of nonlinear analysis which is inherently more difficult than the linear analysis with which we are now so familiar.

If the reader is to understand the computer programs of this text, it is first necessary to understand the computer programs of linear structural analysis. This appendix discusses two of the most simple computer programs of linear structural analysis, the space truss and the plane frame. These programs are listed in full here whereas the other computer programs discussed must be printed by the reader from the disk supplied with the text.

A5.2 Space Trusses

The space truss program listed here has three parts like most linear analysis computer programs which use the displacement or node method: the system matrix is set up, it is solved for displacements given loads, and finally internal forces and stresses are computed.

In more detail, following input, the system matrix `C(I,J)` is zeroed and then formed by adding the contribution of each member to it. In the case of bars with two ends, each member contributes at most four matrix terms to the system matrix. In the case of the truss, each of these terms has the form of

$$\pm \mathbf{n}_i \mathbf{n}_i^{\mathrm{T}} A_i E/L_i$$

and all that remains is to determine exactly where these terms go in the matrix `C`. That is done by the subroutine `INSERT`. In the program shown here, these terms require unit vector components which are provided by the subroutine `UNITV`.

In the center of the listing is a section of computer code which solves the system matrix given the joint loads. This code is simply Gaussian elimination: In the first half, rows of `C` are combined to eliminate terms below the diagonal; in the second half a triangular system of equations (back substitution) is solved. Like most linear equation solvers this code is entered with a description of the system matrix and the right hand side of the equations; at time of exit, the answers (the node displacements) are found in the array which originally contained the right hand side (the joint loads). Note also that this code does not include pivoting. There are two reasons for this: First, a zero along the diagonal of the system matrix would imply a singular (unstable) joint in the structure which can not occur; second, the equations of the node method are typically well-conditioned.

Once the node displacements have been computed, the member forces can be computed, bar by bar as

$$\Delta_i = \mathbf{n}_i . (\delta_{\mathrm{A}} - \delta_{\mathrm{C}}); \qquad F_i = \frac{A_i E}{L_i} \Delta_i$$

(It is left to the reader to determine the details, for example, of where each of the above terms is to be placed in the various arrays.)

A5.3 Plane Frames

Once one computer program for linear structural analysis has been written, others follow directly. The plane frame is a case in point. Like the three-dimensional truss it has three degrees of freedom per node, it requires four matrix terms in the system matrix per member, it uses the same equation solver, but is a little more complex to deal with than the truss.

The difficulties of the plane frame arise from the fact that its members are beams rather than rods. This fact also drives the introduction of the concept of a local coordinate system and the use of the rotation matrix. In terms of programming, the impact of these differences is simply that the contribution of each member to the system matrix is left as a matrix product

$$\mathbf{R}_i^{\mathrm{T}}\mathbf{N}_i^{\pm\mathrm{T}}\mathbf{K}_i\mathbf{N}_i^{\pm}\mathbf{R}_i$$

rather than being written out explicitly. This, incidentally, makes it an easy matter to modify the stiffness matrix which now is set up for straight, uniform elements. The subroutine RNK generates the matrices just mentioned and then PROD3 performs the required matrix multiplications. Otherwise, this program uses much from the space truss program.

A5.4 Listing of TR3D.FOR

```
C      SPACE TRUSS
       DIMENSION NP(100),NM(100),S(100)
       DOUBLE PRECISION R(100),P(100),C(100,100),UVEC(3)
      1,C1,D1,F1,F2,FAC
       MAXC=100
C
C      INITIALIZE PARAMETERS/ARRAYS
C
       E = 30.0D06
100    READ(5,150)NB,NN,NS
150    FORMAT (3(I4,3X))
       WRITE(6,1)NB,NN,NS
     1 FORMAT(I5,'  NO. MEMBERS'/I5,'  NO. NODES'/I5,
      1'  NO.SUPPORTS'//)
       READ(5,156)(R(3*K-2),R(3*K-1),R(3*K),P(3*K-2),
      1P(3*K-1),P(3*K),K=1,NN)
       WRITE(6,157)(K,R(3*K-2),R(3*K-1),R(3*K),P(3*K-2),
      1P(3*K-1),P(3*K),K=1,NN)
157    FORMAT (1H1,25X,11HCOORDINATES,40X,5HLOADS//
      114X,1HX,19X,1HY,19X,1HZ,18X,2HPX,18X,2HPY,
      1 18X,2HPZ//(I4,6D20.8))
       NNN = NN - NS
       N=3*NNN
C
C      SET UP SYSTEM MATRIX
C
       DO 30 I = 1,N
       DO 30 J = 1,N
    30 C(I,J) = 0.
       WRITE(6,159)
       DO 999 L=1,NB
       READ(5,151)NP(L),NM(L),S(L)
       WRITE(6,160)L,NP(L),NM(L),S(L)
151    FORMAT (2I5,8X,E10.6)
160    FORMAT (3I10,E20.8)
       K = 3*NP(L)
       M = 3*NM(L)
       CALL UNITV(K,M,C1,UVEC,R)
       CALL INSERT(C,K,M,UVEC,MAXC,N,E,S(L),C1)
 999   CONTINUE
C
C      SOLVE FOR JOINT DISPLACEMENTS
C
       M = N - 1
       DO 17 I = 1,M
       L = I + 1
```

```
      DO 17 J = L,N
      IF(C(J,I)) 19,17,19
   19 DO 18 K = L,N
   18 C(J,K) = C(J,K) - C(I,K)*C(J,I)/C(I,I)
      P(J) = P(J) - P(I)*C(J,I)/C(I,I)
   17 CONTINUE
      P(N) = P(N)/C(N,N)
      DO 20 I = 1,M
      K = N - I
      L = K + 1
      DO 21 J = L,N
   21 P(K) = P(K) - P(J)*C(K,J)
      P(K) = P(K)/C(K,K)
   20 CONTINUE
      WRITE(6,161)(I,P(3*I-2),P(3*I-1),P(3*I),I=1,NNN)
  161 FORMAT (1H1,13HDISPLACEMENTS/20X,1HX,19X,1HY,19X,1HZ
     1//(I10,3D20.8))
      WRITE(6,162)
  162 FORMAT(1H1,3X,6HMEMBER,9X,2HDL,17X,5HFORCE,
     1 14X,6HSTRESS//)
C
C     COMPUTE MEMBER FORCES AND DISPLACEMENTS
C
      DO 998 I=1,NB
      K = 3*NP(I)
      M = 3*NM(I)
      CALL UNITV(K,M,C1,UVEC,R)
      K1=K
      D1=0.
      FAC=1.
      DO 997 J=1,2
      IF(K1.GT.N) GO TO 996
      D1=D1+FAC*(P(K1-2)*UVEC(1)+P(K1-1)*UVEC(2)+P(K1)*UVEC(3))
  996 FAC=-1.
      K1=M
  997 CONTINUE
      F1=D1*E*S(I)/C1
      F2=F1/S(I)
      WRITE(6,1000) I,D1,F1,F2
  998 CONTINUE
      GO TO 100
  156 FORMAT (8X,6F11.6)
 1000 FORMAT (I10,3D20.8)
  159 FORMAT (1H1,3X,6HMEMBER,5X,5H+ END,5X,5H-
END,6X,4HAREA//)
      END
C
      SUBROUTINE UNITV(K,M,C1,UVEC,R)
      DOUBLE PRECISION R(1),C1,UVEC(3)
      C1=0.
      DO 1 I=1,3
      UVEC(I)=R(K+I-3)-R(M+I-3)
    1 C1=C1+UVEC(I)**2
      C1=DSQRT(C1)
      DO 2 I=1,3
    2 UVEC(I)=UVEC(I)/C1
```

```
      RETURN
      END
C
      SUBROUTINE INSERT(C,K,M,UVEC,MAXC,N,E,S,C1)
      DOUBLE PRECISION C(MAXC,MAXC),UVEC(3),C1
      K1=K
      DO 1 I=1,2
      IF(K1.GT.N) GO TO 1
      M1=K
      DO 2 J=1,2
      IF(M1.GT.N) GO TO 2
      FAC=1.
      IF(I.NE.J) FAC=-1.
      DO 3 L=1,3
      I1=K1-3+L
      DO 3 L1=1,3
      J1=M1-3+L1
    3 C(I1,J1)=C(I1,J1)+UVEC(L)*UVEC(L1)*S*E*FAC/C1
    2 M1=M
    1 K1=M
      RETURN
      END
```

A5.5 Listing of FR2D.FOR

```
C     PLANE FRAMES
      DIMENSION A(100),AL(100),ST(100),TH(100),NP(100),MI(100)
      DOUBLE PRECISION
P(100),C(100,100),R(3,3),SK(3,3),SNP(3,3)
     1,SNM(3,3),AI(3,3),AJ(3,3),NRP(3,3),NRM(3,3),ANG
      MAXC=100
C
C     INITIALIZE PARAMETERS/ARRAYS
C
      DO 6 I=1,3
      DO 6 J=1,3
      R(I,J)=0.
      SK(I,J)=0.
      SNP(I,J)=0.
    6 SNM(I,J)=0.
      R(3,3)=1.
      SNP(1,1)=1.
      SNP(2,3)=1.
      SNM(1,1)=-1.
      SNM(3,3)=1.
      PI=3.14159    /180.
      E=29000000.
    5 READ(5,2,END=999) NB,NN,NS
    2 FORMAT (3I5)
      WRITE(6,1)NB,NN,NS
    1 FORMAT(I5,'  NO. MEMBERS'/I5,'  NO. NODES'/I5,'  NO.
SUPPORTS'//)
      NNS=NN-NS
      N=3*NNS
      READ(5,3) (  P(3*I-2),P(3*I-1),P(3*I),I=1,NNS )
    3 FORMAT(3D10.2)
      WRITE (6,903)(I,P(3*I-2),P(3*I-1),P(3*I),I=1,NNS )
  903 FORMAT(1H1,11HJOINT LOADS /13X,2HPX,18X,2HPY,19X,1HM//
```

```
     1 (I4,3D20.8))
      WRITE            (6,901)
  901 FORMAT(18H1MEMBER PROPERTIES
/11X,4HAREA,15X,6HLENGTH,16X,1HI,
     1 17X,5HANGLE,14X,5H+ END,5X,5H- END)
      DO 904 I=1,N
      DO 904 J=1,N
  904 C(I,J)=0.
      DO 926 K=1,NB
      READ(5,31) A(K),AL(K),SI(K),TH(K),NP(K),MI(K)
   31 FORMAT(4E10.2,2I5)
      WRITE(6,900) K, A(K),AL(K),SI(K),TH(K),NP(K),MI(K)
  900 FORMAT(I4,4E20.8,2I10)
      ANG=TH(K)*PI
      CALL RNK(R,SK,SNP,SNM,NRP,NRM,AL,SI,A,E,ANG,K)
      IF(NP(K).GT.NNS) GO TO 23
      CALL PROD3(NRP,SK,NRP,AI)
      CALL INSERT(C,MAXC, NP(K),NP(K),AI)
      IF(MI(K).GT.NNS) GO TO 926
      CALL PROD3(NRP,SK,NRM,AI)
      CALL INSERT(C,MAXC, NP(K),MI(K),AI)
      DO 14 I=1,3
      DO 14 J=1,3
   14 AJ(I,J)=AI(J,I)
      CALL INSERT(C,MAXC, MI(K),NP(K),AJ)
   23 CALL PROD3(NRM,SK,NRM,AI)
      CALL INSERT(C,MAXC, MI(K),MI(K),AI)
  926 CONTINUE
C
C     SOLVE FOR DISPLACEMENTS
C
  927 M=N-1
      DO 91 I=1,M
      L=I+1
      DO 91 J=L,N
      IF (C(J,I)) 93,91,93
   93  DO 92 K=L,N
   92  C(J,K)=C(J,K)-C(I,K)*C(J,I)/C(I,I)
      P(J)=P(J)-P(I)          *C(J,I)/C(I,I)
   91  CONTINUE
      P (N)=P(N)/C(N,N)
      DO 94 I=1,M
      K=N-I
      L=K+1
      DO 95 J=L,N
   95  P(K)=P(K)-P  (J)*C(K,J)
   94  P  (K)=P(K)/C(K,K)
      WRITE  (6,231)(I,P(3*I-2),P(3*I-1),P(3*I),I=1,NNS )
  231 FORMAT(20H1JOINT DISPLACEMENTS
/13X,2HDX,16X,2HDY,16X,2HTH/
     1 (I3,3X,3(D15.8,3X)))
C
C     COMPUTE MEMBER FORCES AND DISPLACEMENTS
C
      WRITE (6,230)
```

```
  230 FORMAT(28H1MEMBER DISPLACEMENTS-
FORCES/13X,2HDL,15X,3HAL+,16X,
     1 3HAL-,16X,1HT,16X,2HM+,16X,2HM-)
      DO 207 I=1,NB
      ANG=TH(I)*PI
      CALL RNK(R,SK,SNP,SNM,NRP,NRM,AL,SI,A,E,ANG,I)
      DO 222 J=1,3
  222 AI(2,J)=0.
      IF(NP(I).GT.NNS) GO TO 213
      DO 208 J=1,3
      DO 208 K=1,3
      L4=3*NP(I)-3+K
  208 AI(2,J)=AI(2,J)+NRP(J,K)*P(L4)
  213 IF(MI(I).GT.NNS) GO TO 215
      DO 308 J=1,3
      DO 308 K=1,3
      L4=3*MI(I)-3+K
  308 AI(2,J)=AI(2,J)+NRM(J,K)*P(L4)
  215 DO 226 J=1,3
       AI(1,J)=0.
      DO 226 K=1,3
  226  AI(1,J)= AI(1,J)+SK(J,K)*AI(2,K)
  207 WRITE(6,217) I,AI(2,1),AI(2,2),AI(2,3), AI(1,1),
     6 AI(1,2), AI(1,3)
  217 FORMAT (I3,3X,6(D15.8,3X))
      GO TO 5
  999 STOP
      END
C
      SUBROUTINE RNK(R,SK,SNP,SNM,NRP,NRM,AL,SI,A,E,ANG,K)
      DIMENSION AL(1),SI(1),A(1)
      DOUBLE PRECISION
R(3,3),SK(3,3),SNP(3,3),SNM(3,3),NRP(3,3),
     1 NRM(3,3),ANG
      R(1,1)=DCOS(ANG)
      R(2,2)=R(1,1)
      R(1,2)=DSIN(ANG)
      R(2,1)=-R(1,2)
      SK(1,1)=E*A(K)/AL(K)
      SK(2,2)=E*4.*SI(K)/AL(K)
      SK(3,3)=SK(2,2)
      SK(2,3)=SK(2,2)*.5
      SK(3,2)=SK(2,3)
      SNP(2,2)=-1./AL(K)
      SNP(3,2)=SNP(2,2)
      SNM(2,2)=1./AL(K)
      SNM(3,2)=SNM(2,2)
      DO 1 I=1,3
      DO 1 J=1,3
      NRP(I,J)=0.
      NRM(I,J)=0.
      DO 1 L=1,3
      NRP(I,J)=NRP(I,J)+SNP(I,L)*R(L,J)
    1 NRM(I,J)=NRM(I,J)+SNM(I,L)*R(L,J)
      RETURN
      END
C
      SUBROUTINE INSERT(C,MAXC,I1,J1,A)
      DOUBLE PRECISION C(MAXC,MAXC),A(3,3)
      DO 1 I=1,3
```

```
      I2=3*I1-3+I
      DO 1 J=1,3
      J2=3*J1-3+J
    1 C(I2,J2)=C(I2,J2)+A(I,J)
      RETURN
      END
C
      SUBROUTINE PROD3(A,B,C,D)
      DOUBLE PRECISION A(3,3),B(3,3),C(3,3),D(3,3)
      DO 1 I=1,3
      DO 1 J=1,3
      D(I,J)=0.
      DO 1 K=1,3
      DO 1 L=1,3
    1 D(I,J)=D(I,J)+A(K,I)*B(K,L)*C(L,J)
      RETURN
      END
```

□

APPENDIX 6

GRAPHICS ON A PC

A6.1 Introduction

Computer graphics can be fun to use when computing. It can also be very useful when you are attempting to check input data and it can be indispensable when dealing with three-dimensional visualization of the type required for fabric structures and cable nets. Unfortunately, graphics code is the least portable code of all.

In order to get started with graphics, it is instructive to compile and run a simple program on the disk called PLOTDEMO.FOR. That is done on our system using the (Microsoft FORTRAN) commands

```
FL32 PLOTDEMO.FOR
PLOTDEMO
```

It should be noted that the use of graphics requires loading a graphics library in addition to the usual FORTRAN libraries. With the compiler used above, these libraries are loaded automatically. Other versions of FORTRAN require a specific reference to this library. For example, on some compilers the command is

```
FL PLOTDEMO.FOR   /LINK   GRAPHICS.LIB
PLOTDEMO
```

If the graphics libraries have not been loaded properly, the system will be quite vocal about so informing you.

PLOTDEMO.FOR generates a three-dimensional figure which can be rotated using keyboard commands. It generates this figure with and without node numbers.

If this program will not compile, there can be several reasons. The most likely one has to do with the use of fonts when writing letters and numbers. Our program uses the command

```
FONTPATH= '\F32\LIB\COURB.FON'
```

This statement implies that we are running our programs in our root directory and that the fonts reside in a file called 'COURB.FON' which can be reached through the commands 'CD F32' and 'CD LIB'. If PLOTDEMO does not run on your system you may have to change this statement. Otherwise you may have to redo the command

```
DUMMY=SETVIDEOMODE($VRES16COLOR)
```

The program PLOTDEMO.FOR has two pieces. Its main program simply generates the data for the three-dimensional structure which is displayed. (Once the structure has been displayed, it is a straightforward matter to read the code which generated it.) The main program subsequently calls plotting routines.

There are two plotting routines. The main program calls a subroutine SPLOT which in turn calls a subroutine PLOT. SPLOT is a more or less portable subroutine which rotates a three-dimensional object. PLOT is a very system dependent subroutine which produces the actual drawing on the CRT.

A6.2 Plotting in 2-D

Any plot on a flat surface is of course a two-dimensional drawing. Three-dimensional drawings are produced here by first rotating a three-dimensional object to obtain a desired view of it and then plotting the x-y plane in the rotated coordinate system. Only stick figures which can be described by points and lines are considered here.

Figure A.1 gives some of the basic information about creating two-dimensional plots on a PC. First of all the image to be drawn has its own coordinate system which must be transformed so that this image will appear on the screen when drawn using the coordinate system of the screen. In the case of the subroutine PLOT, the transformation is simply a shift and a scaling.

Otherwise there is system overhead to be dealt with. That is, when you have a line to draw or a piece of text to write, in addition to the draw and write commands, many other commands must be given to "set up the system". These are called "overhead" here since they add nothing to the plotting process and only enable the system; they are more or less recognizable do to their otherwise lack of function. The subroutine PLOT scales a two-dimensional picture and draws it on the CRT line by line. The subroutine SPLOT performs the required three-dimensional rotations (see Appendix 3) using the idea of a compound rotation.

With regard to detail, in order to plot on the face of the CRT the main program calls a subroutine SPLOT as

```
CALL SPLOT(NP,MI,NN,NB,R,FOR,IWRITE)
```

with

NP,MI	- arrays which describe member node numbers
NN	- number of nodes
NB	- number of bars or members
R	- coordinate array
FOR	- array of member forces
IWRITE	- describes text features
IWRITE=0	no text
IWRITE=1	shows member forces
IWRITE=2	shows node map

Subroutine SPLOT generates the new x-y coordinates of all points after the coordinate system has been rotated to establish a point of view. It begins by setting the rotation matrix ROT to be the identity matrix which subsequently produces an initial top view of the structure. It then sets the rotation increment DTH to be 10 degrees. It finally forms the rotation matrix as the product of rotations about coordinate axes and computes new plane coordinates RXY which are subsequently plotted by the subroutine PLOT. Hard copy can be obtained using the PRINTSCREEN key.

It may be noted that the code for SPLOT comes primarily from the three-dimensional frame program which also uses a rotation matrix.

PLOTDEMO has utility beyond the truss which is displayed. PLOT3D is a program which is created from it by replacing the main program with read routines. In this form PLOT3D can be used to display data sets. For example, the command

PLOT3D TR2DNL.DAT

will produce a picture of the structure described by the data set TR2DNL.DAT.

A6.3 Drawing Lines in 2-D

Coordinate Systems

- Real coordinates - describe the object to be drawn
- Screen coordinates(draw) - are used to draw the picture on the screen of the CRT

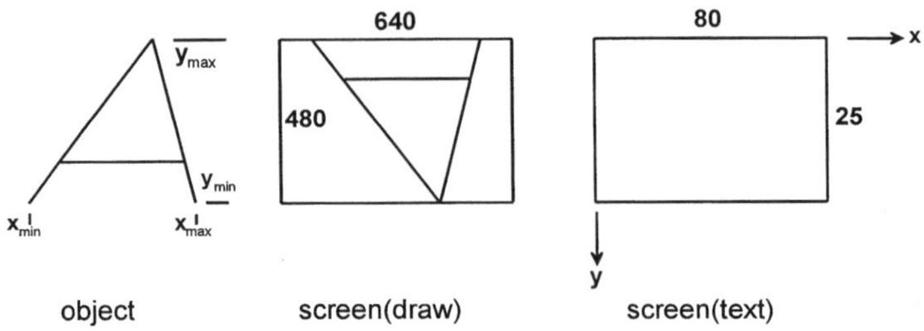

Figure A7.1 Three coordinate systems of a picture.

• Screen coordinates(text) - are used to write text on the screen when fonts are not used

Real coordinates must be transformed so that the picture will appear on the screen:

$$x_{\text{screen}} = (x_{\text{real}} - x_{\text{shift}}) / \text{scale}$$
$$y_{\text{screen}} = (y_{\text{real}} - y_{\text{shift}}) / \text{scale}$$

Pick "scale" to make the object fit screen

$$\text{scale} = \max((x_{\max} - X_{\min})/640; \qquad (y_{\max} - y_{\min})/480)$$

Pick x_{shift}, y_{shift} to center object in screen

$$x_{\text{shift}} = \frac{x_{\max} + x_{\min}}{2} - \frac{640}{2} \times \text{scale}$$
$$y_{\text{shift}} = \frac{y_{\max} + y_{\min}}{2} - \frac{480}{2} \times \text{scale}$$

INDEX